T0140149

Studies in Systems, Decision and Control

Volume 275

Series Editor

Janusz Kacprzyk, Systems Research Institute, Polish Academy of Sciences, Warsaw, Poland

The series "Studies in Systems, Decision and Control" (SSDC) covers both new developments and advances, as well as the state of the art, in the various areas of broadly perceived systems, decision making and control–quickly, up to date and with a high quality. The intent is to cover the theory, applications, and perspectives on the state of the art and future developments relevant to systems, decision making, control, complex processes and related areas, as embedded in the fields of engineering, computer science, physics, economics, social and life sciences, as well as the paradigms and methodologies behind them. The series contains monographs, textbooks, lecture notes and edited volumes in systems, decision making and control spanning the areas of Cyber-Physical Systems, Autonomous Systems, Sensor Networks, Control Systems, Energy Systems, Automotive Systems, Biological Systems, Vehicular Networking and Connected Vehicles, Aerospace Systems, Automation, Manufacturing, Smart Grids, Nonlinear Systems, Power Systems, Robotics, Social Systems, Economic Systems and other. Of particular value to both the contributors and the readership are the short publication timeframe and the world-wide distribution and exposure which enable both a wide and rapid dissemination of research output.

** Indexing: The books of this series are submitted to ISI, SCOPUS, DBLP, Ulrichs, MathSciNet, Current Mathematical Publications, Mathematical Reviews, Zentralblatt Math: MetaPress and Springerlink.

More information about this series at http://www.springer.com/series/13304

Kaizhong Guo · Shiyong Liu

Error Systems: Concepts, Theory and Applications

 Springer

Kaizhong Guo
School of Management
Guangdong University of Technology
Guangzhou, China

Guangzhou Vocational College
of Science and Technology
Guangzhou, China

Shiyong Liu
Research Institute of Economics
and Management
Southwestern University of Finance
and Economics
Chengdu, China

ISSN 2198-4182   ISSN 2198-4190   (electronic)
Studies in Systems, Decision and Control
ISBN 978-3-030-40762-9   ISBN 978-3-030-40760-5   (eBook)
https://doi.org/10.1007/978-3-030-40760-5

© Springer Nature Switzerland AG 2020
This work is subject to copyright. All rights are reserved by the Publisher, whether the whole or part
of the material is concerned, specifically the rights of translation, reprinting, reuse of illustrations,
recitation, broadcasting, reproduction on microfilms or in any other physical way, and transmission
or information storage and retrieval, electronic adaptation, computer software, or by similar or dissimilar
methodology now known or hereafter developed.
The use of general descriptive names, registered names, trademarks, service marks, etc. in this
publication does not imply, even in the absence of a specific statement, that such names are exempt from
the relevant protective laws and regulations and therefore free for general use.
The publisher, the authors and the editors are safe to assume that the advice and information in this
book are believed to be true and accurate at the date of publication. Neither the publisher nor the
authors or the editors give a warranty, expressed or implied, with respect to the material contained
herein or for any errors or omissions that may have been made. The publisher remains neutral with regard
to jurisdictional claims in published maps and institutional affiliations.

This Springer imprint is published by the registered company Springer Nature Switzerland AG
The registered company address is: Gewerbestrasse 11, 6330 Cham, Switzerland

To our family and friends who have been giving unselfish and generous emotional and material supports over the course of writing.

Preface

Jérôme Kerviel, ever a French trader working for Société Générale, was convicted for unauthorized trading resulting in estimated €4.9 billion loss for the company. In 1995, former Daiwa Bank trader Toshihide Iguchi pleaded guilty of covering up his 12-year loss estimated at $1.1 billion due to unauthorized trading. The fraudulent and unauthorized trading activities of Nicholas William Leeson, a former UK derivatives broker, caused the collapse of Baring Bank in 1995. Although there are still arguments on who were really responsible for the huge loss, all those events were caused by decision-making errors either in companies' staffing mechanism and performance evaluation or trading processes. The well-known research institute and think tank RAND Corporation indicated that the collapse of 85% large and medium companies was attributed to errors in decision making.

Errors permeate every corner of the world. The decision-making processes in different entities including state governments, for-profit enterprises, nonprofit organizations, and people are often susceptible to error. The occurrence of error does not differentiate the developed countries or underdeveloped countries, well-tuned organizations or under-performing organizations, those famous or infamous persons in history. Errors were often seen in scientific and technological fields whether they were in developed countries or developing countries. History has witnessed many different errors in either primitive or civilized societies. The causes for an error might result from multiple elements or a single element of the system in which the error is embedded. The consequences of the errors in an object, a case, a decision, or a theoretical system may lead to minor loss, catastrophic casualties, disbandment of organizations, dissolution of countries, or even the ter-mination of human beings. For instance, the mid-air collision on July 1, 2002 between Bashkirian Airlines Flight 2937 (Tu-154 passenger jet plane) and DHL Flight 611 (Boeing757 cargo jet plane) was caused by a series of errors including problems of the arrangement for both personnel and equipment maintenance of SWISS air traffic control service and procedural errors of using Traffic Collision Avoidance System (TCAS). The disaster killed 69 passengers and crew members on Flight 2937 and 2 crew members of the Flight 611, which offered a thought-provoking lesson for relevant organizations and decision makers.

The generation on the idea of developing error logic, error theory, and error systems can be traced back to 1983 while professor Wen Cai and the first author of this book were attending workshop named "Value-mapping Engineering" in former Guangzhou Nitrogen Fertilizer Plant. During a noon nap time, noises coming from the dispute between two housekeeping ladies outside my room woke me up. One of them said "you are wrong." The other one argued that "I am not wrong, it is you who made the mistake." The first one replied "Would you please tell me where my fault is if you think I was wrong?" The other returned very rapidly "and where is my fault if you think I am wrong? On what basis do you judge if I went wrong?" "On what basis do you judge if I went wrong?" struck my mind while I was still in the hypnopompic state and it ignited my inspiration and passion in research of conceptualization, symbolization, and modeling of error over the past 30 years. Yes, indeed! On what basis does one judge if certain thing, matter, decision, action, etc., is erroneous and what caused the error? Is there any condition, measure or way to eliminate the error? More importantly, is there any condition or method to prevent error from happening. If there exist conditions or measures to prevent erring, one must figure out the laws of change, transformation, and transition of errors. Question then arose again. Are there laws in preventing, identifying, and consequently eliminating errors? While pondering these issues, I woke up, like what I used to do, and wrote them down on my notebook. From then on, I started my life-long studies in error-related theories and practices.

Thus far, many papers and books regarding the investigation of errors have been published. For sake of brevity, we mainly review some books and please refer to reference for other publications. Books are listed as follows: (1) *Malfunction Diagnosis*: diagnosing the causes and position or predicting the occurrence of malfunction; (2) *Enterprise Diagnosis*: figuring out discrepancies between standards and the results and corresponding laws for generating such errors by scanning the internal and external environments of organizations, identifying the laws and mechanisms for improving management; (3) *Enterprise Disaster*: risk management; (4) *Rand Diagnosis* and *Rand Decision-making*: conducting in-depth and extensive research on the causes of errors, countermeasures dealing with errors, and approaches in diagnosing errors. Rasmussen (Eds.) (1987) summarized the topics addressing various approaches to human error analysis in different fields, especially in the context of technological development. Reason (1990) identified cognitive processes to gain better understanding on mechanism of human error. Bogner (1994) presented those examples of human errors in medicine collected from various medical fields including psychology, medicine, engineering, cognitive science, human factors, gerontology, and nursing. Dorner (1997) proposed the "logic of failure" and analyzed the roots of catastrophes. Hollnagel (1998) put forward CREAM (Cognitive Reliability and Error Analysis Method) to offer an error taxonomy that incorporates individual, technological and organizational factors based on cognitive engineering principles. Reason and Hobbs (2003) systematically analyzed the types of error in maintenance and provided a well-organized guide to managing maintenance error accordingly. Dekker (2004) assessed two views of human errors with old view of looking at errors as the cause of an incident or accident and new view of deeming error as a symptom of deeper problem within the system. The author preferred to adopt the new view. Dismukes (2009) indicated that *"human errors in aviation industry are not root causes but symptoms of the way industry operates"*. The book edited by Hofmann and Frese (Eds.) (2011) talked about the errors in organizations and future directions of related research. Hagen (2013), using aviation industry as an example, offered insights on how to exercise error management in organizations. Haselton and Galperin (2013) exhibited the application of Error Management Theory (EMT) in relation management for the purpose of reducing biases. Dhillon (2007, 2009, 2012, and 2014), in four books, discussed human errors in transportation systems, engineering maintenance related to aviation and power generation, engineering systems in general, and nuclear power plants, respectively. Blokdyk (2018) provided the essentials of error management theory and extensive criteria summarized from past projects. There are also many theories and methodologies for studying errors and mistakes in different fields and disciplines such as "reductio ad absurdum" in mathematics, clinical "misdiagnosis", "criminal psychology" in forensic science, "abnormal thinking" in thinking, and "theory of fault tolerance" in computer science.

Nevertheless, none of the above-mentioned research has conducted quantitative and holistic studies on errors in general, let alone the studies on the patterns and laws of error transformation and transition. Since 1983, our team started to build the theoretical framework and systems for error elimination where common errors are research objects and causes & mechanisms of erring, patterns and laws of error transformation and transition, and methodologies for preventing and eliminating error are the primary contents. In summary, the research done by our team includes the following 7 aspects:

1. Definition of error
   Suppose that U is the universe of discourse, G is a set of rules for judging error defined within U, if $\exists\, G \nRightarrow a$ (including the cases that a can not be completely or partially obtained by exercising G; or the case that a has nothing to do with G), a is erroneous defined within U under the rule of judging errors G. Based on the definition, we know that the error is relative. Its existence pertains to certain universe of discourse, a group of rules, and the implications manifested in the definition. Note: G is a set of predetermined and qualified rules for judging if an object is correct or not. Otherwise, $G \Rightarrow a$ holds if G is the rule for judging whether an object is erroneous or not.

2. Rules for judging errors
   Research contents in this area contain: (1) conditions for justifying the appropriateness of those chosen rules and their applicable factors such as fields and time; (2) relationships between chosen rules; (3) relationships between chosen rules and the object being investigated; (4) methods for screening and assessing rules.

3. Error function

   In order to quantitatively depict the scale and value of errors, error function is introduced. The particular formulation of error function should hold the principle that the obtained error value should reflect the degree to which the object violates the chosen rules. Scalar and vector functions are two relatively simple error functions.

4. Error set theory

   On the basis of classic and fuzzy set theories, the established error set is composed of classic error set, fuzzy error set, and error set with critical points. In error set theory, we investigate both static and dynamic relationships and interactions between subjects. We especially focus on the study of relationships between change and transformation of subjects. Six basic transformations, namely, similarity or equivalence transformation, displacement transformation, decomposition transformation, addition transformation, destruction transformation, and unit transformation are introduced for the purpose of studying change and transformation of subjects. Dynamic parameters are assigned to elements of error set to capture the change and transformation of subjects. Thereafter, by using 6 basic transformations, operations can be qualitatively and quantitatively conducted on elements of error set.

5. Error logic

   Built on the foundation of classic mathematical logic, fuzzy mathematical logic, and dialectic logic, the concept of error logic is proposed in order to identify the patterns and laws for transition and transformation of errors. In our error logic system, in addition to denotation connective, connotation connective, and individual connective, 6 more transformation connectives are created, namely, similarity transformation connective, displacement transformation connective, decomposition transformation connective, addition transformation connective, destruction transformation connective, and unit transformation connective as well as their inverse transformation connectives. More importantly, research in error logic covers many aspects including but are not limited to (1) principles and laws of operation of each newly created connective, operations between new connectives, and operations between each of the new connective and existing denotation connective, connotation connective, or individual connective; (2) truth table, normal forms, and valid reasoning methods in error logic; (3) concepts and parameters of error predicate logic; (4) semantic structure and interpretation of subject term (the variable), quantifier, predicate, etc.

6. Error system

   For convenience of judging error, concept of error system is introduced. By so doing, the subject being evaluated will be abstracted to be an object system which is composed of issue set, condition set, conclusion set, intrinsic features, objective features, and relation set. Under the influence of certain judging rules, error values are assigned to this object system. In a broad sense, all the object systems and those corresponding judging rules defined within universe of discourse form a general error system.

7. Application of error theory

The motivation of conducting research on error theory originates from issues in the real world and the achievements from the theoretical research should be put into practice to help people to prevent and eliminate errors. Among all actual application cases of error theory, one worth mentioning is a project supported by China National Natural Science Foundation (CNSF) from 1991 through 1993, which was titled as "Research on Expert System Used for Judging and Examining Decision-making Errors in Fixed Asset Investment". Having completed the project, several valid methods for identifying, judging, and investing decision-making errors in fixed asset investment had been found. With appropriate summary, induction, and deduction, the "Fifteen", "Six", and "Three" methods for preventing and eliminating errors were proposed.

Since 1983, although the fundamental architecture for error theory has been established and has gained attention from scholars in different fields, the research on the theory system still remains at a relatively immature level. Many aspects and concerns in this architecture need to be fixed and consummated. Examples include: (1) refinement on particular structure need to done; (2) need to develop a systematic guidelines for defining steps, methods, and formats in building error functions; (3) the conditions and ways of using the 6 basic transformations need more extensive investigations; (4) the research on the laws of error transition and transformation only resides in logical analysis instead of systematic exploration; (5) the developed "Fifteen", "Six", and "Three" methods can only provide a guideline in helping predicting, preventing, and eliminating errors. Therefore, it is very necessary to develop more systematic, concrete, and valid methodologies to enhance the operability, applicability, and practical values in addressing emerging issues in real world.

This monograph, by putting error research in the context of a typical system, investigates the root causes and mechanisms of erring, transition and transformation of errors, and consequently the conditions and methods of preventing and eliminating error in such system.

The essence of systems science resides in the philosophy of holism. When talking about the state that system reaches optimal, it generally refers to global optimum of the whole system with respect to the objective features that are included in the intrinsic features of a system. The attainment of global optimum of a system must rely on the normal deployment of functions (features) of its subsystems.

The word "system" originates from Latin word *systēma* meaning whole made of several parts or members. Many different definitions have been made by scholars for system from different perspectives based on their particular research objectives. Let's give some examples of them. "System is a pre-given set composed of elements and their normal behaviors"; "System is a well-organized wholeness"; "System is an entity made of connected materials and processes"; "System is a body composed of ordered elements/factors working towards a common goal" are some popular examples for the definitions of system.

The development of systems theory and their applications in different fields were mainly attributed to the contributions of scholars such as biologist Ludwig von Bertalanffy (1901–1972), Norbert Wiener (1894–1964), Ross Ashby (1903–1972), John Henry Holland (1929–2015), and Murray Gell-Mann (1929–2019). Bertalanffy pioneered the general system theory by introducing models, principles, and laws that apply to it. Wiener and Ashby used mathematics to study systems. Holland, Gell-Mann, and others proposed the term "complex adaptive system".

General system theory attempts to provide a definition that can capture common properties of various systems. The definition for general system is a body composed of well-organized elements working toward attaining particular goals or features. This definition apparently includes 4 concepts and their relationships, namely, system, element, structure, feature, relationships between elements, relationships between elements and structure, and relationships between system and its external environment.

The purpose of system theory is to investigate form, structure, and laws of general systems, to examine the common properties of those systems, to capture and illustrate their features using mathematical methods, and consequently to identify the mechanisms, rules, laws, principles, and mathematical models that can be applied to general systems. And the ultimate objective of learning system theory is to use the understanding on the system to better manage, control, renovate, or change the current system structures (natural or man-made systems) to align them with the needs of our civilized world. With better understanding on the system structures, we can introduce all kinds of interventions or policies to enable the systems of interest to attain their optimal performance or outcomes. Moreover, by gaining better understanding on the dynamics of a system over time, decision/policy makers and practitioners can prevent policy-resistance (counter-intuitive behaviors). System theory is recognized a discipline that possesses both mathematical and logic characteristics. System theory proclaims that holism, connectedness, hierarchical structure, and dynamic equilibrium, time-dependence are common properties of all systems, which are both philosophies of system thinking and principles of using system approach. As a branch of scientific approaches, system theory helps identify the objective laws on how world is running and also offers human being a way of thinking about the world. Therefore, system theory is also called system approach since it can represent concept, view, model, and mathematical methods as well.

In Bertalanffy's masterpiece titled "General System Theory; Foundations, Development, Applications", he emphasized the concept of holism. System, as an organic body, is not mechanical combination or simple addition of its constituents but an organic combination of its elements working together towards a common goal. The system's features are its emerging behaviors, which can not be found in its individual elements or subsystems. By quoting Aristotle's "*A whole is greater than the sum of its part*", Bertalanffy opposed those mechanical philosophies that the wholeness (system behaviors) can be observed or inferred from the behavior of a particular element of the system. He also stated that each element of the system is in a particular location in the system hierarchy, which is also tightly coupled with other elements. The connectedness among system's elements renders system integral and holistic. The "should-be" function of a system's element will disappear once it is separated from the system structure. For example, having done the hand amputation due to traumatic injury, the removed "hand" would never function as it could have been doing when it is an integral part of a person.

The fundamental thoughts of system theory is to treat the object being investigated as a system and to analyze the structure, function, dynamic relationships between elements, system, and their environment. With better understanding on the dynamics, complexities, and uncertainties associated with the system, the ultimate goal is to find how it attains its optimal target and, consequently provide counter-factual analysis when interventions need to be implemented in the system. Systems are ubiquitous in the universe. From cosmos to the microscopic world, systems exist everywhere such as Milky Way, solar system, earth system, social system, transportation system, production system, human body system, bacterial system, cell system, and atom, etc.

For ease of studying system, many ways were used to categorize system: (1) natural systems and artificial systems (whether designed by human being or not); (2) natural systems, social systems, and thinking systems (according to research subject); (3) macrosystems, mesa system, microsystems, and microscopic systems (scale of the systems); (4) open systems, closed systems (whether there exists interaction with environment); (5) balanced systems (systems having equilibrium), non-equilibrium systems, near-equilibrium systems, and far-from-equilibrium systems (whether there exists equilibrium).

The emergence of system theory brought profound changes on the way how people think about the world. In conventional research practices, Descartes' philosophy of "reductionism" had been dominating the academic fields. Under such influence, the general practice in research is to divide a complicated issue or subject into multiple parts and investigate each part individually. Thereafter, the characteristics of those individual parts are then used to infer the behaviors of the original issue or object. The reductionism approach focuses on local substructures or elements and abides by the unidirectional causal-effect determinism. Although this approach had proved valid for centuries within certain confined ranges and had served as the most popular way of thinking in mainstream research communities, it can only handle simple issues or objects without being able to capture the wholeness, dynamic interactions, and circular causalities of complicated objects (i.e., systems in the language of system theory). With accelerated development in economy, technology, and society, human beings with traditional analytical thinking become incompetent in dealing with issues/objects with thousands or even millions of variables connected/networked in various ways. However, the emergence of system theory, cybernetics, and informatics paved the way for human beings to drive the rapid advancement of modern science and technologies. The widespread applications of system theory make it become basis for developing new theories in handling complicated system in the fields of politics, economy, military, culture, science, and society, etc.

Regarding the trend of system theory, the authors think it is moving towards the formation of unified framework that summarizes the achievements obtained from the empirical and theoretical research in different fields. System thinking ensued by system theory has become very powerful force to overturn the ingrained singular causation thinking.

In many occasions, people like to use the formula "system function $\neq \sum$ {functions of all subsystems}" to represent the relationship between system features and subsystem features, which is not a precise expression. First of all, we are not sure if there exists additivity among subsystem features. Secondly, the system features could be larger, equal, or smaller than the sum of its subsystem features since the complicated synergistic effects caused by interactions between subsystem renders the causality prohibitively difficult to understand. In reality, there are tons of examples to illustrate this phenomenon, which brought troubles for researchers to examine the relationships between global optimization in system and local optimization in subsystems. In the following paragraphs, we provide some examples.

Example 1 MiG-25, a supersonic interceptor designed by the Soviet Union's Mikoyan-Gurevich Bureau and built mainly using stainless steel, can reach an altitude of 35,000 meters, top speed of Mach 3.2, and an average speed of 2,319.12 km/h over a 1,000 km circuit (on 16 March 1965) (Gordon 2008). It is a highly maneuverable fighter. Although the airplane primarily used nickel-steel alloy (80%) and aluminum (11%) and only 9% of titanium, it attained super performance through advanced design and synergistic effects of its subsystem structures. This case became a classical example for illustrating how system theory (systems engineering) helped engineer to achieve a better performance in system design and operations.

Example 2 A common commercial airplane is primarily composed of 6 subsystems, i.e., wings, fuselage, tail wings and rudder, landing gears, engines, and control. The major function of the commercial airplane is to safely deliver passengers from departure location to destination, which must be achieved through the synergistic interaction of the 6 subsystems. During flying process, airflow on the top of the wing is faster than that of airflow underneath the wing, which generates lower air pressure on the top of the wing and higher air pressure underneath the wing. Therefore, the difference in air pressure produces the lift. Moreover, the spoilers, ailerons, and winglets on the wings offer the function of lift, drag, and roll actions. The wings also provide installation locations for turbine engines (2 or 4). The fuselage of an airplane is the backbone to install wing, tail, landing gears, and cockpit (control circuit and equipment) and carry passengers and payload as well. The tail wing is used as horizontal stabilizer and to change and control pitch. While the tail rudder is used as vertical stabilizer and to change yaw. Landing gears are used to support airplane on the ground and allow airplane to take off, land, and taxi without damage. Turbine engines generate thrust and also provide power supply for control equipment, air conditioning, and lighting for normal airplane operation. Given the above-mentioned descriptions, the features of subsystems have essential difference with the system's features of "safely deliver passengers from departure location to destination" and they do not have additivity towards the realization of the whole system's features. The global optimum of this kind of system can not be obtained by simply optimizing its subsystems.

Example 3 In a factory with three identical production lines, each production line is a subsystem of this factory. If daily output of the factory is the feature (goal) of this factory, the daily throughput (goal) of each production line apparently has additivity. The total daily throughput is the sum of three production lines' daily output. Under such circumstance, the global optimum of the system and local optimum of each subsystem can be simultaneously achieved.

Example 4 The prisoner's dilemma is a typical non-zero-sum game in game theory where the players can be win-win or lose-lose depending on how they play the game. In this game, the game is a system and the total years in prison of the two suspects is system's goal. Each member is a subsystem and the year this player serves is the feature of this subsystem. Suppose that both members want to optimize their features (betraying the other), it turns out that the feature of the whole system gets a total of 4 years instead of the global optimum of 2 years. Therefore, in this case, the global optimum of the system can not be guaranteed when each subsystem of the system reaches its optimum and vice versa.

From the above discussion and examples, we know that there exist certain relationships between the subsystem optimum and system optimum. What are those relationships? In this monograph, we will conduct in-depth research in this respect. Two major relationships are the major themes of our research: (1) relationship between the optimum of certain intrinsic feature GY_j of system S and the feature optimum of certain subsystem in this system; (2) relationship between the optimum of certain objective feature GY_j of system S and the feature optimum of certain subsystem in this system. One issue worth noting here is that the optimum of certain intrinsic feature GY_j of system S is different from the global optimum of the whole system. Global optimum of a system is attained when all intrinsic features having consistency with all corresponding objective features of the system reach optimum. The consistency between intrinsic features and objective features dictates the alignment of system objectives, which not only requires the consistency of objectives over time but also needs the alignment the subsystems' objectives with the system's objective as well as the dynamic coordination between subsystems' goals. The system's goal is divided into sub-goals and then assigned to different subsystems or elements. Before joining a system, subsystem, as an individual self-sustained entity, also has its own goal and particular interests. Nevertheless, once employed by a system, the subsystem must be subject to the goal of the whole system and should align its individual goal or interests with the system's goal. In other words, in the process of achieving global optimization of a system, the subsystem must provide the features required by the system even with the price of sacrificing individual interests. Or the subsystem can not negatively affect the whole's features when it exercises its own features or interests.

In social system, local interests should be subject to the whole organization's interests and lower hierarchy's tactical goals should be aligned with the organizational strategy. All resources are moving towards the realization for goals of the whole organization by which the optimization of the system can be achieved. Otherwise, if there is no alignment between local goals and global goal and each element acts on its own will, the system will become a chaotic mass. Similarly, in an engineering system, the characteristics and features of constituent parts of the system must be compatible and match with each other and they must conform to the requirements of the whole system.

The combination pattern of different elements in a system plays a critical role in determining what kind of characteristics and features a system might have. For instance, due to different structures that carbon atoms can build, multiple distinct materials (e.g., diamond, graphite, and graphene) can be formed, which even have absolutely distinct characteristics. Therefore, system structure determines system behavior and characteristics.

When examining the issues regarding optimization and relationship of subsystem and system, one must consider the cost of attaining both local goals and global optimum. In China's history, there was a very famous game theory example called "Tian Ji Horse Racing". In this story, the emperor of Qi Kingdom (Emperor Qiwei) wanted to conduct horse racing with Tian Ji (370 BC-313 BC) a general of Qi Kingdom. Both of them had three groups of horse categorized as fast, medium, and slow. In the first round, the stratagem of Tian Ji advised by Sun Bin (author of Sun Bin's Art of War) was to use slow horse to race with Emperor Qiwei's fast horse. Of course, there was no doubt that Emperor Qiwei won in this round with an overwhelming advantage. However, the cost of Emperor Qiwei was that he already used his fastest horse. Tian Ji won the second and third rounds since he used his fastest horse to race with Emperor Qiwei's medium speed horse in the second and used his medium speed horse to race with Emperor Qiwei's slow horse. Finally, Emperor Qiwei lost the whole game since he consumed too much resources in realizing local goal and negatively affected the realization of the goal of the whole system.

A well-functioned system demands coordination of its elements and subsystems. In a system with feature additivity, incentives or mechanisms must be designed to simultaneously reach local optimization and align individual's interests with the system goal to attain global optimization. In "Jingyezi Anecdote", a story illustrated an old Chinese proverb: "make the best possible use of a person according to his/her strengths just like selecting tools based on what they are capable of." A family had five sons which had different merits or defects with one rustic (unsophisticated), one smart, one blind, one hunchback, and one lame. According to their specific situations, the father helped them arrange their career: with rustic one working as a farmer, smart one being a businessman, blind one offering massage service, the one with hunchback stranding rope, and lame one hand-spinning cotton. In this family system, every member used their potentials and enabled the whole family to live a bountiful life.

While in a complicated system beyond the simple aggregation of elements/subsystems, the system design must handle the trade-off between sacrificing local optimization and attaining global optimization. More attentions should be paid to the interface between subsystems. For purpose of achieving global optimization of a system, dynamic interactions with other system constituent play more important role than attaining individual optimum. Examples can found in cases like "the Dilemmas of Prisoners" and "Tian Ji Horse Racing".

A successful entrepreneur ever wrote this equation "$100-1 = 0$". The meaning of this equation is that one-time bad service experience could offset the positive image created by 100 times excellent services. In 2000, a well-known real property management company in Shanghai lost the chance in managing a luxury community due to its poor service and ensuing boycotts from property owners, which rendered its built brand loyalty and image worthless. In many well-recognized companies, they deem quality as the life of the company and "either 100 or 0" is used to define the quality acceptance standard, which means 0 tolerance toward defect, failure, mistake, or error. One can imagine the operation of a nuclear power station. A minor traffic violation could cause catastrophic property damage and casualty. A 1% chance of maloperation in medical surgeries will exert 100% impact on the patients involved. Taking another example, in the process of inputting password for a bank account, you can not log in as long as you have one typo.

Based on the above analysis, in order to investigate system optimization, we must first build a series of concepts: (1) critical subsystems, major subsystems, and important subsystems; (2) critical structures, major structures, and important structures; (3) critical elements, major elements, important elements, and subsystem independence. Next, we discuss the theory and method for conducting system optimization. The steps of system optimization are listed as follows: (1) determine if a system has error; (2) error should be eliminated if the system has error. In some instances, for error occurred in a system, one can start addressing errors from the bottom layer of a system and then move up to higher hierarchies according to acting forces of different subsystems. The method for investigating system error is then confirmed and used to eliminate error in the whole system. (3) for a system without errors (or a system tolerating errors), a programming model is built in which: the expected overall intrinsic features are the objectives; each intrinsic feature contained in objective features is defined as an independent variable; and the system's conditions are constraints. By solving the programming model, the values or the value ranges of each intrinsic feature are obtained. (4) feature-based system optimization is then conducted given the values or the value ranges of each intrinsic feature.

Except for the investigation on the relationship between system structure and system features, we also explore the strength, spatial dynamics, direction, and sequence of acting of mutual interacting forces of system structure and how they affect the attainment of system's global optimum. For the sake of gaining better understanding on how interactions of system (or subsystem) structures affect system features, we propose the concepts of system structural acting force ("Shi" or "potential" or "quan"), chained structural acting force, and accumulated acting force.

Thereafter, transformations of system structural acting forces, relationship between system structural acting forces and system features, and relationship between system structural acting forces and optimization of error system are examined. The relevant issues, contents, and examples of system acting forces are provided accordingly.

Systems are categorized into two types when studying the relationship between subsystem optimization and system optimization.

1. Additivity of features

    (1) Complete additivity:

$$GY_j = \sum_{i=1}^{n} GY_{ji}$$

    (2) Partial additivity:

$$GY_j \neq \sum_{i=1}^{n} GY_{ji}$$

2. Non-additivity of features
    $S(GY_j) = S(S_1(GY_{j1}(a_1, b_1)), S_2(GY_{j2}$
    $(a_2, b_2)), \ldots, S_i(GY_{ji}(a_i, b_i)), \ldots, S_n(GY_{jn}(a_n, b_n)))$, where (a_i, b_i) denotes the range for the feature GY_{ji} of ith subsystem S_i (i = 1, 2,..., n).

In system design and operation, we should investigate the states of system and its subsystems based on the system structure when the system's global optimum is reached. Then, one should consider all critical subsystems and important subsystems. And then one should consider to choose the least-costly scenario of allocating resources among all the subsystems s_i when the system global optimum is attained if there are one more scenarios.

A typical error system S has 4 major factors, namely, domain, system, time, and judging rules. We can conduct transformation on each factor or their different combinations thereof. Then we have

    (1) Domain transformation;
    (2) System transformation;
    (3) Temporal transformation;
    (4) Rules transformation;
    (5) Simultaneous transformation on domain and system;
    (6) Simultaneous transformation on domain and time;
    (7) Simultaneous transformation on domain and rules;
    (8) Simultaneous transformation on system and time;
    (9) Simultaneous transformation on system and rules;

(10) Simultaneous transformation on time and rules;
(11) Simultaneous transformation on system, domain, and time;
(12) Simultaneous transformation on system, domain, and rules;
(13) Simultaneous transformation on domain, time, and rules;
(14) Simultaneous transformation on system, time, and rules;
(15) Simultaneous transformation on system, domain, time, and rules.

Additionally, 6 basic transformations can be conducted on each factor.

1. Similarity or equivalence transformation $T_n \subseteq \{T_{xly}, T_{xsw}, T_{xlj}, T_{xtz}, T_{xlz}, T_{xcz}, T_{xhs}, T_{xsj}, T_{xgz}, T_{xzh}\}$ (similarity or equivalence), similarity or equivalence transformation includes similarity in domains, similarity in factors of conditions, similarity in structures of conditions, similarity in rules of conditions, similarity in factors of conclusions, similarity in structures of conclusions, similarity in rules of conclusions, similarity in factors of intrinsic features, similarity in structures of intrinsic features, similarity in rules of intrinsic features, similarity in factors of objective features, similarity in structures of objective features, similarity in rules of objective features, similarity in factors of structures, similarity in structures, and spatial similarity.

2. Displacement transformation $T_n \subseteq \{T_{zly}, T_{zsw}, T_{zlj}, T_{ztz}, T_{zlz}, T_{zcz}, T_{zhs}, T_{zsj}, T_{zgz}, T_{zzh}\}$ (displacement), displacement transformation includes displacement in domains, displacement in factors of conditions, displacement in structures of conditions, displacement in rules of conditions, displacement in factors of conclusions, displacement in structures of conclusions, displacement in rules of conclusions, displacement in factors of intrinsic features, displacement in structures of intrinsic features, displacement in rules of intrinsic features, displacement in factors of objective features, displacement in structures of objective features, displacement in rules of objective features, displacement in factors of structures, displacement in structures, and spatial displacement.

3. Addition transformation $T_n \subseteq \{T_{znly}, T_{znsw}, T_{znlj}, T_{zntz}, T_{znlz}, T_{zncz}, T_{zncz}, T_{znsj}, T_{zngz}, T_{znzh}\}$ (addition), addition transformation includes addition in domains, addition in factors of conditions, addition in structures of conditions, addition in rules of conditions, addition in factors of conclusions, addition in structures of conclusions, addition in rules of conclusions, addition in factors of intrinsic features, addition in structures of intrinsic features, addition in rules of intrinsic features, addition in factors of objective features, addition in structures of objective features, addition in rules of objective features, addition in factors of structures, addition in structures, and spatial addition.

4. Decomposition transformation $T_n \subseteq \{T_{fly}, T_{fsw}, T_{flj}, T_{ftz}, T_{flz}, T_{fcz}, T_{fhs}, T_{fsj}, T_{fgz}, T_{fzh}\}$ (decomposition), decomposition transformation includes decomposition in domains, decomposition in factors of conditions, decomposition in structures of conditions, decomposition in rules of conditions, decomposition in factors of conclusions, decomposition in structures of conclusions, decomposition in rules of conclusions, decomposition in factors of intrinsic features, decomposition in structures of intrinsic features,

decomposition in rules of intrinsic features, decomposition in factors of objective features, decomposition in structures of objective features, decomposition in rules of objective features, decomposition in factors of structures, decomposition in structures, and spatial decomposition.

5. Destruction transformation $T_n \subseteq \{T_{hly}, T_{hsw}, T_{hlj}, T_{htz}, T_{hlz}, T_{hcz}, T_{hhs}, T_{hsj}, T_{hgz}, T_{hzh}\}$ (destruction), destruction transformation includes destruction in domains, destruction in factors of conditions, destruction in structures of conditions, destruction in rules of conditions, destruction in factors of conclusions, destruction in structures of conclusions, destruction in rules of conclusions, destruction in factors of intrinsic features, destruction in structures of intrinsic features, destruction in rules of intrinsic features, destruction in factors of objective features, destruction in structures of objective features, destruction in rules of objective features, destruction in factors of structures, destruction in structures, and spatial destruction.

6. Unit transformation T_d (unit) and its corresponding inverse unit transformation T_d^{-1}.

7. Quantifier for error logic, there are another three combinations with respect to 6 basic transformation:

(1) Conjunction of transformations;
(2) Disjunction of transformations;
(3) Inverse transformation.

In brevity, we employ 15 paths, 6 basic transformations, and 3 combinations to figure out solutions to prevent and eliminate errors. This is the core of our research, i.e., employing "15, 6, 3" methodology system to prevent and eliminate errors.

The book is arranged as follows. Chapter 1 proposes the basic concepts of error system. In Chap. 2, methods for identifying errors are discussed. In order to identify errors, the first thing is to define universe of discourse (research domain). And the next step is to define a group of rules G for judging errors. The conditions, fields, and time for qualifying G must be clearly defined in advance. The relationship among rules and relationship between rules and object of interest are then examined. Having done that, the interacting mechanism and laws are obtained. Step 3 is to find method for identifying error given that tools and processes are well mastered. Examples are provided in the end of this chapter. Five basic structures of system and their brief descriptions are presented in Chap. 3. They are: (a) series structure, (b) parallel structure, (c) feedback structure, (d) expanding and shrinking structure, (e) inclusion structure, and other types. Types (a), (b), and (d) can be represented by $m \times n$ form where m stands for the number of parallel series structures from starting point (observing from left to right) and n stands for the number of parallel series structures at the ending point (observing from left to right). Type (a) is $m = n = 1$; (b) is $m = n \leq 2$; (d) expanding is the case of $m = 1$ and $n \leq 2$ and shrinking is the case of $m \leq 2$ and $n = 1$. Therefore, the basic types of system are (1) $m \times n$ type, (2) feedback structure, (3) inclusion structure, and (4) other types. Chapter 4 provides the basic structures of error systems which are:

(a) series structure, (b) parallel structure, (c) feedback structure, (d) expanding and shrinking structure, (e) inclusion structure, and other types. Chapter 5 talks about the conditions and modes for conducting 6 basic transformations on 5 basic system structures. In Chap. 6, a new concept in system theory is proposed here, i.e., System Acting Force (abbreviated as **AF**, i.e., system's "shi" or potential). Structural AF of system, element AF of system, and the relationship between them are discussed at length. Chapter 7 mainly addresses the transformations on universe of discourse (domain) and rules of error systems. In Chap. 8, concepts and forms of error function[1] are presented. This chapter also explores the relationship between error function and judging rules. Chapter 9 provides the detailed application of error system. This chapter also presents practical methods and principles in avoiding and eliminating errors. Some examples are offered as well.

Guangzhou, China   Kaizhong Guo
Chengdu, China   Shiyong Liu
October 2019

---

[1]Function: it serves two purposes in this book with one denoting mathematical relationship between sets that maps each element in one set to exactly one element in another set and the other meaning is the action for which a person or thing is specially fitted or used or for which a thing exists: purpose (Merriam-Webster). Therefore, in order to avoid confusion, we prefer to use "feature" when second meaning is needed.

Acknowledgements

We are grateful to financial and administrative supports from the Southwestern University of Finance and Economics, Fundamental Research Funds for the Central Universities (JBK1904005), China's National Natural Science Foundation (CNSF), Guangdong University of Technology, Guangdong Vocational College of Technology. We are genuinely indebted to my students and friends Yungang Bian, Qiwei Guo, Haoran Huang, Jize Huang, Hongbing Liu, Zhenghuang Liu, Xilin Min, Jia Shi, Haiou Xiong, Qixin Ye, Xiaoping Zhou, Hongli Zhu, Weiwei Zhang, and Lan Zhang for their dedicated help in idea generating, writing, editing, and revising this manuscript. The authors evenly contributed to the completion of this manuscript.

Guangzhou, China   Kaizhong Guo
Chengdu, China   Shiyong Liu
October 2019

Contents

Chapter 1
Brief Introduction of Error Systems

Abstract In this chapter, we put error events and issues in the context of a system and treat $X = X(\{W_i\}, T(t_1, t_2), J, GY, MG, R)$ as an object system, where $\{W_i\}$ is the set composed of all the issues associated with research objects; T is the set of conditions with t_1 stands for constrained conditions and the t_2 represents conditions borne by the issue; J is the set for all conclusions of research objects; GY is the set of intrinsic features of the system formed by all research objects; MG is the set of objective features of research objects; R is the set containing all relationships connecting all research objects. In the object system, concept and classification of error, operation and transformations of error system are investigated accordingly. Within the context of object system, we examine the critical subsystems, major subsystems, important subsystems, critical structures, major structures, important structures, critical elements, major elements, important elements, and independence of subsystems. Next, we discuss the theory and method for conducting system optimization which should be carried on a non-erroneous system. For a system without errors (or a system tolerating errors), a programming model is built where the expected overall intrinsic features are the objectives; each intrinsic feature contained in objective features is defined as an independent variable; and the system's conditions are constraints. By solving the programming model, the value or the range of values of each intrinsic feature is obtained. Feature-based system optimization is then conducted given the value or the range of values of each intrinsic feature.

1.1 Concepts of Error System

1.1.1 Definition of Error System

In history, when commemorating famous scientists and inventors, people seldom mentioned the failures and errors they experienced on their way to their success but their achievements, wisdom, and proven correct thinking process and methods. People have neither systematic traces nor holistic analysis, thinking, and studying on failures and errors made by those scientists and inventors. Error actually played very

© Springer Nature Switzerland AG 2020

K. Guo and S. Liu, *Error Systems: Concepts, Theory and Applications*,
Studies in Systems, Decision and Control 275,
https://doi.org/10.1007/978-3-030-40760-5_1

important roles in the cognitive development of human being. Error and correctness always exist in the same context. Error sometimes became the road sign or guidance for the followers, which drove people to initiate inverse thinking and find the right way to success. For example, James Prescott Joule, with many failures in his experiments, found the Law of Conservation of Energy. In the process of finding the appropriate materials for light bulb, Thomas Edison failed thousands of times before he found the ideal one.

People like correctness and hope to act correctly. However, in order to obtain correct results or act correctly, one must learn to prevent and eliminate errors, which demands researchers to examine the causes and laws in causing errors. Most importantly, a system, a proposition, or a decision free of errors might be transformed to be the one having errors with the change in time and natural environment as well as the development of science and technology. Therefore, error elimination is the continual theme in the evolution of human society. This chapter presents concept and properties of error system and preliminary exploration on causes and laws in generating errors.

Definition 1.1 Suppose that U is the universe of discourse, $a \in U$, G is a set of rules for judging error defined within U, if $\exists\, G \nRightarrow a$ (including the cases that a can not be completely or partially obtained by exercising G; or the case that a has nothing to do with G.), a is erroneous defined within U under the rule of judging errors G.

Definition 1.2 The set composed of decisions, arguments, propositions, and things is called an issue set noted by $W_0, W_1, W_2, \ldots \ldots W_n$. For example, both $W_1 = \{$the sum of the three angles of triangle is equal to $180°$; conducting field experiment in the afternoon on 25th February 1985; sun rises in the east$\}$ and $W_2 = \{$the technologies in the future will be more advanced than current ones$\}$ are issue sets.

Definition 1.3 For certain issue set, the total impacts on society exerted by all the conclusions obtained under all conditions are called the intrinsic features of this issue set. While certain impact on society of this issue set is called an intrinsic feature.

Definition 1.4 The features necessary for realizing objectives of a system are called objective features. In the process of investigating issue set, it is very difficult to unfold all the intrinsic features. For example, due to limitations in the development of science and technology, some natural laws have not been discovered and explored at all. There also exist some natural phenomena for which sufficient and necessary explanations have not been provided to prove the faultlessness. Moreover, when studying certain issue set, it is not necessary to list or find out all intrinsic features because it is enough to clarify whether the objective features for the issue of interest are equal to or included in intrinsic features or not. For example, $W_9 = \{$calculator is a good product$\}$, the provision of certain computing capability is the intrinsic feature of W_9. In another example, $W_{11} = \{$food$\}$, the intrinsic feature of W_{11} is that it can be eaten and digested and safely provide human with necessary energy.

Definition 1.5 The system composed of the condition T, conclusion J, intrinsic feature GY, objective feature MG, and the relationship (structure) R of an issue set is called an object system (simplified as system in this chapter), noted by ($\{W_i\}$, $T(t_1, t_2)$, J, GY, MG, R). Where, $\{W_i\}$ is certain issue set; $T(t_1, t_2)$ is the condition set of $\{W_i\}$; t_1 stands for constrained conditions and t_2 represents conditions borne by the issue; J is the conclusion set of $\{W_i\}$; GY is the intrinsic feature set of $\{W_i\}$; MG is the objective feature set of $\{W_i\}$, R is the set for relationships of interests in $\{W_i\}$. For the sake of simplicity, they can be noted by: $X(T(t_1, t_2), J, GY, MG, R)$, or $X(T, J, GY, MG, R)$ or x, y, ...

Definition 1.6 In an object system, it is called an error system as long as at least one of its constituent factors has error. For example, suppose that $W_{10} = \{$if paper is red then it can be used for writing couplets, building spacecraft, and constructing foundations for piers of a bridge$\}$, then ($\{W_{10}\}$, T, J, GY, MG, R) is an error system.

1.1.2 Relationships and Operations Between Object Systems

Definition 1.7 Suppose that $X_1 = (\{W_i\}_1, T_1, J_1, GY_1, MG_1, R_1)$, $X_2 = (\{W_i\}_2, T_2, J_2, GY_2, MG_2, R_2)$, where $\{W_i\}_1 = \{W_i\}_2$, $T_1 = T_2$, $J_1 = J_2$, $GY_1 = GY_2$, $MG_1 = MG_2$, and $R_1 = R_2$ hold, then X_1 is said to be equal to X_2 noted by $X_1 = X_2$.

Proposition 1.1 *Suppose that X_1, X_2, and X_3 are object systems, (a) if $X_1 = X_2$, then $X_2 = X_1$; (b) if $X_1 = X_2$, $X_2 = X_3$, then $X_1 = X_3$.*

Proof Proof is omitted here.

Definition 1.8 Suppose that X_1 and X_2 are error systems, if $\{W_i\}_1 \Rightarrow \{W_i\}_2$, $T_1 \Rightarrow T_2$, $J_1 \Rightarrow J_2$, $GY_1 \Rightarrow GY_2$, $MG_1 \Rightarrow MG_2$, and $R_1 \Rightarrow R_2$ hold, then X_1 is said to contain X_2 noted by $X_1 \supset X_2$, or $X_2 \subset X_1$, where $\Rightarrow$ represents the right side can be deduced from left side.

Proposition 1.2 *Suppose that X_1, X_2, and X_3 are object systems, if there exist $X_1 \supset X_2$ and $X_2 \supset X_3$, then $X_1 \supset X_3$ holds.*

Proof Proof is omitted here.

Definition 1.9 Suppose in object systems X_1 and X_2, if $\{W_i\}_1 \Rightarrow \{W_i\}_2$, then the issues in X_1 contain the issues in X_2 noted by $X_1 \overset{\cdot}{\supset} X_2$.

Proposition 1.3 *Suppose that X_1 and X_2 are object systems, (a) if $X_1 \supset X_2$, then $X_1 \overset{\cdot}{\supset} X_2$ holds; (b) if $X_1 \overset{\cdot}{\supset} X_2$, $X_1 \supset X_2$ may not hold.*

Proof (a) If $X_1 \supset X_2$, then $\{W_i\}_1 \Rightarrow \{W_i\}_2$ holds (Definition 1.8), $\therefore X_1 \overset{\cdot}{\supset} X_2$.
(b) If $X_1 \overset{\cdot}{\supset} X_2$, at least $MG_1 \Rightarrow MG_2$ may not hold because even though there exists the relationship of $\{W_i\}_1 \Rightarrow \{W_i\}_2$. The reason is that the objectives of the two issue sets could have the relationship of equivalence, non-equivalence, or inclusion.
Proof is completed.

Proposition 1.4 *Suppose that X_1 and X_2 are object systems, if both $X_1 \dot{\supset} X_2$ and $R_1 \Rightarrow R_2$ hold, then the relationships $T_1 \Rightarrow T_2$, $J_1 \Rightarrow J_2$, and $GY_1 \Rightarrow GY_2$ hold.*

Proof $\because X_1 \dot{\supset} X_2$
$\therefore \{W_i\}_1 \Rightarrow \{W_i\}_2$
$\because R_1 \Rightarrow R_2$, the issue set and relationships set for both X_1 and X_2 have inclusion relationship, then all elements but objective features in these two error systems have inclusion relationships. On the contrary, suppose that $t \in T_2$ and $t \notin T_1$:
$\because T_1$ is the condition set of $\{W_i\}_1$ and T_2 is the condition set of $\{W_i\}_2$, this contradicts the fact $\{W_i\}_1 \Rightarrow \{W_i\}_2$.
$\therefore T_1 \Rightarrow T_2$.
Similarly, $J_1 \Rightarrow J_2$ and $GY_1 \Rightarrow GY_2$ can be successfully proven.
  Proof is completed.

Definition 1.10 Suppose that X_1 and X_2 are two object systems, if both $X_1 \supset X_2$ and $X_2 \supset X_1$ hold, then the X_1 is said to be equivalent to X_2 noted by $X_1 \Leftrightarrow X_2$.

Proposition 1.5 *Suppose that X_1, X_2, and X_3 are object systems, (a)if $X_1 \Leftrightarrow X_2$, then $X_2 \Leftrightarrow X_1$ holds; (b) if both $X_1 \Leftrightarrow X_2$ and $X_2 \Leftrightarrow X_3$ hold, then $X_1 \Leftrightarrow X_3$ holds.*

Proof Proof is omitted.

Definition 1.11 Suppose that x and y are two arbitrary object systems, if $x \supset y$, then y is called a subsystem of x.

Definition 1.12 Suppose that x and y are two arbitrary object systems, if there does not exist equivalent non-ordinary subsystems in both x and y, then x is said to be independent of y noted by $x \, d \, y$.

Proposition 1.6 *Suppose that x and y are two object systems, then $x \, d \, y = y \, d \, x$ holds.*

Proof Proof is omitted.

Definition 1.13 Suppose that A_1 and A_2 are two arbitrary object sets, if A_3 is the union of all equivalent subsets in both A_1 and A_2, A_3 is called the maximum equivalent subsets of both A_1 and A_2 noted by $A_3 = Z\ (A_1, A_2)$.

Definition 1.14 Suppose that X_1 and X_2 are two object systems, if $\{W_i\}_3 = Z\ (\{W_i\}_1, \{W_i\}_2)$, $T_3 = Z\ (T_1, T_2)$, $J_3 = Z\ (J_1, J_2)$, $GY_3 = Z\ (GY_1, GY_2)$, $MG_3 = Z\ (MG_1, MG_2)$, $R_3 = Z\ (R_1, R_2)$, then $X_3 = (\{W_i\}_3, T_3, J_3, GY_3, MG_3, R_3)$ is called the intersection of X_1 and X_2 noted by $X_3 = X_1 \cap X_2$.

Proposition 1.7 *Suppose that X_1, X_2, and X_3 are three object systems, (1) if $X_1 \supset X_2$, then $X_1 \cap X_2 = X_2$; (2)$X_1 \cap X_1 = X_1$; (3) $X_1 \cap X_2 = X_2 \cap X_1$; (4)$(X_1 \cap X_2) \cap X_3 = X_1 \cap (X_2 \cap X_3)$.*

Proof Proof is omitted.

Definition 1.15 Suppose that X_1 and X_2 are two object systems, if $\{W_i\}_3 = Z$
$(\{W_i\}_1, \{W_i\}_2) \cup (\{W_i\}_1 - Z (\{W_i\}_1, \{W_i\}_2)) \cup (\{W_i\}_2 - Z (\{W_i\}_1, \{W_i\}_2)); T_3 =$
$Z (T_1, T_2) \cup (T_1 - Z (T_1, T_2)) \cup (T_2 - Z (T_1, T_2)); J_3 = Z (J_1, J_2) \cup (J_1 - Z (J_1,$
$J_2)) \cup (J_2 - Z (J_1, J_2)); GY_3 = Z (GY_1, GY_2) \cup (GY_1 - Z (GY_1, GY_2)) \cup (GY_2$
$- Z (GY_1, GY_2)); MG_3 = Z (MG_1, MG_2) \cup (MG_1 - Z (MG_1, MG_2)) \cup (MG_2$
$- Z (MG_1, MG_2)); R_3 = Z (R_1, R_2) \cup (R_1 - Z (R_1, R_2)) \cup (R_2 - Z (R_1, R_2)).$
Then $X_3 = (\{W_i\}_3, T_3, J_3, GY_3, MG_3, R_3)$ is called the union of X_1 and X_2 noted
by $X_3 = X_1 \cup X_2$.

Proposition 1.8 *Suppose that X_1, X_2, and X_3 are three object systems, (a) $X_1 \cup X_1$
$= X_1$; (b) $X_1 \cup X_2 = X_2 \cup X_1$; (c) if $X_1 \supset X_2$, then $X_1 \cup X_2 = X_1$; (d) $(X_1 \cup X_2)$
$\cup X_3 = X_1 \cup (X_2 \cup X_3)$.*

Proof Proof is omitted.

Proposition 1.9 *Suppose that X_1, X_2, and X_3 are three object systems, (a) $X_1 \cup$
$(X_2 \cap X_3) = (X_1 \cup X_2) \cap (X_1 \cup X_3)$; (b) $X_1 \cap (X_2 \cup X_3) = (X_1 \cap X_2) \cup (X_1 \cap$
$X_3)$.*

Proof Proof is omitted.

Proposition 1.10 *Suppose that X is an arbitrary object system, there exists a mechanism that can represent X by the union of n derived independent subsystems by
X-i.e., $X = X_1 \cup X_2 \cup \ldots \ldots \cup X_n$, where X_i is independent of X_j ($i, j = 1, 2,$
$\ldots, n, i \neq j$).*

Proof Proof is omitted.

Definition 1.16 Suppose that X_1 and X_2 are two object systems, if $\{W_i\}_3 = Z$
$(\{W_i\}_1, \{W_i\}_2) \cup (\{W_i\}_1 - Z (\{W_i\}_1, \{W_i\}_2)) \cup (\{W_i\}_2 - Z (\{W_i\}_1, \{W_i\}_2)); MG_3$
$= Z (MG_1, MG_2) \cup (MG_1 - Z (MG_1, MG_2)) \cup (MG_2 - Z (MG_1, MG_2)); R_3$
$= Z (R_1, R_2) \cup (R_1 - Z (R_1, R_2)) \cup (R_2 - Z (R_1, R_2))$. Then $X_3 = (\{W_i\}_3, T_3, J_3,$
$GY_3, MG_3, R_3)$ is called the approx-union of X_1 and X_2 noted by $X_3 = X_1 h X_2$,
where T_3, J_3, GY_3 are derived from $\{W_i\}_3$.

Proposition 1.11 *Suppose that X_1, X_2, and X_3 are three object systems, (a) $X_1 h$
$X_1 = X_1$; (b) $X_1 h X_2 = X_2 h X_1$; (c) $(X_1 h X_2) h X_3 = X_1 h (X_2 h X_3)$.*

Proof Proof is omitted.

Definition 1.17 Suppose that X_1 and X_2 are two object systems, if $\{W_i\}_3 = Z$
$(\{W_i\}_1, \{W_i\}_2)$; $MG_3 = Z (MG_1, MG_2); R_3 = Z (R_1, R_2)$. Then $X_3 = (\{W_i\}_3, T_3,$
$J_3, GY_3, MG_3, R_3)$ is called the approx-intersection of X_1 and X_2 noted by $X_3 =$
$X_1 t X_2$, where $T_3, J_3,$ and GY_3 are derived from $\{W_i\}_3$.

Proposition 1.12 *Suppose that X_1, X_2, and X_3 are three object systems, (a) $X_1 t$
$X_1 = X_1$; (b)$X_1 t X_2 = X_2 t X_1$; (c) $(X_1 t X_2) t X_3 = X_1 t (X_2 t X_3)$.*

Proof Proof is omitted.

Proposition 1.13 *Suppose that X_1, X_2, and X_3 are three object systems, (a) $X_1 h$
$(X_2 t X_3) = (X_1 h X_2) t (X_1 h X_3)$; (b) $X_1 t (X_2 h X_3) = (X_1 t X_2) h (X_1 t X_3)$.*

Proof Proof is omitted.

1.1.3   Categorization of Error Systems

Definition 1.18 An object system is called a complete error system if all the elements and factors that compose this system are wrong. For example: if $W_3 = \{$if sun rises in the west, then it sets in the south$\}$, then $T = \{$sun rises in the west$\}$, $J = \{$sun sets in the south$\}$, $GY = \{\ldots \ldots\}$, $MG = \{$studying plane geometry$\}$, $R = \{$axioms for studying plane geometry, theorems, rules, $\ldots \ldots\}$, therefore, $X = (\{W_3\}\, T, J, GY, MG, R)$ becomes a complete error system (under normal conditions).

Definition 1.19 An error system is called a pseudo error system if the elements and factors attributing to errors in this system are not necessary for achieving the objective of this system. For example: if $W_{10} = \{$the sum of all angels in a triangle is $180°$, all plants grow downward$\}$, $T = \{\angle A, \angle B, \angle C$ are three angles of triangle $\triangle ABC$ and $\angle A + \angle B + \angle C = 180^0; P_1, P_2, P_3, \ldots, P_n \ldots$ grow downward.$\}$, $GY = \{\ldots\}$, $MG = \{$studying plane geometry$\}$, $R = \{$axioms for studying plane geometry, theorems, rules, $\ldots\}$, therefore, $X = (\{W_{10}\}\, T, J, GY, MG, R)$ is a pseudo error system.

Definition 1.20 Suppose that X is an error system, under the conditions of guaranteeing the realization of objective features, X can be transformed into a system free of errors through reasonable and achievable transformations. Then X is called an extendable error system.

Definition 1.21 Suppose that X is an error system, under a group of operable conditions, X is called a transformable error system if all the errors in X can be eliminated while keeping the objective features unchanged. Both extendable error system and pseudo error system are transformable error systems.

Definition 1.22 Suppose that X is an object system, it is called an error system with risk (or risky error system) if at least one of the elements or factors that compose this system is stochastic. For example: $W_{20} = \{$a house will be built if we have bumper crop year$\}$, $Y = (\{W_{20}\}, T, J, GY, MG, R)$ is a risky error system.

Definition 1.23 Suppose that $X = (\{\ \}\, \Phi, \Phi, \Phi, \Phi, \Phi)$, Φ is empty set, X is called an ordinary error system.

Definition 1.24 In an object system, if and only if certain conditions are free of errors and these conditions are not necessary for achieving the objective of this system; other elements or factors of this system are free of error, then this system is called a waste error system.

According to the properties of errors, error systems can be divided into the following types (Fig. 1.1).

In general, different error systems can be obtained if other type of categorizations are adopted.

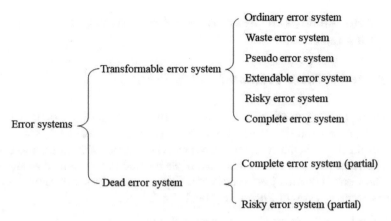

Fig. 1.1 Categorization of error systems

1.1.4 Relation Between Error System and Object System

In Definition 1.1, it is said that an object system is called an error system as long as there exits error in at least one element or factor. In other words, suppose that X is an object system, U is the universe of discourse, G is a set of rules for judging error defined within U, if $\exists u \in U$ in error function $f(x) = f(G \nRightarrow u)$ makes $f(G \nRightarrow u)$ > 0 hold, then the object system X is called an error system. The error system defined here is called a special error system-i.e., the special error system belongs to object system.

In practice, when investigating the errors in system S, the following steps are employed: (1)the S is generally converted to be a corresponding object system X; (2) finding out the universe of discourse U; (3) establishing a set of rules for judging errors under U; (4) constructing error function $f(x) = f(G \nRightarrow u)$ to capture error X; (5) calculating the value of $f(x) = f(G \nRightarrow u)$ in U. And $f(x) = f(G \nRightarrow u)$ is called the error function defined within U no matter what value the $f(x)$ might take. If $\exists u \in U$, $f(G \nRightarrow u) > 0$ holds, then X is called an error system. If $\exists u \in U$, $f(G \nRightarrow u) = 0$ holds, then X is called a critical error system. And if $\exists u \in U$, $f(G \nRightarrow u) \leq 0$ holds, then X is called a system free of errors. That is to say, when exploring errors in a system, we should use corresponding error system as research object. Sometimes, it is also necessary to study the methods and laws for transforming a system free of error X into an error system (when studying the prevention of errors). Therefore, we treat the object system as a general error system. Without particular notification, both special and general error systems are called error systems. Moreover, if the subject set in object system is a system, the object system is a common system.

1.2 Relationships Between System, Subsystems, and Elements

1.2.1 100% Error Produced by 1% Error

The concept of "100% failure was caused by 1% error" indicates that $100-1 = 99$ does not hold but $100-1 = 0$. Many examples exist such as broken windows theory, domino effect, and butterfly effect (chaos theory) in which 100% failure was caused by 1% error. In the first part of this section, we provide some practical examples to show how error occurred just because of a failure or mistake in a system's element or subsystem. A series of stories are provided as follows.

1. "Great failure" of many private companies in China

In many organizations, no employee cared about trivial things as the company was getting too large. Even though trivial things were taken care of, they were not handled with discretion because their were seemingly insignificant. The failure of a business originated from the accumulation of many ill-treated trivial things.

2. Account 88888 sank Barings Bank

In February 1995, due to financial decision error, Nick Leeson caused £ 860 million of loss for Barings Bank which consequently led to the collapse of the over two-century-old bank. The cruel fact tells us that even minor error or failure in administrative process will incur catastrophic disaster to a company.

3. Return of exported frozen shrimp

In 2002, a batch of frozen shrimp exported to Europe was returned because local inspection and quarantine authority found out 0.2 g chloramphenicol sampled from the 1000 metric ton shrimp. This event was caused by an incident in which some employees applied the disinfectant containing chloramphenicol to their wounded hands and contaminated the goods while they were handling the shrimp package. As indicated by Murphy's law, "anything that can go wrong will go wrong no matter how small the initial error."

4. T28 cellphone failed Ericsson in Chinese market

In 2001, due to the ignorance of its product quality and service for T28 model cellphone, Ericsson was boycotted by many medias and users in Chinese Market. Like a running car on the highway, 1% diversion, if maladjusted, can cause severe accident.

5. Argentina vs United Kingdom: Falklands

In Falklands war between Argentina and the United Kingdom in 1982, the ARA San Luis from Argentina had at least three times to breach the anti-submarine patrolling areas and tried to launch six SST-4 wired torpedoes. Unfortunately, none of them struck valid target and missed the excellent striking chances. In the post-war investigation done by two engineers coming from Germany and Netherlands, it was found that the maintenance engineer reversely connected electrodes in the submarine torpedo tube when he conducted maintenance on the fire control system of submarine, which caused the launched torpedoes to loss heading reference. Otherwise, the UK's naval fleet could have suffered heavy losses and Falklands war could be a different result.

6. $100 - 1 = 0$

A successful entrepreneur ever wrote the above equation. The meaning of this equation is that one-time bad service experience could offset the positive image created by 100 times excellent service. In 2000, a well-known real property management company in Shanghai lost its business in managing a luxury community due to its poor service and ensuing boycotts from property owners, which rendered its built brand loyalty and image worthless. In many well-recognized company, they deem quality as life of the company and "either 100 or 0" is used to define the quality acceptance standard, which means 0 tolerance toward defect, failure, mistake, or error. One may still have vivid memory of the nuclear accident of No.4 nuclear reactor in the Chernobyl Nuclear Power Plant in 1986 and the Fukushima Daiichi nuclear disaster in Japan in 2011. In the Chernobyl disaster, during the simulation test for developing a safety procedure to maintain cooling water circulation during power outage, there were a series errors, namely (1) the previous three times tests failed to find solution to take care of the one-minute operating gap between power outage and power generation from backup generator; (2) the fourth test was delayed about 10 hours and operating shift was not present; (3) the test supervisor failed to follow procedure and created unstable operating conditions; (4)RBMK(*Reaktor Bolshoy Moshchnosti Kanalnyy*) reactor design flaws; (5) intentional disabling of several nuclear reactor safety systems(Eden,1999). A minor traffic violation could cause catastrophic property damage and casualties. A 1% chance of maloperation in a transportation system will exert 100% impact on the victim. Taking another example, in the process of inputting password for a bank account, you can not log in as long as you have one typo.

7. A solid levee collapsed due to termite dens

There was a very old Chinese story telling about the origin of the above proverb. In a village adjacent to Yellow river, villagers built sturdy long levee to prevent flooding. One day, one villager found that termite dens had increased significantly recently. The villager had doubt if the increased termites can affect the safety of the levee. He met his son when he went back to the village to report this. He son said the sturdy levee definitely could be not be broken by termites. Then, they together went to their land to work. After a few days, at a heavy-raining night, the water in Yellow river rose rapidly and started to leak through the small holes produced by termite. Before long, the raging flood gushed out of that hole and washed the sturdy levee away. The flood inundated their village and crops. For many years, I thought this was just a proverb for helping people from ignoring trivial things. Having read so many records and stories in this respect, I started to believe they were true. Since 1970s, in Qingyuan, Guangdong province, 13 levees and nine dams collapsed where nine levees and five dams were caused by termites. In July 1986, in Meizhou city,

Guangdong province, the flood breached 62 points on the Mei river levee in which 55 were caused by termites; on September, 1981, in Yangjiang, Guangdong province, 18 breached holes appeared at the Moyang embankment section where six were caused by termites. In the summer of 2003, Yangtze river suffered heavy flooding which caused piping at Nanping section of Jingjiang levee, Gongan county, Hubei Province. With extensive investigation by experts in hydraulic engineering, it was confirmed that the piping was caused by termites. Measures were quickly taken to fix the problem and the dangers were ultimately eliminated.

Termites are gregarious species living in gigantic underground nest built by worker termites. Generally, there are more than millions of termites living in a community which has very strict social division of labors. Termite king and queen are on the top of the hierarchy in the community, which are responsible for producing offspring for the community; termite workers are taking charge of building nest and collecting food; while termite soldiers are in charge of supervising the job of termite workers and safeguarding the nest. The place where termite king and queen live is the main nest having the size of one to several square meters. Many tunnels radiate from the main nest to other nests and sometimes perforate the interior and exterior sides of the levees, which have a diameter from six to 12 cm.

1.2.2 Chaos

Butterfly effect, coined by Edward Lorenz, is obtained from a metaphor that the formation of a tornado was influenced by the minor perturbation of the air flow such as the flapping of a butterfly in a very far distance. Dr. Lorenz used this term to demonstrate an important feature of the chaotic system where the behavior of the system is very sensitive to the initial conditions. Here there is a very famous quotation from Benjamin Franklin *"For the want of a nail the shoe was lost, for the want of a shoe the horse was lost, for the want of a horse the rider was lost, for the want of a rider the battle was lost, for the want of a battle the kingdom was lost, and all for the want of a horseshoe-nail."* The lost nail is a trivial thing at the beginning and it causes the collapse of a kingdom through cascading actions starting from initial conditions.

1.2.3 Non-linearity

Linearity dictates that two variables have proportional relationship demonstrated by a straight line. The linearity represents that a physical part is moving in regular and smooth manner over time in a certain space. However, nonlinearity stands for the phenomenon that the relationship of two variables is not proportional and can not be represented by a straight line. In reality, it represents the irregular and abrupt movement over time in certain space. For example, when asking how many times two eyes' visual acuity is that of one eye, the answer is 6–10 times instead of twice, which exists nonlinarity. That is to say $1 + 1 \neq 2$. The generation of laser is a nonlinear process. When voltage is relatively small, the laser generator just produces normal scattering light. However, when the voltage exceeds the threshold value, a powerful coherent light is formed, where temporal coherence, spectral coherence, spatial coherence, and polarization coherence must be met.

1.2.4 Almost-There Phenomenon

Two individual stockholders were talking about the experience of investing in the stock market. One person said "I got stuck in the stocks in which I invested". The other person said "my stocks were almost sold at the top and I could have escaped successfully." Although they were using different descriptions, both of them actually were stuck in their invested stocks. It is not uncommon for people talking about almost-there phenomenon such as some writer almost got Nobel Prize, professor A almost cracked the "Goldbach's conjecture", the national soccer team almost won the match, certain technology almost surpassed other similar technologies in the world. Almost-there is not there and "almost succeeded" was failure. It is a pity for having reached so close to the target but it is still not the target. The reality does not lie and never cares about how people might feel. Although "almost there" can be attributed to bad luck sometimes, the fact is that the gap was not filled or some error happened and was not fixed, which could be subjective or objective reasons.

In the work manual of McDonald, uneven round bread and rough cutout can not be used; milkshake must be kept under $4°C$ at delivery and the supplies will be returned if actual storage temperature at delivery is higher than $5°C$; lettuce must be discarded two hours after it was taken out of refrigerator; the beef patty must go through more than 40 quality checks. The standardized food preparation is conducted and monitored by computerized control. The ready-to-serve food and time label are placed on the insulated food containers. French fries exceeding seven minutes and hamburger exceeding 10 minutes will be discarded. There are 560 pages in the work manual and even the instructions on roasting beef patties account for 20 pages. The employer and the employees are well aware of the proverb: "1% error leads to 100% failure"; "*A miss is as good as a mile*"; "*Heaven and hell both reside in the details.*" The genuine greatness lies in the accumulation of details. Most of time, we do not lack the wisdom and diligence to succeed but the spirit of making details perfect. Finishing each simple thing or job in a spirit of craftsmanship is not simple!

1.2.5 Concept of Error Systems

This section gives a brief introduction regarding the concepts and characteristics of error system as well as the causes and laws for generating error in a system.

Definition 1.25 In a typical system S, the total effects that a system can exerts on the society are called the intrinsic features of this system noted by GY; certain effect that a system can exert on the society is called an intrinsic feature noted by GY_i.

Definition 1.26 In a typical system S, for purpose of achieving system objectives, the features that a system must have are called objective features MD.

Definition 1.27 In a typical system S, it is called an error system as long as there is at least one error in the elements or subsystems that construct the system.

Figure 1.2 provides graph that depicts the structure for a general system.

1.2.6 Relationship Between Systems, Subsystems, and Elements

1. Relationship between system and subsystems in general system

   (1) In general, features of system $\neq \sum$ features of all subsystems

   (2) Due to the external forces on system such as physical, chemical or other actions, creation or destruction transformation might happen in some of the

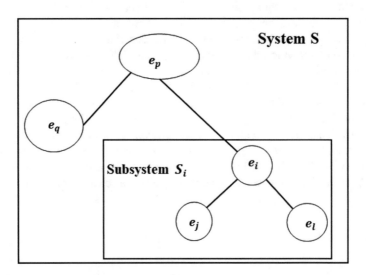

Fig. 1.2 Diagram depicting general system structure

system's features even though some features of the subsystems of the system might still function well.

(3) The relationship between the system optimization and subsystem optimization.

In order to realize the global optimization of a system, the constituent subsystems must provide necessary and sufficient support.

Definition 1.28 Suppose that the feature GY_j of system S belongs to the same type as the feature of subsystem GY_{ji}, $i = 1, 2, \ldots, \ldots, n$, then the feature GY_j of system S has the property of additivity. Suppose that features of system $S = \sum$ features of subsystems S_i, then the feature GY_j of system S has the property of complete additivity with respect to the features of subsystem GY_{ji}, $i = 1, 2, \ldots, \ldots, n$.

For example, it is assumed that the whole nation of China is system S and the constituent provinces, cities, and autonomous regions are subsystems S_i, as one of the features (GY_j)-GDP has the additivity. In another example, suppose that a complete exam sheet is treated as system S, all questions in the exam sheet are subsystem S_i, for the feature GY_j of final grade a student can get from the exam, it has the property of complete additivity with respect to the features (score from each question) of subsystem GY_{ji}, $i = 1, 2, \ldots, \ldots, n$.

Theorem 1.1 *Suppose that the features GY_j of system S have the property of complete additivity with respect to the corresponding features of subsystem GY_{ji}, if the the feature GY_j has the optimal value, then all corresponding features GY_{ji}, $i = 1, 2, \ldots, \ldots, n$ of system can also attain the optimal value.*

(4) Relationship between system optimization and subsystems optimization

Under the circumstance of optimal subsystems, due to the situation of suboptimization, the features, properties, or the range of features can not provide necessary and sufficient conditions that enable the achievement of optimal state of the system S. In other words, the features for subsystems required by optimal system are not consistent with the actual features of constituent subsystems.

(a) Investigation on the values of the features of subsystems and their range.

(b) In the investigation of subsystems, for the purpose of realizing objective features of system, it is necessary to identify: which subsystems are important (without such subsystems, the objective features of system can only be achieved within the range of $[0, a]$, the value of a is determined by actual requirements; they also include the smallest important subsystems, important elements, and important structures); which subsystems are critical(without this subsystem, the objective features of system by no means can be achieved; they also contain smallest critical subsystems, critical subsystems, critical subsystems; critical subsystem is a special case of important subsystem); which are major subsystems(the major subsystems have the largest contribution to the realization of objective features of the system; major elements, major structures). Having identified the important, critical, and major subsystems, further research will be conducted on them to pinpoint laws associated with them. In the same system, for different objective features, there exist different important, critical, and major subsystems.

The study on the pivotal connections between subsystems can help decision makers gain better understanding on why either "$1\% = 100\%$" or "$1\% \neq 100\%$" appears. The research can provide theoretic basis for establishing practical and operable methods to prevent and eliminate errors.

(c) It is also imperative to examine the changes of different elements i.e., dynamics (non-critical subsystems are converted to critical subsystems, e.g., "butterfly effect" and "nesting effect".) and the laws explaining the changes in system objective features.

(5) The symbiosis of system and subsystems

A system is composed of interacting subsystems which provide the foundation and architecture for the system to realize its objective features. A system must also provide the context and necessary material support for its corresponding subsystems. For example, China consists of provinces, special administrative regions, autonomous regions, and direct-controlled municipalities which provide the necessary materials such as food, and industrial materials for the whole country to function well. The nation provides necessary national security, social security, fair competition environment, and law, etc. for the constituent subsystems.

1.2.7 Important Concepts of Error Subsystems

Definition 1.29 Suppose that subsystem S_i in a system S is removed, the intrinsic features GY_j of S can only be partially realized, i.e., $[0, a\%]$, where $0 \leq a \leq 100$, then subsystem S_i has the contribution of $100 - a$ to the intrinsic features GY_j of the whole system. Especially, when $a = 0$, subsystem S_i is the critical subsystem to the intrinsic features GY_j of the whole system. And when $a = 100$, subsystem S_i is an unnecessary (or redundant)subsystem to the intrinsic features GY_j of the whole system.

For instance, for a diesel engine system S, it is composed of fueling subsystem S_1, cooling subsystem S_2, transmission subsystem S_3, and $\ldots, \ldots$, and engine body subsystem S_n. The fueling subsystem is a critical subsystem with respect to the power function GY_j of the engine system. The power generation feature of the engine system can not function if the fueling subsystem is removed from the system. While the cooling subsystem is an important subsystem with respect to the power function GY_j of the engine system. The engine system can work for a period until overheating makes the system stop if the cooling subsystem is removed.

Definition 1.30 Suppose that element y_i in a system S is removed, the intrinsic features GY_j of S can only be partially realized i.e., $[0, a\%]$, where $0 \le a \le 100$, system element y_i has the contribution of $100 - a$ to the intrinsic features GY_j of the whole system S. Especially, when $a = 0$, system element y_i is the critical element to the intrinsic features GY_j of the whole system S. And when $a = 100$, y_i is a surplus(redundant) element to the intrinsic features GY_j of the whole system S.

For example, for a human system S, it consists of metabolism subsystem S_1, blood circulation subsystem S_2, nervous subsystem $S_3, \ldots\ldots$, bone subsystem S_n, etc. For blood circulation subsystem S_2, $S_2 = \{$heart S_{21}, cerebral arteries S_{22}, heart arteries $S_{23}, \ldots\ldots$, arm arteries S_{2i}, cerebral veins S_{2i+1}, heart veins $S_{2i+2}, \ldots\ldots$, arm veins $S_{2i+k} \ldots\ldots \}$. The heart S_{21} is a critical system element that facilitates blood circulation GY_j. The human body system will loss the function of blood circulation if heart S_{21} is removed from the system. While arm artery S_{2i} is an important system element for blood circulation GY_j because GY_j can only be partially achieved $[0, a\%]$ $(0 < a < 100)$ if S_{2i} is removed from the human body system.

Definition 1.31 Suppose that structure r_i in a system S is removed, the intrinsic features GY_j of S can only be partially realized i.e., $[0, a\%]$, where $0 \le a \le 100$, system structures r_i have the contribution of $100 - a$ to the intrinsic features GY_j of the whole system. Especially, when $a = 0$, system element r_i is the critical structure to the intrinsic features GY_j of the whole system. And when $a = 100$, y_i is a surplus(redundant) system structure to the intrinsic features GY_j of the whole system.

System structure indicates the relationship, connections, and interactions among subsystems and elements and it also dictates the interacting forms/sequences or the arrangement/combination of subsystems and elements over time and space.

For a flashlight system S, it is composed of forefront battery subsystem S_1, rear battery subsystem S_2, bulb subsystem S_3, housing subsystem $S_4, \ldots\ldots$, control subsystem S_n. The system structure is illustrated in Fig. 1.3. Suppose that $R = \{$the negative electrode of forefront battery subsystem S_1 is connected to the positive electrode of rear battery subsystem S_2 represented by $r_1 \ldots\ldots$, the relative position

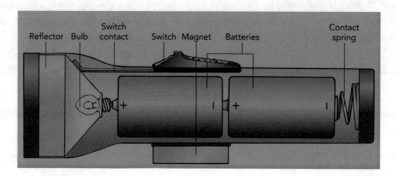

Fig. 1.3 Flashlight system S

between reflector and light bulb represented by r_m}, r_1 here is a critical structure for illumination feature GY_j because the system S will lose GY_j if r_1 is removed from the system structure. However, r_m is an important structure because the illumination feature of the system S can only be achieved with a range of $[0, a\%]$ if the structure r_m is changed.

Definition 1.32 Critical series structure of multiple subsystems: suppose that GY_j are the intrinsic features of system S, there exists a series structure of multiple subsystems $\{S_1, S_2, \ldots, S_n\}$, the intrinsic features GY_j of this system by no means can be achieved if any subsystem S_i (i = 1,2, ..., n) in this series structure is removed. Then the series subsystem structure $\{S_1, S_2, \ldots, S_n\}$ is a critical subsystem structure for the intrinsic features GY_j of the system S.

Definition 1.33 Critical complete series structure of multiple subsystems: suppose that GY_j are the intrinsic features of system S, there exists a series structure of multiple subsystems $\{S_1, S_2, \ldots, S_n\}$ that is equivalent to the total structure of the system S, the intrinsic features GY_j of this system by no means can be achieved if any subsystem S_i in this series structure is removed. Then the series subsystem structure $\{S_1, S_2, \ldots, S_n\}$ is a critical complete series structure of multiple subsystems for the intrinsic features GY_j of the system S.

Theorem 1.2 *Suppose a system S has a critical subsystem S_i, the features GY_j of system S by no means can be achieved if the S_i generates error which makes GY_{ji} unattainable.*

Proof Because S_i is a critical subsystem, GY_{ji} can not be achieved as error occurs in S_i. From definition on critical subsystem, it is easy to know that the GY_j of system S can by no means be attained. Proof is over!

Theorem 1.3 *Suppose that S is a system without critical subsystem, although any subsystem S_i in system S has error that makes its features GY_i as a whole unattainable, partial features GY_j in system S can still be attained.*

Proof From the definition on the critical subsystem, for a system S without critical subsystem, if error arises in any subsystem S_i which makes its features GY_i as a whole unattainable, this can not prevent the case that partial features GY_j of system S can still be achieved. A typical case of this is the exam sheet(One still get certain score to get decent GPA even he/she made mistakes in one question). Proof is over!   □

Theorem 1.4 *For a system S, there may exist multiple critical series structures (subsystems).*

For example, the human body system S is to sustain life function GY_j (refer to Fig. 1.4).

Fig. 1.4 Human system S

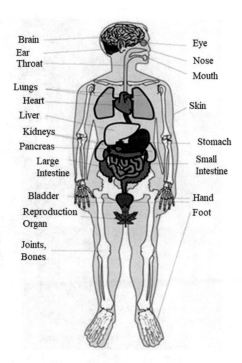

Brain
Ear
Throat
Lungs
Heart
Liver
Kidneys
Pancreas
Large
Intestine
Bladder
Reproduction
Organ
Joints,
Bones

Eye
Nose
Mouth
Skin
Stomach
Small
Intestine
Hand
Foot

Human respiratory system S_1 (Fig. 1.5) is composed of respiratory tract and lungs. Respiratory tract consists of nasal cavity, nasal conchae, nasal vestibule, larynx, pharynx, trachea, and bronchi. During the course of metabolism, body has to consume oxygen and generate carbon dioxide. The process of air exchange between body and external environment is called respiration. Air exchange is done at two positions with one happening between ambient environment and lung called external respiration and the other occurring between the cells of the body and the blood by way of fluid-bathing the cells. The common characteristics of respiration organ and organism are large surface area combined with the thin-walled nature of the alveolar cells which have direct contact with capillaries in the circulatory system.

Digestive system consists of digestive tract and digestive gland. Digestive tract is a "pipeline" connecting mouth, pharynx, larynx, esophagus, stomach, small intestine(duodenum, jejunum, ileumi, and large intestine(caecum, colon, rectum). Digestive gland is composed of small and large digestive glands. Small digestive glands are located in the wall of digestive tract and large digestive glands lie in liver, pancreas, and three salivary glands (parotid gland, submandibular gland, and sublingual gland). They all excretes digestive enzymes and other materials to digestive tract. The basic function of digestive system is to provide necessary nutrients and energy for the organism through digestion and absorption of food. The vitamins, water, and inorganic salts in food can be absorbed directly by the organism without further processing. And the carbohydrates, proteins, and fats in food can not be absorbed

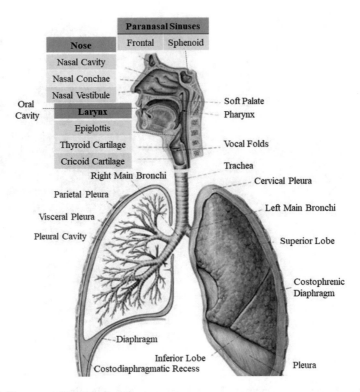

Fig. 1.5 Human respiratory system S_1

by the organism before they are broken into smaller molecules by bacteria and fungi
in digestive tract. The process of small food molecules enters into blood and lymph
through mucosal epithelial cells in digestive tract is called digestion. Those unab-
sorbed residues are excreted in the form of feces. Respiratory system S_1 and digestive
system S_2 are two different critical series structures(subsystems) (Fig. 1.6).

In an ordinary error system $X=(S, G)$, the intrinsic features GY_j can not be
realized if error occurs in its critical subsystems. Under such circumstance, the GY_j
can not be achieved at all no matter how much percentage $a\%$ does the critical
subsystem account for in the system $X = (S, G)$ i.e., "$a\% = 100\%$". However,
the realization of intrinsic featured GY_j can not be affected if error occurs in its
non-critical subsystems. Under this circumstance, the realization of GY_j can not be
influenced no matter how much percentage $a\%$ does the subsystem account for in
the system $X=(S, G)$ i.e., "$a\% \neq 100\%$".

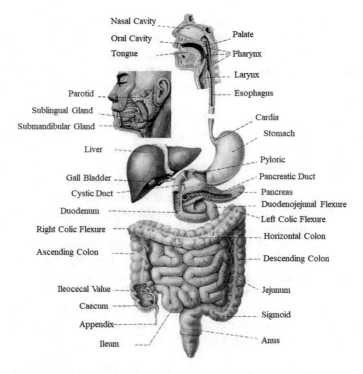

Nasal Cavity

Oral Cavity

Tongue

Palate

Pharynx

Larynx

Esophagus

Parotid

Sublingual Gland

Submandibular Gland

Cardia

Stomach

Liver

Pyloric

Gall Bladder

Cystic Duct

Duodenum

Right Colic Flexure

Ascending Colon

Pancreatic Duct

Pancreas

Duodenojejunal Flexure

Left Colic Flexure

Horizontal Colon

Descending Colon

Ileocecal Value

Caecum

Appendix

Ileum

Jejunum

Sigmoid

Anus

Fig. 1.6   Human digesting system S_2

1.3   Feature-Based System Optimization

1.3.1   Background and Related Problems

The essence of systems science resides in the holism and integrity. The optimization of a system dictates the global and holistic optimum of the system with respect to certain objective features contained in the system intrinsic features. System thinking provides people with a paradigm of looking at the world and dealing with issues with holistic perspective and global optimization. The optimization of a system is the synergistic functioning and interacting of its subsystems and elements.

In many occasions, people like to use the formula "system function $\neq \sum$ {functions of all subsystems}" to represent the relation between system function and subsystem function, which is not a precise expression. First of all, we are not sure if there exists additivity among subsystem features. Secondly, the system features could be larger, equal, or smaller than the sum of its subsystem features since the complicated synergistic effects caused by interactions between subsystem renders the causality prohibitively difficult to understand. In reality, there are tons of examples to illustrate this phenomenon, which brought troubles for researchers to examine the relationship between global optimization in system and local optimization in subsystems.

With the proposed challenge, we need to examine: (1) what relationships exist between system optimization and the subsystem optimization; (2)how to address their relationships?

With many years of research, we basically think system optimization must start from a non-erroneous system. Regarding the system of interest, first of all, it is necessary to delineate the universe of discourse U; secondly, the rules for judging error defined on the U need to be constructed. The steps involve: (a)to establish error system: $S = (S, G)$; (b)to judge if the established system has error; (c)to define the measures that are taken to remove the identified errors; (d)in the non-erroneous system, all the intrinsic features contained in the objective features are analyzed. Thereafter, a programming model (PM) is established with: each intrinsic feature is defined as a decision variable (independent variable); attaining all the intrinsic features is the objective of the system; conditions (resources) binding system are constraints. By solving the PM, the feature value or value range is obtained for achieving optimal system state.

Suppose that x_j, $j = 1, 2, \ldots, n$ is the jth intrinsic feature contained in the objective features, z is overall intrinsic feature that the system attempts to achieve, $b_i, i = 1, 2, \ldots, m$ is the ith condition (resource) then:

$$Max(min)z = c_1x_1 + c_2x_2 +, \ldots, +c_nx_n, (or\ f(x_1, x_2, \ldots\ldots, x_n))$$
$$a_{11}x_1 + a_{12}x_2 +, \ldots, +a_{1n}x_n \leq b_1$$
$$a_{21}x_1 + a_{22}x_2 +, \ldots, +a_{2n}x_n \leq b_2$$
$$\ldots\ldots\ldots\ldots\ldots\ldots\ldots\ldots\ldots\ldots$$
$$a_{m1}x_1 + a_{m2}x_2 +, \ldots, +a_{mn}x_n \leq b_m$$

is a linear programming model for optimizing system feature, where $a_{ij}, i = 1, 2, \ldots, m, j = 1, 2, \ldots, n$ are the technical coefficients.

For the undesired intrinsic features (e.g., pollution) of a system, a separate linear programming model is established. In the PM, the objective function is to minimize total pollution; decision variables are the undesired intrinsic features of system; conditions are the overall intrinsic features of system when it is optimized. On the other hand, an overall system optimization model is constructed by considering the decision variables for undesired intrinsic features of system in the system optimization model. Finally, feature-based system optimization is conducted under the values or value range that each intrinsic feature should have when system is optimal.

1.3.2   Relationship Between System Optimization and Subsystems Optimization

For the sake of convenience, suppose that certain intrinsic feature of any system S is denoted by GY_j, the subsystems that the system contains are denoted by S_i ($i = 1, 2,\ldots, n$), in order to realize the feature GY_j of system S, the features provided by all S_i ($i = 1, 2,\ldots, n$) are denoted by GY_{ij} ($i = 1, 2,\ldots, n$).

Definition 1.34 Suppose that the intrinsic feature GY_j of system S and the intrinsic feature GY_{ij} ($i = 1, 2,\ldots, n$) of subsystem S_i ($i = 1, 2,\ldots, n$) belong to the same type of feature, then the GY_j have additivity with respect to the GY_{ij} ($i = 1, 2,\ldots, n$). If $GY_j = \sum_{i=1}^{n} GY_{ji}$, then the GY_j have complete additivity with respect to the GY_{ij} ($i = 1, 2,\ldots, n$).

Definition 1.35 Suppose that the realization of feature GY_j of system S needs its subsystems S_i to provide feature GY_{ji} ($i = 1, 2,\ldots, n$), the change in the feature GY_{ji} of any subsystem S_i does not lead to the change in the feature GY_{jk} ($k = 1, 2,\ldots, i - 1, i + 1, n$) of other subsystems S_k, then the features GY_{ji} ($i = 1, 2,\ldots, n$) of subsystem S_i are mutually independent.

1. The optimum of certain feature GY_j of subsystem S has additivity with respect to the optimum of GY_{ji} of system S

    (a) Complete additivity

$$GY_j = \sum GY_{ji}$$

    In this case, the optimization of system and subsystems is generally realized simultaneously.
    (b) Partial additivity

$$GY_j \neq \sum GY_{ji}$$

    The features GY_j of system S have partial additivity with respect to the features GY_{ij} of subsystem S_i. There exists the relationship of $GY_j < \sum GY_{ji}$ if the system structure does not provide creation/generation action on the feature. Under such case, the optimization for both system and subsystems is very complicated. Due to the difference in system structure, there exist three cases: (1) creation (emergence) of system features; (2) destruction in system features; (3) overlap in system features.

In China, there was an old story telling about three monks and their anecdotes. Once upon a time, a monk found a very good location at the summit of a hill with water flowing around both sides of the hill, which was a scenery place for hermit. With years of work, he built a temple and settled there. He climbed down and up every day to carry water by himself. A few years later, another monk joined him. They started to carry water together every day. After another two years, they became a team of three members. However, they started to face water shortage problem because none of them would like to fetch water. In the system composed of three monks, suppose that the feature considered is the provision of sufficient water, then this system has partial additivity. This system can have following structures: (1) the first monk works as team leader and the other two monks need to report to the team leader; (2) a loosely coupled system in which each monk does not have to report to anyone else; (3) a system with rotating leader and total involvement is needed in decision-making process. Obviously, due to the difference in system structure, the features are also different.

2. The optimum of certain feature GY_j of system S have no additivity with respect to the optimum of features GY_{ji} of subsystem S_j.

In this case, one needs to know what features GY_{ji} the subsystem S_i should possess and what the values or value range of them when the feature GY_j of system S reaches optimum. If the features GY_{ij} of subsystem S_i have no additivity towards the optimum of GY_j of system S, then it can be denoted by $S(GY_{ji}) = S((S_1(GY_{j1}(a_1, b_1)), S_2(GY_{j2}(a_2, b_2)), \ldots \ldots, S_n(GY_{jn}(a_n, b_n)))$.

For example, for the flashlight system S, it consists of forefront battery subsystem S_1, rear battery subsystem S_2, light bulb subsystem S_3, flashlight housing subsystem $S_4, \ldots \ldots$, control subsystem S_n, etc. Regarding the illumination feature of flashlight system S, it can be expressed by: illumination feature $S(GY_{ji}) = S($ forefront battery subsystem $S_1(GY_{j1}(a_1, b_1))$, rear battery subsystem $S_2(GY_{j2}(a_2, b_2)), \ldots \ldots$, light bulb subsystem $S_i(GY_{ji}(a_i, b_i)), \ldots \ldots$, control subsystem $S_n(GY_{jn}(a_n, b_n))$.

Due to the existence of certain external forces (certain structure), system might have transformation in feature creation or destruction although some features are kept. As suboptimization at subsystem is achieved, it generally can not meet value or value ranges that renders the system optimal or the features in subsystem required by optimal system are not consistent with the features provided the current subsystem. In reality, it is critical to investigate the relationship between the value range of the features provided by subsystem (element) and the criteria generated by the subsystem (element). During the course of identifying errors, it is also imperative to find if there exists overlap between the value ranges of features in different subsystems. Moreover, the value or value range of the features that a subsystem should provide must also be examined.

1.3.3   Optimization of Feature-Additive Systems

From the aforementioned analysis, system can be categorized into two types, namely feature-additive system and feature-non-additive system. For the system having no feature-additive property, system might go through transformation in feature creation or destruction under certain external forces because the subsystem features are not aligned with system features. In the process of transformation, some features are kept.

Theorem 1.5   *Suppose that the the feature GY_j of system S has the property of complete additivity with respect to subsystem features GY_{ji} and the subsystem features GY_{ji} are mutually independent, then the subsystem features GY_{ji} ($i = 1, 2, \ldots, \ldots, n$) and features GY_j of system S can reach optimum simultaneously.*

Proof   From the definition of the complete additivity of subsystem features, GY_{ji} ($i = 1, 2, \ldots, \ldots, n$) and GY_j are same types of features, they meet $GY_j = \sum GY_{ji}$, therefore the optimal value of them might be either maximum or minimum at the same time. It is assumed that both of them have reached maximum. Suppose that the GY_j of system S has achieved maximum while at least one of the subsystem features GY_{ji} ($i = 1, 2, \ldots, \ldots, n$) is not at their maximum. Without loss of generality, it is assumed that the feature GY_{jk} of S_k did not reach maximum and the value of GY_{jk} has the gap of $\Delta GY_{jk} > 0$. As the subsystem features GY_{ji} ($i = 1, 2, \ldots, \ldots, n$) are mutually independent, the improvement on GY_{jk} will not lead to the change of other subsystem features, therefore $GY'_j = GY_{j1} + GY_{j2} + GY_{j3} +, \ldots, + GY_{j(k-1)} + (GY_{jk} + \Delta GY_{jk}) + GY_{j(k+1)} +, \ldots, + GY_{jn}$. Apparently, $GY_j < GY'_j$ holds, which contradicts the hypotheses "GY_j has reached maximum". Therefore, all features GY_{ji} ($i = 1, 2, \ldots, \ldots, n$) of subsystem S_i have reached optimum when the feature GY_j of system S achieves optimum. Similarly, as $GY_j = \sum GY_{ji}$, the feature GY_j of system S has achieved optimum as all features GY_{ji} ($i = 1, 2, \ldots, \ldots, n$) of subsystem S_i reach optimum. Proof can also be conducted in the case of minimum.
Proof is over!

Theorem 1.6   *Suppose that the features GY_j of system S has the property of complete additivity with respect to the subsystem features GY_{ji} and the subsystem features GY_{ji} are not mutually independent, then the optimum of feature GY_j of system S is not better than the sum of the optimum of subsystem features GY_{ji} ($i = 1, 2, \ldots, \ldots, n$).*

Proof   From the definition of the complete additivity of subsystem features, GY_{ji} ($i = 1, 2, \ldots, \ldots, n$) and GY_j are same types of features, they meet $GY_j = \sum GY_{ji}$.
Suppose that the optimum attained is maximum. Since the subsystem features GY_{ji}

are not mutually independent, it is possible that not all features GY_{ji} ($i = 1, 2, \ldots,$ $\ldots, n$) of all subsystems S_i can reach maximum at the same time i.e., $Max \sum\limits_{i=1}^{n} GY_{ji}$ $\leq \sum\limits_{i=1}^{n} Max(GY_{ji})$ and $GY_j = \sum GY_{ji}$, so $MaxGY_j \leq \sum\limits_{i=1}^{n} Max(GY_{ji})$. Similarly, it can be proved when the optimum is the minimum.

Proof is over!

Theorem 1.7 *Suppose that the features GY_j of system S has the property of partial additivity with respect to the subsystem features GY_{ji}, if the system structure does not have creation transformation on the feature GY_j, then when GY_j has the maximum, there exist the following relationship between the features GY_j of system S and the features of subsystem GY_{ji}:*

$$if\ GY_j < \sum_{i=1}^{n} GY_{ji}$$

$$then\ Max(GY_j) < \sum_{i=1}^{n} Max(GY_{ji})\ holds$$

Proof Because the features GY_j of system S have the property of partial additivity with respect to the features GY_{ji}, so $GY_j \neq \sum GY_{ji}$. Because the system structure does not have creation transformation on the feature GY_j,

$$\therefore\ GY_j < \sum_{i=1}^{n} GY_{ji}$$

$$\therefore\ Max(GY_j) < Max(\sum_{i=1}^{n} GY_{ji})$$

$$\because\ Max(\sum_{i=1}^{n} GY_{ji}) \leq \sum_{i=1}^{n} Max(GY_{ji})$$

$$\therefore\ Max(GY_j) < \sum_{i=1}^{n} Max(GY_{ji})$$

Proof is over!

The system optimization means global optimization where all intrinsic features aligned with objective features of the system are optimized. However, the optimization of certain feature in a system needs to be discussed in two cases-i.e., feature-additive and feature-non-additive categories. In the feature-additive case, we investigate the laws of system feature optimization and three theorems are proposed. In the non-feature-additive case, when conducting system design and operation, we need to know the states of system and subsystems when they reach optimum. Thereafter, we consider all critical and important subsystems, especially all critical subsystems. System optimization must take the cost into consideration. When there exist alternatives in realizing optimization of different subsystems, it is necessary to consider the minimum possible cost as the system optimum is achieved.

Chapter 2
Identification of Error

Abstract This chapter discusses the identification of errors. In order to identify error, three steps must be followed: (1) defining the universe of discourse; (2) determining a group of meta-rules G as the conditions for evaluating the rules for judging errors, figuring out the pertinent fields and time, and exploring relationships among rules and relationships between rules and object of interest; (3) finding out methods for identifying error given that tools and processes for identifying errors are well mastered. This chapter systematically presents the above-mentioned steps and provides applications accordingly.

2.1 Necessity of Studying the Rules for Judging Errors

The universe of discourse is determined by problems of interests and stakeholders in the discussion. This chapter mainly discusses the rules for judging errors. Rules are terms, regulations, statutes, law, and constitutions drawn up by organizations and legislatures for the purpose of regulating people's behaviors or providing evaluation criteria in decision making process.

2.1.1 Objective Existence of Rules for Judging Errors

During the course of grading homework, teachers always put some symbols or written comments on the homework sheet, which were used to judge if some questions or the steps of certain question were correct or not. How are those symbols or comments obtained? Of course, they are a group of rules obtained from solution sheets, axioms, theorems, and laws. In court, judge decided if a defendant was guilty and what kind of penalty needed to be put on criminal. In an organization, the under-performing employees were punished based on the company's regulations or key performance indicators. Parents disciplined children according to the social norms or values, law, regulations, and disciplines in a family. In summary, rules are needed to determine

© Springer Nature Switzerland AG 2020

K. Guo and S. Liu, *Error Systems: Concepts, Theory and Applications*,
Studies in Systems, Decision and Control 275,
https://doi.org/10.1007/978-3-030-40760-5_2

whether the political system, policies, decisions, theoretic systems in a country are correct or appropriate or not. Therefore, a group of rules need to be defined before one makes judgment.

2.1.2   Theoretical Foundations of Studying Rules for Judging Errors

What conditions does rule G meet? What relationship do G_i and G_j $(i \neq j)$ have? What laws does this relationship abide by? What principles do we obey when establishing the rules for judging errors? Suppose that G is known, a is proven to have error. Can this conclusion be generalized to other situations? In many cases, the same question may have conflicting results judged under different rules in different situations. All of them are questions we need to address in this chapter.

2.2   Properties of Rules for Judging Errors

2.2.1   Changeability of Rules for Judging Errors

1. The necessity and inevitability of changes in rules for judging errors
   In order to meet the very basic physiological needs, human beings must have some actions and activities. Regarding the results of certain activity, how should we know if they are wrong or right? This is easier said than done. The following items must be clarified before answering the question: (1) where did this activity happen? (2) when did this activity happen? (3) which field did this question belong to? (4)what was the purpose of addressing this question? (5)what kinds of knowledge, skills and techniques were employed to solve this question? In some occasion, item (5) is included in (1) and (3). Nevertheless, when judging the results of a question, the above items must be clearly understood.

Example 2.1 It is right or wrong if a driver is driving his car on the right side of the road. In China, according to the traffic rules and regulations, his action has no problem. However, he is wrong by breaking the rules if he is driving in the UK because all vehicles must keep left side of the road.

Example 2.2 Someone did a calculation $1 + 1 = 1$, is this correct? According to laws in binary computing (Table 2.1):
   The result is incorrect if this person is performing binary computation. On the contrary, the computing result is correct if this person is doing Boolean calculation. Because the Boolean calculation is as (Table 2.2):

Table 2.1 Computation of binary numbers

Binary number 1	Binary number 2	Sum
0	0	0
0	1	1
1	1	10

Table 2.2 The computation in Boolean algebra

Boolean number 1	Boolean number 2	Sum
0	0	1
0	1	0
1	1	1

Example 2.3 Tying knots and scratching marks on stone are advanced tally mark and numerical system. Is this proposition correct? In 5000 years ago, as tally mark, tying knots, and scratching marks on stones were advanced and the above proposition is correct in that sense. However, many advanced numerical systems have been developed to handle this since then. Comparatively, the primitive techniques are outdated and the proposition is false nowadays.

Example 2.4 In evaluating the functionality of a bicycle, residents in urban areas require a light and aesthetic style while farmers in village use it as transport tool demanding heavy-duty and all-terrain use style.

By analyzing the above four examples, example 1 tells us that rules for judging errors change with respect to the location (or spatial attributes). And example 2 exhibits that different rules in different fields are needed to evaluate errors in corresponding areas or fields. While example 3 shows that rules change over time and are contingent on the development of economy, technology, and society. Moreover, rules for judging errors must also change with various purposes of using them as demonstrated in example 4.

2. The laws in the changes of rules for judging errors

    (1) Function for rules of judging errors

        Suppose that $S = [S_0, S_1]$, $K = \{(x, y, z) \mid a \leq x \leq b, c \leq y \leq d, e \leq z \leq f\}$, $Z = \{z_1, z_2, \ldots \ldots, z_n\}$, $M = \{m_1, m_2, \ldots \ldots, m_i\}$, where S is the temporal set; K is the spatial set; Z is the set composed of different fields; M is the purpose set. In general, technological level is a function with respect to time.

(i) Definition

Definition 2.1 Suppose that G is a group of rules, $D = S \times Z \times M \times K$, if $f : D \to G$, then f is called a function for rules of judging errors simplified as rule function, noted by $G = f(D)$, $G = f(s, k, z, m)$, $G = G(s, k, z, m)$.

(ii) Special forms of rule function

  (a) When space, field, and objective are given, the rule is the function with respect to time, i.e., if $k = \{k_0\}$, $z = \{z_0\}$, and $m = \{m_0\}$, $G = f(s, k_0, z_0, m_0) = f(s)$ holds.

  (b) When time, field, and objective are given, the rule is the function with respect to space, i.e., if $s = \{s_0\}$, $z = \{z_0\}$, and $m = \{m_0\}$, $G = f(s_0, k, z_0, m_0) = f(k)$ holds. Similarly, the following relationships hold.

  (c) If $s = \{s_0\}$, $k = \{k_0\}$, and $m = \{m_0\}$, $G = f(s_0, k_0, z, m_0) = f(z)$ holds.

  (d) If $s = \{s_0\}$, $k = \{k_0\}$, and $z = \{z_0\}$, $G = f(s_0, k_0, z_0, m) = f(m)$ holds.

  (e) If $s = \{s_0\}$ and $k = \{k_0\}$, $G = f(s_0, k_0, z, m) = f(z, m)$ holds.

  (f) If $k = \{k_0\}$ and $z = \{z_0\}$, $G = f(s, k_0, z_0, m) = f(s, m)$ holds.

  (g) If $m = \{m_0\}$ and $z = \{z_0\}$, $G = f(s, k, z_0, m_0) = f(s, k)$ holds.

  (h) If $s = \{s_0\}$ and $m = \{m_0\}$, $G = f(s_0, k, z, m_0) = f(k, z)$ holds.

  (i) If $s = \{s_0\}$ and $z = \{z_0\}$, $G = f(s_0, k, z_0, m) = f(k, m)$ holds.

  (j) If $k = \{k_0\}$ and $m = \{m_0\}$, $G = f(s, k_0, z, m_0) = f(s, z)$ holds.

  (k) If $s = \{s_0\}$, $G = f(s_0, k, z, m) = f(k, z, m)$ holds.

  (l) If $k = \{k_0\}$, $G = f(s, k_0, z, m) = f(s, z, m)$ holds.

  (m) If $z = \{z_0\}$, $G = f(s, k, z_0, m) = f(s, k, m)$ holds.

  (n) If $m = \{m_0\}$, $G = f(s, k, z, m_0) = f(s, k, z)$ holds.

  (o) If $s = \{s_0\}$, $k = \{k_0\}$, $z = \{z_0\}$, and $m = \{m_0\}$, $G = f(s_0, k_0, z_0, m_0)$ holds. It is called a constant rule function noted by $G = f_0$ or $G = G_0$ or G_0.

(2) Operations of rule functions

  (a) Difference and union operations on rule functions

  Given that we have two data sets regarding the law in country C with one 1980–2010 and the other 1990–2018, one is able to study all law in C over period of 1980-2018; similarly, given that the axioms, theorems, and theories in different sub-disciplines of math have been known, we can study the holistic axioms, theorems, and theories of the math system. Both examples in the above need a union operation of rule functions.

Definition 2.2 Suppose that $G_1(s_0, k_0, z_0, m_0)$ and $G_2(s_0, k_0, z_0, m_0)$ are two constant rule functions, if $G_3(s_0, k_0, z_0, m_0) = \{x \mid x \in G_1(s_0, k_0, z_0, m_0)$ and $x \bar{\in} G_2(s_0, k_0, z_0, m_0)\}$, then G_3 is called the difference between G_1 and G_2 noted by $G_3 = (G_1 - G_2)$.

Definition 2.3 Suppose that $G_1(s_1, k_1, z_1, m_1)$ and $G_2(s_2, k_2, z_2, m_2)$ are two rule functions, respectively defined on $U_1 = S_1 \times Z_1 \times M_1 \times K_1$ and $U_2 = S_2 \times Z_2 \times M_2 \times K_2$, if $U = U_1 \cup U_2$, $\forall (s_1, k_1, z_1, m_1), (s_2, k_2, z_2, m_2) \in U$, $G_3 = (G_1 - G_2)$ holds, then $G_3(s_3, k_3, z_3, m_3)$ is the difference between $G_1(s_1, k_1, z_1, m_1)$ and $G_2(s_2, k_2, z_2, m_2)$ noted by $G_3 = (G_1 - G_2)$.

Definition 2.4 Suppose that $G_1(s_0, k_0, z_0, m_0)$ and $G_2(s_0, k_0, z_0, m_0)$ are two constant rule functions, and $G = Z(G_1, G_2)$, if $G_3(s_0, k_0, z_0, m_0) = \{X \mid X \in G_1$ or $X \in (G_1 - G_2)\}$, then $G_3(s_0, k_0, z_0, m_0)$ is called the union of G_1 and G_2 noted by $G_3 = G_1 \cup G_2$.

Definition 2.5 Suppose that $G_1(s_1, k_1, z_1, m_1)$ and $G_2(s_2, k_2, z_2, m_2)$ are two rule functions defined on $U_1 = S_1 \times Z_1 \times M_1 \times K_1$ and $U_2 = S_2 \times Z_2 \times M_2 \times K_2$, respectively, if $U = U_1 \cup U_2$, $\forall (s_1, k_1, z_1, m_1)$, $(s_2, k_2, z_2, m_2) \in U$, $G_3 = G_1 \cup G_2$ holds, then $G_3(s_3, k_3, z_3, m_3)$ is the union of $G_1(s_1, k_1, z_1, m_1)$ and $G_2(s_2, k_2, z_2, m_2)$ noted by $G_3 = G_1 \cup G_2$.

Proposition 2.1 *The union of rule functions meets the commutative law, i.e., $G_1 \cup G_2 = G_2 \cup G_1$.*

Proof Proof is omitted here.

(b) Intersection of rule functions

Given that the respective rules for judging errors in physics and math have been known, it is necessary to take the conjunction on those rule functions in both disciplines when addressing an issue or error involving both subjects.

Definition 2.6 Suppose that $G_1(s_0, k_0, z_0, m_0)$ and $G_2(s_0, k_0, z_0, m_0)$ are two constant rule functions, if $G_3(s_0, k_0, z_0, m_0) = Z(G_1(s_0, k_0, z_0, m_0), G_2(s_0, k_0, z_0, m_0))$, then $G_3(s_0, k_0, z_0, m_0)$ is called the intersection of G_1 and G_2 noted by $G_3 = G_1 \wedge G_2$.

Definition 2.7 Suppose that $G_1(s_1, k_1, z_1, m_1)$ and $G_2(s_2, k_2, z_2, m_2)$ are two rule functions, respectively defined on $U_1 = S_1 \times Z_1 \times M_1 \times K_1$ and $U_2 = S_2 \times Z_2 \times M_2 \times K_2$, if $U = U_1 \cup U_2$, $\forall (s_1, k_1, z_1, m_1)$, $(s_2, k_2, z_2, m_2) \in U$, $G_3 = G_1 \cap G_2$ holds, then $G_3(s_3, k_3, z_3, m_3)$ is the intersection of $G_1(s_1, k_1, z_1, m_1)$ and $G_2(s_2, k_2, z_2, m_2)$ noted by $G_3 = G_1 \cap G_2$.

Proposition 2.2 *The intersection of rule functions meets the commutative law, i.e., $G_1 \cap G_2 = G_2 \cap G_1$.*

Proof Proof is omitted here.

(c) Interactions of rule functions

Definition 2.8 Suppose that the universe of discourse for G_1 and G_2 are U_1 and U_2, respectively, $U_3 = U_1 \cup U_2$, G_3 is a rule function defined on U_3 and $G_3 = G_1 \cup G_2$, if $\forall a \in U_3$, the error values of a under rules G_1, G_2, and G_3 are z_1, z_2, and z_3, respectively, and z_3 meets the condition of $z_3 = min[z_1, z_2]$, then for universe of discourse U_3, G_1 and G_2 have no error interaction, otherwise, G_1 and G_2 have certain error interactions. From the definition, it means that $\forall a \in U_3$, if a can be judged using G_1 and G_2, then a can also be judged using G_3 and the derived results are identical. At the same time, if a can be judged using G_3, then a can also be judged using G_1 and G_2 (if $a \in U_1$, G_1 is used; if $a \in U_2$, G_2 is used; when both $a \in U_1$ and $a \in U_2$ hold, the rule for deriving the minimum error value is used) and derived results are identical, then G_1 and G_2 are, in U_3, called rules without error interactions.

Proposition 2.3 *Suppose that G_1, G_2, G_3 are three rule functions having no error interactions, then*

(a) $(G_1 \cap G_2) \cup G_3 = (G_1 \cup G_3) \cap (G_2 \cup G_3)$
(b) $(G_1 \cup G_2) \cap G_3 = (G_1 \cap G_3) \cup (G_2 \cap G_3)$.

Proof

(a) If $a \in (G_1 \cap G_2) \cup G_3 \Rightarrow a \in Z(G_1, G_3)$ or $a \in Z(G_2, G_3) \Rightarrow a \in Z((G_1 \cup G_3), (G_2 \cup G_3)) \Rightarrow a \in (G_1 \cup G_3) \cap (G_2 \cup G_3)$; on the other hand, if $a \in (G_1 \cup G_3) \cap (G_2 \cup G_3) \Rightarrow a \in Z((G_1 \cup G_3), (G_2 \cup G_3)) \Rightarrow a \in Z(G_1, G_3)$ or $a \in Z(G_2, G_3) \Rightarrow a \in (G_1 \cap G_2) \cup G_3$
(b) Similarly, $(G_1 \cup G_2) \cap G_3 = (G_1 \cap G_3) \cup (G_2 \cap G_3)$ can also be proven.

Proof is completed.

Definition 2.9 Suppose that p is a proposition, then $\overline{p}$ is called the negative of p, i.e., $\overline{p} = -p$, if $G_1 = \{g_1, g_2, g_2, \ldots \ldots, g_n\}$, $G_2 = \{-g_1, -g_2, -g_2, \ldots \ldots, -g_n\}$, then G_2 is called negative of G_1 noted by $G_2 = -G_1$.

Proposition 2.4 *The negative of rule meets* $-(-G) = G$.

2.2.2   The Hierarchy of Rules for Judging Errors

1. Hierarchy of system

System is the unification of structure and features. The structure is the order of permutation and combination of those system elements and the feature is the order of the activities of system elements. The structure and elements are relatively independent from each other. Therefore, system has hierarchy. The hierarchies in system structure and elements determine the hierarchy of system. The hierarchy of system is the fundamental property of general systems. It means that each functioning system element can be regarded as a system. While a system can be thought as a constituent of a system of systems (SoS).

For a particular system, its different hierarchical structure can be demonstrated in two different directions.

(1) **element** → 1st order subsystem → 2nd order subsystem → ... ... → $(n-1)^{th}$ order subsystem → n^{th} order subsystem → **current system**;
(2) **current system** → 1st order ultrasystem → 2nd order ultrasystem → ... ... → $(n-1)^{th}$ order ultrasystem → holistic system.

System structure has not only vertical layers of subsystems and/or elements but also horizontal interacting elements.

Example 2.5 AAAA University of Technology. The vertical and horizontal structures are demonstrated in Table 2.3 and Fig. 2.1.

Table 2.3 Demonstration of the system vertical and horizontal structure

Vertical\Horizontal	H_1	H_2	......	H_m
V_1 subsystem	a_{11}	a_{12}	...	a_{1m}
V_2 subsystem	a_{21}	a_{22}	...	a_{2m}
V_3 subsystem	a_{31}	a_{32}	...	a_{11}
...	...	...	...	...
...	...	...	...	...
V_n subsystem	a_{n1}	a_{n2}	...	a_{nm}
Element	$a_{(n+1),1}$	$a_{(n+1),2}$	...	$a_{(n+1),m}$

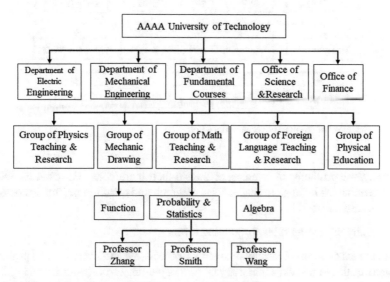

Fig. 2.1 System structure for AAAA University of Technology

2. Hierarchy of objective world

(1) Hierarchy of social systems

Example 2.6 Figure 2.2 provides the structure of national governance of a country.

(2) Hierarchy of other systems
     Hierarchy widely exists in the physical, chemical, geological, astronautical, production, and social thinking aspects.

3. Hierarchy in practical requirements and issues of interests

In modern science, the totality of universe is divided into three hierarchies, namely, universal, macro, and microscopic levels. The range beyond solar system belongs to universal hierarchy. The particles under the level of molecule include atom, nuclei, and quark belong to microscopic level. As both system and science research has

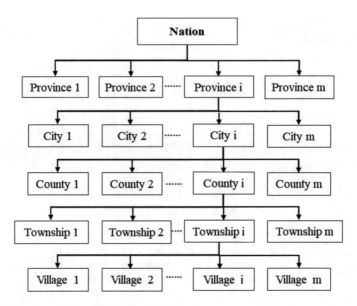

Fig. 2.2 Structure of national governance in China

hierarchy, the problems of interests also have their hierarchies. For example, New-tonian mechanics belongs to macro level research and quantum mechanics belongs to microscopic research.

4. Hierarchy in judging rules for judging errors

Due to the existence of hierarchy in objective objects, system structure, and problems of interests, there must exist hierarchy in rules of judging errors (referring to Fig. 2.3).

For example, in the study of law system, the grand rule is constitution under which there are different hierarchies of law and relevant articles (marriage law and contract law) as sub-rules. General math theories have a set of theoretical laws and rules and each discipline in math also has its relevant rules. The problems in each discipline also have their corresponding rules and laws.

2.2.3   Completeness of Rules for Judging Errors

1. Definition on the completeness of rules for judging errors

Definition 2.10 Suppose that G is a group of non-erroneous rules defined on U, if $\forall\, a \in U$, the error value of a under G is t, and the error value is t_i with applying any group of non-erroneous rules G_i, $t \le t_i$ holds, then G is called a group of rules with completeness noted by $\omega U\,(G)$.

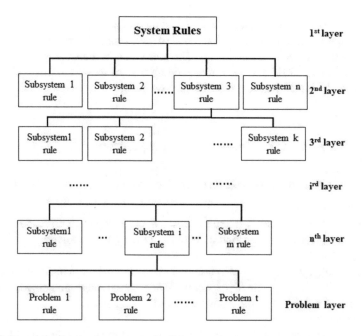

Fig. 2.3 Structure of national governance in China

2. Practical meaning of rule completeness

Suppose that U is the universe of discourse composed of formulations of addition and multiplication of binary numbers, then judging rule $G_1 = \{1 + 1 = 10, 1 + 0 = 1, 0 + 1 = 1, 0 + 0 = 0\}$ is an incomplete rule defined on U because $f(a, G_1) = 1$ when $a = (1 + 0) \times 1 \ldots 1$; $ch(a) = \{0, 1\} \not\subset 0$ under judging rule $G_2 = \{1 + 1 = 10, 1 + 0 = 1, 0 + 1 = 1, 0 + 0 = 0, 0 \times 0 = 0, 1 \times 0 = 0, 0 \times 1 = 0, 1 \times 1 = 1\}$. As both G_1 and G_2 are rules defined on U, therefore G_1 is an incomplete rule defined on U. In this example, the rules qualified for judging errors within U must be complete rules. Otherwise, $\exists a \in U$, a is non-erroneous under the complete rules defined on U and it is erroneous under rule G, which causes confusion in the real application.

3. Objective relativity of rule completeness

(1) Existence of relativity in the objective world
    With the ever-accelerated development of science and technology, the boundaries for different research fields are continuously changing. For instance, the development in math has witnessed ever-expanding boundaries.

(2) Relativity of human beings' understanding on the world
    The understanding of the fundamental particles started from molecule to atom,
    proton, neutron, ..., quark, ... Therefore, due to the existence of relativity
    in objective world and the understanding toward the world, there also exists
    relativity in the complete rules defined on U.

2.2.4 Scientificity of Rules for Judging Errors

The scientificity of rules for judging error dictates that:

1. Accuracy
   For the problem evaluated, the rules chosen must be aligned with academic fields,
   temporal and spatial range, and objectives of the system involved.
2. Completeness
   Theoretically, the rules chosen should possess completeness regarding the problem of interests, academic fields, temporal and spatial range, and objectives. Otherwise, the reliability of the results can not be guaranteed.
3. Conciseness
   The rules for judging error should be as precise as possible. Among all the complete rules within the same academic field, temporal and spatial range, a group of
   simple rules are favorable. It is possible to make mistakes for using too complicated rules.
4. Infallibility
   The credibility of the results obtained using rules with error is doubtful. However,
   due to the limitations in the scientific level or measurement precision, it is not
   possible or necessary to require absolute correctness in the rules for judging errors.
   Nevertheless, what rules are deemed as the non-erroneous rules? In our definition,
   the original rules G_0 that have been scientifically proven correct or tested and
   verified in practical applications are qualified for being the non-erroneous rules.

2.2.5 Fuzziness of Rules for Judging Errors

1. Fuzziness in the objective existence
   The expressions such as "A is high", "he is obese", "she is young", and "she is
   beautiful" have fuzziness[1] in delivering their meaning. In reality, for linguistic
   expressions, there always exist somewhat ambiguity or vagueness (fuzziness).
   Both "they resemble each other" and "he is fluent in speaking English" have

---

[1] If X is a group of objects denoted by x, then a fuzzy set $\tilde{A}$ in X is a set of ordered pair $\tilde{A} = \{(x, \mu_{\tilde{A}}(x)) \mid x \in X\}$, $\mu_{\tilde{A}}(x))$ is called the membership function of x in $\tilde{A}$ that maps X to the membership space M.

fuzziness in their meaning. The degree of truth that A belongs to the set S (A resembles B) is "1" if A is exactly the same as B. While the degree of membership for expression "the boy resembles his father" can only be represented by the number in the range of [0, 1]. For the expression "he is fluent in speaking English", it does not indicate the two polarized points- i.e., "he is an English literature expert" and "he knows nothing about English" but the case that he is at the point between two extremities. If the above two sentences are treated as two logical propositions, then the logical propositions have fuzziness. Given that detailed investigations have made to the phenomenon exhibited by the above concepts and objective facts involved in the above-mentioned two sentences, we know that the fuzziness widely exists in the objective world.

2. Fuzziness in human beings' understanding

   In different historic periods of human society, the degree to which human beings understood the world was distinct due to the limitation in knowledge of science and technology. In the primitive period, surviving with very limited substances, the number larger than two digits can only be quantified using fuzzy expressions such as "many, plenty, abundant". Nowadays, the development of science and technology enabled the research on number theory to reach an unprecedentedly high level by which very large numbers have been investigated. Nevertheless, the understanding for the range from super large number to infinity is still fuzzy. Moreover, human beings are still not very clear about how human brain works and stores memories. Therefore, the understanding of human beings toward the world is fuzzy.

3. Fuzziness in the ways of handling problems

   When judging who is walking up to you, it is not hard to draw a conclusion by comparing the features such as height, body shape, walking gait, and facial appearance with the data stored in your brain. However, if this task is assigned to artificial intelligence, the machine has to measure and calculate data such as height, weight, angle and frequency of arm swinging, walking velocity and acceleration. At the same time, it has to adopt more than 8 decimal number to achieve necessary precision level, which makes the task very complicated and sometimes even draws the wrong conclusion because the critical data profiling a human being have been changing. In another example, the question is to judge if a triangle is a right triangle. By using fuzzy handling, human being can easily draw the "correct" conclusion. On the other hand, if it is assigned to an AI to handle this, it is possible for the computer not being able to find the right triangle because in reality the chance of having an angel that is exactly equal to $90°$ is infinitesimal.

4. Fuzziness in the rules for judging errors

   In fuzzy mathematics, fuzzy reasoning (fuzzy induction and deduction toward fuzzy proposition) and fuzzy recognition (both individual and group recognition) have been explored, where rules for fuzzy reasoning and fuzzy recognition are adopted. When using a group of rules to judge if a proposition is right or wrong, the judging rules G must have certain level of ambiguity since there exists fuzziness in understanding and handling problem of interests.

2.2.6 Multi-objective Features of Rules for Judging Errors

Multi-objective problems are very common in our contemporary world in which multiple conflicting objectives must be evaluated and justified at the same time. Thus, for constructing and implementing rules for judging errors in the case of multi-objective problems, it is necessary to hold the principles that the rules used have no internal contradiction and the derived results have no negative impacts.

2.2.7 Non-contradiction of Rules for Judging Errors

With certain temporal and spatial range, academic field, and research objective, a qualified set of rules for judging errors working at the fundamental level possess has not only the feature of non-contradiction within the same acting domain but also the property of scientific correctness and completeness or at least relative completeness. On the universe of discourse U: $G = \{g_1, g_2, \ldots, g_i, \ldots, g_j, \ldots g_n\}$, $G_1 = \{g_1, g_2, \ldots, g_{i-1}, g_{i+1}, \ldots, g_j, \ldots g_n\}$, $G_2 = \{g_1, g_2, \ldots, g_i, \ldots, g_{j-1}, g_{j+1} \cdots g_n\}$, where g_i and g_j ($i \neq j$ are contradicting rules acting in the same domain. For $\exists a \in U$, $ch_1(a) = 0$ under rule G_1, $ch_1(a) \neq 0$ under rule G_2, and $G_1 \subset G$, $G_2 \subset G$, a can not be judged under rule G. In reality, supposed that G is a set of rules defined with U, then $\exists a \subset U$, a can not be judged under G. This leads to the ambiguity in the results obtained in domain U, which might cause detrimental effects or disastrous impacts.

In a multi-objective decision making process, the objectives of multiple decisions are contradicting. Does this mean that decision makers must choose one objective by dropping the other? In reality, it is possible to choose another path to circumvent those barriers. For example, in a decision making process, the first objective is to prevent horse from eating too much, and the second objective is to have it run faster and longer than ever. Apparently, two objectives are contradicting if they together serve as evaluating criteria. Per the requirements for defining rules of judging errors, they are not qualified for being the rules for judging this type of error. However, it is necessary to make trade-off between the two contradicting objectives. Hereby, we can address this problem by defining a relatively reasonable objective value. On one hand, by fixing the feeding quantity, the faster and longer the horse can run the closer it is to approach the objective. On the other hand, when fixing the velocity or distance that horse can run, the less the horse could eat the closer it is to approach the objective. Moreover, one can choose to maximize an objective function which is subject to certain amount of food and certain velocity or distance.

2.2.8 Unchangeability of Rules for Judging Errors Under Certain Conditions

Under certain conditions or within certain periods, the rules defined are relatively stable. For example, the constitution of a country can not be unceremoniously changed before the legislature modifies it. The proven theorems serving as rules for judging errors in a discipline do not change randomly before major theoretical breakthroughs have been made. Furthermore, due to the limitation in the technology and natural conditions, some objective conditions can not be changed during certain period of time.

2.2.9 Parallelism of Rules for Judging Errors

In general, on the universe of discourse U applicable for issue of interests, there may exist parallel qualified rules for judging errors. For example, suppose that U is the set composed of all the faculty and staffs in AAAA University of Technology, there exist parallel rules for judging employee performance in different departments and judging the performance of an individual in distinct features. In evaluating a trans-disciplinary research, it is not uncommon that multiple experts were invited to provide assessment on the project in which parallel rules defined on the pertinent universe of discourse U were employed.

2.2.10 Acting Forces of Rules for Judging Errors

In the section of 1.2.9, there exist multiple qualified rules defined on the same universe of discourse U. It is possible that the parallel rules do not have even impacts. Therefore, it is necessary to assign a weight to different rules to reflect the differences in their impact and we name it as acting forces of rule. Suppose that an event happened in AAAA University of Technology on January 1st 1999, it was necessary to make judgment whether the event was correct or not. In this event, rules and regulations from the levels of People's Republic of China, AAAA province, BB city, and AAAA University of Technology can all have impacts on the results of judgment. Given that so many rules were acting here, how to obtain a reasonable result? In this example, a weight of 1 was assigned to the national law, and the weight of 0 was assigned to the law and regulations issued by AAAA province, BB city, and AAAA University of Technology. In the process of evaluating research project, it was acceptable when all invited experts gave the identical result. However, what one can do if it is hard to obtain a unified conclusion. In this case, for each of the invited experts $z_1, z_2, \ldots, z_n$, weights $p_1, p_2, \ldots, p_n$ ($p_1 + p_2 + \cdots + p_n = 1$) are assigned to the rules they used to make the judgment. Finally, a score is calculated to confirm the assessment result.

2.3   Methods of Building Rules for Judging Errors

This section introduces the principles, processes, steps, and methods of building rules for judging errors.

2.3.1   Principles for Building Rules for Judging Errors

1. Scientific principles

    (1) Scientifically proven correctness
        Establishing rules for judging errors must abide by the scientific principles. For example, rules used by professor for grading homework and exams of students must be based on pertinent axiom, theorem, and laws. The professor's impression on certain students should not be used as rules for judging the quality of their task.

    (2) Completeness
        The rules should be complete regarding the universe of discourse U to which the problems of interest belong. Otherwise, many issues can not be judged with an incomplete set of rules or even some correct results are deemed as wrong due to the inability of the selected rules to make judgment, which renders the judging process meaningless and consequently causes loss.

    (3) Practicality and conciseness
        Suppose that G_1 and G_2 are two group of rules defined on universe of discourse U, $G_1 \sim G_2$, and they are scientifically proven correct, complete, and having no intrinsic contradiction, G_1 has no practical value for evaluating certain problems defined within U when the complexity of G_1 makes it unable to be implemented to judge if $a \in U$ has error. And if G_2 is simple and feasible in fulfilling the above task, it is said the G_2 has practical value for assessing problems within U.

    (4) Without intrinsic contradictions
        When acting within the same domain, the principles of non-contradiction must be held to make sure that the rules do not generate confusing results in the judging process, which might bring negative impacts on the decision makers or even disastrous consequences. It is possible to have some contradicting rules coexisting in a rule system if those rules are used in different acting domains.

2. Practical principles

    (1) Scientifically proven correctness
        Regarding the problem being evaluated, the chosen judging rules should be accordance with the academic fields, temporal and spatial scope, and objective conditions.

(2) Sufficient flexibility

Theoretically, the rules G must be complete regarding the universe of discourse U. However, owing to the limitations in techniques or the requirements for research objective and measurement precision, although G may not be complete regarding the overall universe of discourse, it is sufficient as long as the selected rules are complete regarding the particular domain that contains the problem of interests. For example, suppose that the overall universe of discourse U is the set composed of all mathematical problems, the rules used for judging if the sum of two integer numbers is correct can be incomplete regarding U. Here, it is sufficient to consider the domain U_a containing problem related to addition computation in which the rules for judging errors are complete.

(3) Principle of minimum applicable domain

From (2), we know that, when constructing a set of rules for judging certain problem, it is sufficient to establish the complete rules defined within the minimum applicable domain containing the problem of interests. For the example in (2), only are the rules applicable to addition of integer numbers enough to make the judgment. Even though most of rules defined in the whole mathematical field (excluding the rules for defining operation of complex numbers) can be used to make judgment and they are complete, scientifically proven correct, and having no intrinsic contradiction, they are not concise and actually waste unnecessary resources.

(4) Implementability

(5) Conciseness

(6) Without intrinsic contradictions.

2.3.2   Process of Building Rules of Judging Errors

Figure 2.4 provides the flowchart for demonstrating the process of building rules for judging errors.

2.3.3   Steps for Building Rules of Judging Errors

1. Problem analysis

(1) The temporal and spatial characteristics and academic disciplines associated with the object

In general, the rules for judging errors vary with time, space, and academic disciplines associated with the problem. Therefore, it is necessary to analyze the temporal and spacial characteristics, academic fields that are related to the object of interests.

Fig. 2.4 Process for building rules of judging errors

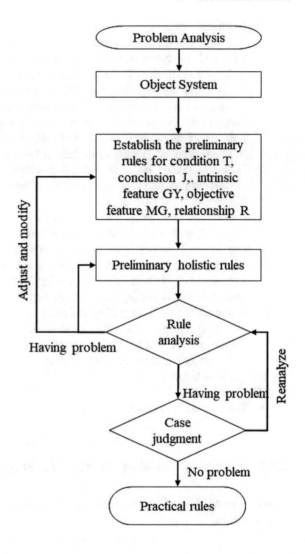

(2) Investigating the objective of the object of interest

In the case of evaluating a pair of shoes, not only is it necessary to evaluate the quality, appearance, and the proper fitting to feet but also to understand the objective of buying this pair of shoes. If this pair of shoes is used specially in raining weather, we think the decision of buying this pair of shoes has no error if it has good water-proof quality, aesthetic appearance, and proper fitting to feet. Otherwise, we think the decision of buying water-proof shoes has error if they are used to walk long way. Thus, it is pivotal to include the objective of the object of interests when defining the rules for judging errors.

2. Establishing object system

   The object system $X = X(\{W_i\}, T(t_1, t_2), J, GY, MG, R)$ contains condition T, conclusion J, intrinsic feature GY, objective feature MG, and relationship R. Therefore, in order to establish the object system of a problem, one must examine the following elements in the object system:

   (1) Conditions
   (2) Conclusions
   (3) Intrinsic feature
   (4) Objective feature
   (5) Relationship
   (6) Object system

3. Establishing rules for judging errors for T, J, GY, MG, and R

   (1) Collect data
       (i) Collect axiom, theory, laws, propositions, criteria, and formulas in the academic fields that are related to the problem of interests;
       (ii) Collect law, articles, policies, moral standards, and social norms and values associated with the problem of interests;
       (iii) Scan environments surrounding the problem of interests;
             (a) Social environment;
             (b) Natural environment;
             (c) Economic environment;
       (iv) Identify conditions the object has
       (v) Confirm research objectives of the object
   (2) Establish preliminary rules for T, J, GY, MG, and R
       (i) Establish preliminary rules for judging errors in conditions T;
       (ii) Establish preliminary rules for judging errors in conclusions J;
       (iii) Establish preliminary rules for judging errors in intrinsic feature GY;
       (iv) Establish preliminary rules for judging errors in objective feature MG;
       (v) Establish preliminary rules for judging errors in relationship R.

       Due to the difference in the object of interests, it is very hard to list a unified concrete rules for the whole object system. Nevertheless, for the five elements T, J, GY, MG, and R, we can list the relevant contents in each element and analyze time, space, academic discipline, and research objective associated with each of them. Then, based on the collected data for each element (e.g., propositions, formulas, and functions), the rules are established for each element and then they are codified in to a rule system.

4. Establish total preliminary rules

   This stage involves building an organic rule system using the rules established for condition T, conclusion J, intrinsic feature GY, objective feature MG, and relationship R, respectively.

5. Rule analysis
   When the total rules are formulated, one needs to analyze if the following aspects are met:

   (1) Scientifically proven correctness
   (2) Relative completeness
   (3) Implementability
   (4) Conciseness
   (5) Having non-contradiction
   (6) Minimum applicable domain

   In the analysis, if the total preliminary rules have problems in one or more than one of the above six items, it is necessary to return to the process for identifying the preliminary rules for T, J, GY, MG, and R and process of formulating the total preliminary rules until the problems are solved. It is a recurrent process.

6. Validating through actual cases
   Even though the total rules already meet the 6 principles for establishing error rules, the theoretical rules often have some discrepancies with actual situations. Therefore, it is necessary to validate the established rules with actual cases and data. The rules must be reexamined unless its theoretical rules are conforming with the actual results. Hitherto, a group of rules have been successfully established and can be used in practical applications.

2.4 Relationship Between Rules for Judging Errors and the Object Being Judged

2.4.1 Axioms

In order to judge if there exist errors in the rules for judging errors, it is necessary to introduce a group of rules. However, for the introduced rules, another group of rules needs to be adopted to make judgment on the correctness of introduced rules. This leads to infinite cycle of introducing rules where one can never get a definite conclusion toward the object of interests. For solving this issue, an axiom system is established and any rule that meets the conditions set by the axiom system is regarded as "non-erroneous".

Axiom: The rules evaluated and confirmed by a group of meta rules (or original rules) G_0 composed of (1) propositions that have been scientifically proven correct, (2) non-erroneous propositions obtained from(1)through correct logical reasoning, (3) non-erroneous propositions tested, verified and validated in practical applications, are qualified for being the non-erroneous rules noted by G_1. Notations: regarding the classification of the fields for G_0:

1. Macro classification of G_0

    (1) Natural science: the G_0 in natural science field are natural laws;

    (2) Social science: the G_0 in social science field are social laws;

    (3) Social activities: the G_0 in social activities are natural law, policies, agreements, contracts, regulations, doctrines, code of ethics, and the original features of the object (form, connotation, . . .).

    (4) Political science: the G_0 in political fields are experiences, wisdom, knowledge, mental models, or collective wisdom of politicians.

    (5) Religion: the G_0 in religion are creeds and doctrines;

    (6) Other: the G_0 in other field are norms and regulations.

2. Micro classification of G_0

According to the needs of researchers or a particular rule, a special range is defined. The definition of "non-erroneous" rules in the special range should be based on the original rules G_0 macro-classification to which they belong.

2.4.2 Equivalence in the Rules for Judging Errors

1. Definition of equivalence

Definition 2.11 Suppose that G_1 and G_2 are two group of rules defined within U, if (a) $G_1 \rightarrow G_2$, (b)$G_2 \rightarrow G_1$, then G_1 and G_2 are equivalent noted by $G_1 \sim G_2$.

2. Characteristics of equivalence

Proposition 2.5 *(a) If $G_1 \sim G_2$, then $G_2 \sim G_1$;*
(b) If $G_1 \sim G_2$ and $G_2 \sim G_3$, then $G_1 \sim G_3$.

Proposition 2.6 *If $G_1 \sim G_2$, $\forall$ a, the error values of a under G_1 and G_2 are equivalent.*

3. Application of equivalence

From Proposition 2.2, the rules for judging errors can be simplified by using the equivalence of rules. For example, in mathematical logic (Referring to Table 2.4), the nine operation rules for the logical expression formed by nine connectives are equivalent to the five operation rules for the logical expression formed by the five connectives ($\neg P$ or $\neg Q$, $P \wedge Q$, $P \vee Q$, $\rightarrow$, and $P \leftrightarrow Q$), which is also equivalent to the three operation rules for the logical expression formed by the three connectives ($\neg P$ or $\neg Q$, $P \wedge Q$, and $P \vee Q$). It is equivalent to the operation rule for the logical expression formed by the connective ($P \downarrow Q$). In this perspective, which rule is simple? Theoretically, operation rule for the logical expression formed by one connective is the simplest one. In practice, it is better to use operation rules for the logical expression formed by the five connectives (1, 2, 4, 5, and 8 in Table 2.4). The choice of operation rules is contingent on the actual situation.

Table 2.4 Operation rules for the logical expression formed by nine connectives

P	Q	1		2	3	4	5	6	7	8	9
		$\neg$		$\wedge$	$\uparrow$	$\vee$	$\downarrow$	$\rightarrow$	$\leftarrow$	$\leftrightarrow$	$\oplus$
		$\neg P$	$\neg Q$	$P \wedge Q$	$P \uparrow Q$	$P \vee Q$	$P \downarrow Q$	$P \rightarrow Q$	$P \leftarrow Q$	$P \leftrightarrow Q$	$P \oplus Q$
T	T	F	F	T	F	T	F	T	F	T	F
T	F	F	T	T	F	T	F	F	F	F	T
F	T	T	F	F	T	T	T	T	T	F	T
F	F	T	T	F	T	F	T	T	F	T	F

4. Changeability of equivalence

The only thing in the world does not change is change. Therefore, change is absolute and constant is relevant. The rules for judging errors change with time, space, academic fields, and objectives. Suppose that DK represents the equivalence relationship between rules G_1 and G_2, DK is a binary relationship between G_1 and G_2 with respect to time, space, academic fields, and objective.

2.4.3   Correlation Between the Rules for Judging Errors

1. Inclusion of rules

Definition 2.12 Suppose that G_1 and G_2 are two group of rules defined within U, if $G_1 \rightarrow G_2$, then G_1 is said to include G_2 noted by $G_1 \supset G_2$ or $G_2 \subset G_1$, G_2 is the subrule of G_1.

Proposition 2.7 *The inclusion has transitive property-i.e., $G_1 \subset G_2$, $G_2 \subset G_3$, then $G_1 \subset G_3$.*

Proposition 2.8 *If $G_1 \subset G_2$ and $G_1 \supset G_2$, then G_1 and G_2 are equivalent; if $G_1 \supset G_2$ and $G_2 \supset G_1 \rightarrow G_1 \sim G_2$.*

Proof
From definition 2.2, if $G_1 \supset G_2$ then $G_1 \rightarrow G_2$; if $G_2 \supset G_1$ then $G_2 \rightarrow G_1$. From definition 2.1, we know that $G_1 \sim G_2$.
Proof is completed.

2. Correlation between the rules for judging errors

Definition 2.13 Suppose that G_1 and G_2 are two group of arbitrary rules defined within U, if $\exists\ g_1 \neq \Phi$, $g_2 \neq \Phi$, $g_1 \subset G_1$, $g_2 \subset G_2$, $g_1 \sim g_2$ holds, then G_1 are correlated with G_2 noted by $G_1 \times G_2$. Otherwise, G_1 is independent of G_2 noted by $G_1\ D\ G_2$.

Proposition 2.9 *Correlation has symmetric property-i.e., $G_1 \times G_2 = G_2 \times G_1$.*

Proposition 2.10 *Independence has symmetric property-i.e., $G_1 \, D \, G_2 = G_2 \, D \, G_1$.*

Proposition 2.11 *Correlation does not have transitive property-i.e., $G_1 \times G_2$ and $G_2 \times G_3 \not\Rightarrow G_1 \times G_3$.*

Proof

Suppose that $G_1 = \{a_1, a_2, a_3\}$, $G_2 = \{a_1, a_4 \, a_5\}$, $G_3 = \{a_5, a_6 \, a_7\}$, where a_1, a_2, a_3, $a_4 \, a_5$, a_6, and a_7 are mutually independent, then $G_1 \times G_2$ and $G_2 \times G_3 \not\Rightarrow G_1 \times G_3$ hold.
Proof is completed.

Proposition 2.12 *Independence does not have transitive property, that is $G_1 \, D \, G_2$ and $G_2 \, D \, G_3 \not\Rightarrow G_1 \, D \, G_3$*

Proof Suppose that $G_1 = \{a_1, a_2, a_3\}$, $G_2 = \{g_1, g_2\}$, $G_3 = \{a_3, a_4 \, a_5\}$, where a_1, a_2, g_1, $g_2 \, a_3$, a_4, and a_5 are mutually independent, then $G_1 \, D \, G_2$ and $G_2 \, D \, G_3 \not\Rightarrow G_1 \, D \, G_3$ hold.
Proof is completed.

Proposition 2.13 *Equivalence is the special case of correlation*

Proof If $G_1 \times G_2$, $g_1 = G_1$, and $g_2 = G_2$, then $g_1 \sim g_2$ holds, and then $G_1 \sim G_2$.
Proof is completed.

3. Background related to the correlation of rules for judging errors

In reality, when evaluating the performance of a research project, a committee composed of multiple experts with different disciplinary background is used to improve the reliability and infallibility of the conclusion. Admittedly, this is one of the methods that can improve the robustness of the conclusion. However, it can not guarantee the reliability and infallibility of the conclusion being evaluated. The major reason is that each of the invited experts involved is treated as a group of rules for judging errors. Suppose that conclusion a is non-erroneous under rule g_i, g_i is a group of erroneous subrules and $g_i \subset G_i$ $(i = 1, 2, \ldots, n)$, it can not guarantee a is correct and infallible. From the above description, although a is non-erroneous if n rules $G_1 \times G_2 \times \ldots \times G_n$ are simultaneously used to judge certain object, the conclusion that a is non-erroneous is still not reliable. Therefore, we need to study the correlation between rules for judging errors.

4. Changeability of correlation

Due to existence of changeability in rules, the correlation between rules is dynamic and the relevant correlation forms a binary relationship between G_1 and G_2.

2.4.4 Good and Bad Rules for Judging Errors

Definition 2.14 Suppose that a and b are two piece of rules, if $a \sim b$, a is said to be better than b if a is more scientific and complete, simpler and easier to implement than that of b noted by $a > b$ (or $b < a$).

Definition 2.15 Suppose that $G_1 = \{a_1, a_2, \ldots, a_n\}$, $G_2 = \{b_1, b_2, \ldots, b_n\}$, if $\exists$ certain order that makes $G_1 = \{a_{j1}, a_{j2}, \ldots, a_{jn}\}$, $G_2 = \{b_{j1}, b_{j2}, \ldots, b_{jn}\}$, where $j_1, j_2, \ldots, j_n$ is certain permutation of $1, 2, \ldots, n$, then

(a) If $a_{j1} > b_{j1}, a_{j2} > b_{j2}, \ldots, a_{jk} > b_{jk}, 1 \leq k \leq n, a_{jk+1} = b_{jk+1}, a_{jn} = b_{jn}$ hold, then G_1 is better than G_2, or G_2 is worse than G_1 noted by $G_1 \geq G_2$ or $G_2 \leq G_1$;

(b) For $k = n$ in (a), G_1 is absolutely better than G_2 or G_2 is absolutely worse than G_1 noted by $G_1 > G_2$ or $G_2 < G_1$.

Proposition 2.14 *Transitive property of good and bad rules for judging errors*

(a) *If $G_1 \geq G_2$ and $G_2 \geq G_3 \Rightarrow G_1 \geq G_3$;*
(b) *If $G_1 \leq G_2$ and $G_2 \leq G_3 \Rightarrow G_1 \leq G_3$;*
(c) *If $G_1 > G_2$ and $G_2 > G_3 \Rightarrow G_1 > G_3$;*
(d) *If $G_1 < G_2$ and $G_2 < G_3 \Rightarrow G_1 < G_3$;*

2.4.5 Comparison Between Rules for Judging Errors

1. Necessity for studying comparability of rules for judging errors
   The axiom systems G_1 and G_2 composed of 15 axioms and five axioms, respectively have relationship of $G_1 \sim G_2$. Which axiom system is more favorable? For a "Phoenix model 26" bicycle, bad and good conclusions can be derived when using two different rules-i.e., $G_1 = \{$bicycle should be light, convenient, and aesthetically appealing$\}$ and $G_2 = \{$bicycle should be heavy-duty, durable, and reliable$\}$. Which rule is better and more scientifically valid? In short, it is common to investigate the pros and cons of two different rules, which makes it very necessary to examine the comparability of different rules.
2. Conditions for comparison
   In general, it is possible to compare the rules in the same category. It is meaningless to compare two distinct items which have no property to be compared. Therefore, when comparing two different rules for judging errors G_1 and G_2, they must be conducted in the same academic fields, spatial and temporal range, and research objective.
3. Foundations for comparison

   (1) If $G_1 \sim G_2$, and G_1 is the subset of G_2, then G_1 is better than G_2;
   (2) If $G_1 \sim G_2$, G_1 is easier implemented than G_2, then G_1 is better than G_2;
   (3) If G_1 is complete within U, and G_2 is incomplete within U, then G_1 is better than G_2 on U;
   (4) If G_1 is scientifically valid within U, and G_2 is scientifically invalid within U, then G_1 is better than G_2 on U.

   The indicator used for comparing two rules is $G = \{$one rule is the true subset of the other; one rule is easier implemented than the other; one rule is scientifically more valid than the other; and one rule is more complete than the other$\}$.

4. Implementation methods for comparison

   (1) Calculate the implementation cost of G_1 and G_2;
   (2) Evaluate the judgment results derived from two group rules;
   (3) Evaluate and appraisal rules based on expert opinion;
   (4) Deduction approach;
   (5) Comprehensive approach: combining (1) through (4).

5. Execution of comparison

   (1) Manual execution;
   (2) Calculating using computer;
   (3) Combining human and machine by using theory system or publicly accepted rules.

6. Changeability of comparison

Although comparison is conducted within the same fields, spatial and temporal range, and objectives, the comparison needs to be contingent on the change of rules for judging error since the rules have been changing with the academic field, temporal and spatial range, and research objectives.

2.4.6 Relationship Between Rules for Judging Errors and Object Being Judged

1. Four propositions

2. Is the value of a really erroneous in reality if it is judged as erroneous under rule G. For example, in the badminton competition at the 1988 Asian Games, a referee tendentiously favored the Korean player and intentionally made the wrong judgment for the Chinese player. In this case, the same results were given two different judgments based on different rules. In the above problem, two opposite conclusions- i.e., "shuttlecock fell out of bound" and "shuttlecock fell inside" when two rules G_1 = {the referee makes judgment based on where the shuttlecock actually fell} and G_2 = {the referee made judgment based on what he thought or he "wanted"} are used to make judgment. How should one explain the above conclusions? Hereby, the above-mentioned axioms (2.4.1) are used to obtain the following 4 propositions.

Proposition 2.15 *Suppose that a has error under rule G on universe of discourse U and G is non-erroneous, the conclusion that a has error under rule G is valid.*

Proposition 2.16 *Suppose that G_1 and G_2 are two different rules, $a \in U$ generally has two different error values under G_1 and G_2.*

Proposition 2.17 *For $a \in U$, $G = G_1 \cap G_2$, only is G actually used when a is judged under rules G_1 and G_2, then a has the same error value under G_1 and G_2.*

Proposition 2.18 *For a ∈ U, a has no error under rule G_1 and it is erroneous under rule G_2, either at least one of the two rules G_1 and G_2 is erroneous or G_1 and G_2 have different applicable scopes.*

2.5 Method for Identifying Errors

This section introduces the methods for identifying errors.

2.5.1 Flow Chart Used for Identifying Errors

Please refer to Fig. 2.5 for the flowchart used to identify errors.

2.5.2 Implementation Steps for Judging Errors

1. Preparation
   Preliminary research is conducted toward the problem of interests regarding the fields, academic range, and temporal and spatial characteristics to which the

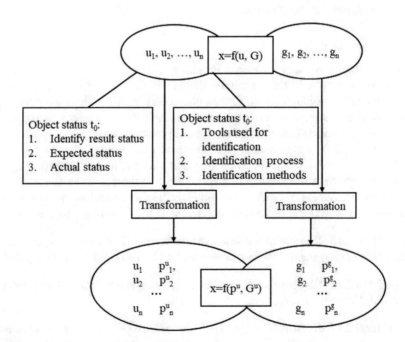

Fig. 2.5 Flow chart used for identifying errors

problem belongs. And the sequence, priority, route, emerging issues in investigation and associated solutions need to be clarified and well planned.

2. Investigation

   Investigation on the problem of interests can be conducted through:

   (1) Mail or email survey
   (2) Structured or unstructured brainstorming
   (3) Delphi method
   (4) Online survey
   (5) Field visit
   (6) Conference call (video or audio) or messaging
   (7) Or a combination thereof

   Each of the above methods undoubtedly has its strengths and weakness. Therefore, the choice of method and specific implementation and actions should be contingent on the dynamics of the problem of interests as well as the context in which the problem is embedded.

3. Comprehensive analysis

   Based on the investigation, comprehensive analysis is initiated to find out the pivotal factors and indicators for judging the problem of interests. At the same time, the factors and indicators must be accurate and accordance with the rules for judging errors.

2.5.3   Implementation Methods for Judging Errors

In the above steps, through comprehensive analysis, method for judging the problem of interests should be proposed and evaluated. Generally, selection of method is critical for the whole process, which directly affects if the judgment can be executed or the validity of the conclusion. The judgment methods can be categorized into the following five types:

1. Expert review Here three forms of expert evaluation are listed as follows.

   (1) Experts are invited to have a review meeting. Based on preset evaluation rules or emerging rules formed during evaluation process, experts provide their view, critics, and evaluation on relevant questions and the expert committee votes for the candidate scenarios or alternatives.
   (2) Delphi method: the organizer mails the problem or issue or topic needing evaluation to relevant experts and follows up and collects feedback from experts. Those reviews and critics are evaluated accordingly. In order to have reliable and convincing results, it might be necessary to repeat this process several times. One thing that needs to emphasize is the design of questions and descriptions, which must be concise and meaningful to save experts' time in comprehending them.

(3) One can take a combination thereof. This means the flexibility and cost-effectiveness of high time value and unavailability of long travel time slot of those well-recognized experts, the organizer can arrange either the mail/email evaluation or conference call to collect feedback from them. Or a representative can be sent to the expert's site to elicit their opinions or views.

2. Mathematical model

For those problems that can be quantified, variables are extracted to formulate mathematical model to gain better understanding of the problem of interests. In general, for an evaluation rule and the problem of interests, both distributed and centralized ways will be employed (if they do not have impacts on the evaluation results) when using mathematical models, which means that factors for the problem of interests are evaluated individually first and then problem is evaluated at the aggregate level. This operation brings more convenience to the evaluation than purely centralized evaluation and also facilitates the remote evaluation by experts located in different places. For example, when assessing the pros and cons of a decision, the aspects such as cost/benefit, value analysis, return on investment, social benefits, and loss can be evaluated using correlation matrix method.

3. Logical reasoning

Fig. 2.6 provides the steps for conducting logical reasoning.

Logical reasoning, starting from evaluation rules, is the process that compares the problems with the results derived using deductive and inductive approaches. Or logical reasoning is conducted on the conditions or relationships between the problem of interests and the derived results are compared with the conclusion obtained using rule system. This method is apt to evaluate the philosophical problem, certain theories, or certain problems in natural science.

4. Computer-aided expert system

For certain given field, spatial and temporal characteristics, having established a group of rules and one or multiple mathematical models, an expert system can be built to judge the errors that happen in the above-given context. This is our major research objective for next stage.

5. Artificial intelligence

Having finished collecting enough big data related to errors, rules, and relevant contexts in which errors happened, machine learning algorithm can be used to train the system, which helps the expert system to provide intelligent judgment for certain problem and make prediction for potential errors.

2.5.4 Analysis on the Implementation Effectiveness in Judging Errors

For a specific problem, it is necessary to analyze the implementation effectiveness after the processes of identifying and judging errors have been completed. One needs

Fig. 2.6 Flow chart used for identifying errors

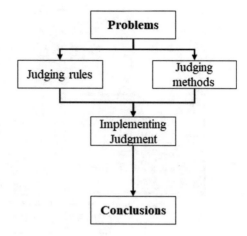

to check the validity of the rules, models, and the process used if the derived results have obvious conflict with reality.

2.6 Case Analysis for Error Identification

2.6.1 Addition of Binary Numbers

Figure 2.7 provides an example of identifying error process using addition of binary numbers. In this example, object of interest is $U = \{u_1 = (1 + 1 = 2), u_2 = (1 + 1 = 10), u_3 = (1 + 0 = 1)\}$; rule is defined as $G = \{g_1 = (0 + 0 = 0), g_2 = (0 + 1 = 1), g_3 = (1 + 0 = 1), g_4 = (1 + 1 = 10)\}$. As the object has been given, (1) the states of derived results; (2) should-be states; (3) true states are all given at time t_0.

Expert review
Based on the definition for error, the expert gives the following result $f = f(u_1 = (1 + 1 = 2), G = \{g_1 = (0 + 0 = 0), g_2 = (0 + 1 = 1), g_3 = (1 + 0 = 1), g_4 = (1 + 1 = 10)\}) = 1$. As $u_1 = (1 + 1 = 2)$ is in contradiction to $g_4 = (1 + 1 = 10)$, the object $u_1 = (1 + 1 = 2)$ is erroneous under rule $G = \{g_1 = (0 + 0 = 0), g_2 = (0 + 1 = 1), g_3 = (1 + 0 = 1), g_4 = (1 + 1 = 10)\}$, where $f = f(u, G)$ is classic error function.

Similarly, $f = f(u_2 = (1 + 1 = 10), G = \{g_1 = (0 + 0 = 0), g_2 = (0 + 1 = 1), g_3 = (1 + 0 = 1), g_4 = (1 + 1 = 10)\}) = 0$, so the object $u_2 = (1 + 1 = 10)$ is correct under rule $G = \{g_1 = (0 + 0 = 0), g_2 = (0 + 1 = 1), g_3 = (1 + 0 = 1), g_4 = (1 + 1 = 10)\}$;

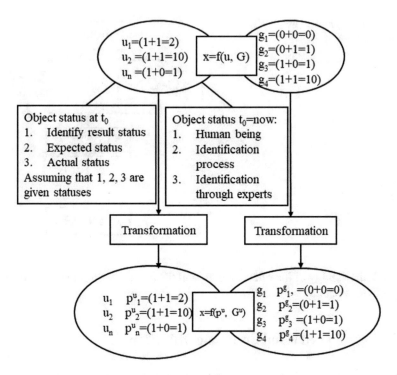

Fig. 2.7 Example of identifying errors-addition of binary numbers

$f = f(u_3 = (1 + 0 = 1), G = \{g_1 = (0 + 0 = 0), g_2 = (0 + 1 = 1), g_3 = (1 + 0 = 1), g_4 = (1 + 1 = 10)\}) = 0$, so the object $u_3 = (1 + 0 = 1)$ is correct under rule $G = \{g_1 = (0 + 0 = 0), g_2 = (0 + 1 = 1), g_3 = (1 + 0 = 1), g_4 = (1 + 1 = 10)\}$.

Logical reasoning
Based on the definition for error, $f = f(u_1 = (1 + 1 = 2), G = \{g_1 = (0 + 0 = 0), g_2 = (0 + 1 = 1), g_3 = (1 + 0 = 1), g_4 = (1 + 1 = 10)\}) = (1$, if $p^{u_i} = \neg\ p^{g_i}$, $(i = 1, 2, 3, 4)$, 0 otherwise). Because $u_1 = (1 + 1 = 2)$ and $g_4 = (1 + 1 = 10)$ in G are propositions with mutual incompatibility (Antinomy), so $u_1 = (1 + 1 = 2)$ is erroneous under rule $G = \{g_1 = (0 + 0 = 0), g_2 = (0 + 1 = 1), g_3 = (1 + 0 = 1), g_4 = (1 + 1 = 10)\}$.

Similarly, $f = f(u_2 = (1 + 1 = 10), G = \{g_1 = (0 + 0 = 0), g_2 = (0 + 1 = 1), g_3 = (1 + 0 = 1), g_4 = (1 + 1 = 10)\}) = 0$, so the object $u_2 = (1 + 1 = 10)$ is correct under rule $G = \{g_1 = (0 + 0 = 0), g_2 = (0 + 1 = 1), g_3 = (1 + 0 = 1), g_4 = (1 + 1 = 10)\}$; $f = f(u_3 = (1 + 0 = 1)\ G = \{g_1 = (0 + 0 = 0), g_2 = (0 + 1 = 1), g_3 = (1 + 0 = 1), g_4 = (1 + 1 = 10)\}) = 0$, so the object $u_3 = (1 + 0 = 1)$ is correct under rule $G = \{g_1 = (0 + 0 = 0), g_2 = (0 + 1 = 1), g_3 = (1 + 0 = 1), g_4 = (1 + 1 = 10)\}$. Because G_1 and G_2 are scientifically proven correct on the universe of discourse U, the above two conclusions are true.

This example tells us: (1) for the same problem, different error values (conclusions) can be derived under different rules-i.e., the same problem has different explanations in different fields; (2) if a problem is erroneous under certain rule, it does not mean that the problem is always erroneous or completely erroneous in reality. For many issues and facts in the history of human being, they were just "so-called" facts measured and collected by people during the period that relevant events happened, which to great extent reflected the subjective (or biased) understanding, explanation, and abstraction from the people involved. Due to the constraints in measurement tool and availability of necessary data and information, many rules and laws in certain historical stage deemed correct are proven invalid or even absurd with the development of history and the availability of more advanced technologies and measurement techniques. Therefore, error always has relativity. An object might be erroneous in realistic world even though it has no error under rule G. An object can also change from having no error under rule G to the situation that error emerges with change in time and conditions under the same rule.

2.6.2 Case in Identifying Trademark Error

Figure 2.8 presents the process for identifying errors related to trademark. In this example, $U = \{u_1 = $ Chongqing (toothpaste), $u_2 = $ Flying pigeon (bicycle), $u_n = $ Trademark $n\}$. As this object is not in the form of proposition, it is necessary to convert all the features corresponding to the object (under associated rules) into propositions. They are presented as follows:

$u_1 = $ Chongqing(toothpaste): $p_1 = $ Chongqing trademark is concise; $p_2 = $ Chongqing trademark is simple; $p_3 = $ Chongqing trademark is easy to write; $p_4 = $ Chongqing trademark is easy to read; $p_5 = $ Chongqing trademark is legible; $p_6 = $ Chongqing trademark is easy to memorize; $p_7 = $ Chongqing trademark has good feeling; $p_8 = $ Chongqing trademark has no unpleasant expression; $p_9 = $ Chongqing trademark is easy to pronounce; $p_{10} = $ Chongqing trademark has good feeling when reading; $p_{11} = $ Chongqing trademark has multiple pronunciation; $p_{12} = $ Chongqing trademark is eligible for packaging or labeling; $p_{13} = $ Chongqing trademark has no cliche; $p_{14} = $ Chongqing trademark has no confusion with registered trademarks; $p_{15} = $ for exported products, Chongqing trademark's pronunciation should be eligible; $p_{16} = $ Chongqing trademark has no unpleasant, dirty and pessimistic meaning; $p_{17} = $ Chongqing trademark has no confusion with other language; $p_{18} = $ Chongqing trademark provides hint for the purpose of the product. As the object has been given, (1) the states of derived results; (2) should-be states; (3) true states are all given at time t_0.

Expert review
Based on the definition for error, the expert gives the following result $f = f(u_1 = $ Chongqing(toothpaste), $G = \{g_1, g_2, g_3, \ldots \ldots, g_{18}\}) = 2$. As in $u_1 = $ Chongqing (toothpaste), $p_{11} = $ "Chongqing trademark has multiple pronunciation" is in

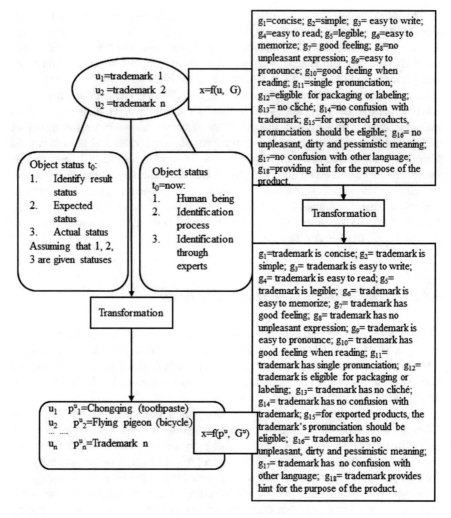

Fig. 2.8 Structure demonstrating the brand error recognition

contradiction to $g_{11}=$ "Chongqing trademark has single pronunciation"; $p_{18}=$ "Chongqing trademark does not provide hint for the purpose of the product" is in contradiction to $g_{18}=$ "Chongqing trademark provides hint for the purpose of the product". In conclusion, the object $u_1 =$ Chongqing(toothpaste) is erroneous under rule $G = \{g_1, g_2, g_3, \ldots \ldots, g_{18}\}$ and there exist two errors here.

Similarly, $f = f(u_2 =$ Flying pigeon (bicycle), $G = \{g_1, g_2, g_3, \ldots \ldots, g_{18}\}) = 0$. For object $u_2 =$ Flying pigeon (bicycle), it is correct under rule G.

Logical reasoning

Based on the definition for error, $f = f(u_1 = $ Chongqing(toothpaste), $G = \{g_1, g_2, g_3, \ldots \ldots, g_{18}\}) = (1$, if $p^{u_i} = \neg\, p^{g_i}, (i=1, 2, \ldots \ldots, 4), 0$ otherwise$) = 2$. As in $u_1 = $ Chongqing(toothpaste), $p_{11} = $ "Chongqing trademark has multiple pronunciation" is in contradiction to $g_{11} = $ "Chongqing trademark has single pronunciation"; $p_{18} = $ "Chongqing trademark does not provide hint for the purpose of the product" is in contradiction to $g_{18} = $ "Chongqing trademark provides hint for the purpose of the product", i.e., $p_{11} == \neg\, g_{11}$ and $p_{18} == \neg\, g_{18}$. Therefore, the object $u_1 = $ Chongqing(toothpaste) is erroneous under rule $G = \{g_1, g_2, g_3, \ldots \ldots, g_{18}\}$ and there exist two errors here.

Similarly, $f = f(u_2 = $ Flying pigeon (bicycle), $G = \{g_1, g_2, g_3, \ldots \ldots, g_{18}\}) = 0$. For object $u_2 = $ Flying pigeon (bicycle) under $G = \{g_1, g_2, g_3, \ldots \ldots, g_{18}\}$, there does not exist $p_i == \neg\, g_i, (i = 1, 2, \ldots \ldots, 18)$. So, $u_2 = $ Flying pigeon (bicycle) has no error under rule G.

Chapter 3
Basic Structures of Systems

Abstract The essence of system science resides in the philosophy of holism. When talking about the state that system reaches optimal, it generally refers to global optimum of the whole system with respect to the objective features that are included in the intrinsic features of a system. The attainment of global optimum of a system must rely on the normal deployment of functions (features) of its subsystems. The word 'system' originates from Latin word *systēma* meaning that a whole is made of several parts or members. Many different definitions had been made by scholars for system from different perspectives based on their particular research objectives. Let's give some examples of them. "System is a pre-given set composed of elements and their normal behaviors"; "System is a well-organized wholeness"; "System is an entity made of connected materials and processes;" "System is a body composed of ordered elements/factors working towards a common goal" are some popular examples for the definition of system. The development of system theory and their applications in different field were mainly attributed to the contributions of scholars such as biologist Ludwig von Bertalanffy (1901–1972), Norbert Wiener (1894–1964), Ross Ashby (1903–1972), John Henry Holland (1929–2015), and Murray Gell-Mann (1929–2019). Bertalanffy pioneered the general system theory by introducing models, principles, and laws that apply to it. Wiener and Ashby used mathematics to study systems. Holland, Gell-Mann, and others proposed the term "complex adaptive system". General system theory attempts to provide a definition that can capture common properties of various systems. The definition for general system is: a body composed of well-organized elements working toward attaining particular goals or features. This definition apparently includes 4 concepts and their relationships, namely system, element, structure, feature, relationships between elements, relationships between elements and structure, and relationships between system and external environment. The purpose of system theory is to investigate form, structure, and laws of general systems, to examine the common properties of those systems, to capture and illustrate their features using mathematical methods, and consequently to identify the mechanisms, rules, laws, principles, and mathematical models that can be applied to general systems. And the ultimate objective of learning system theory is to use the understanding on the system to better manage, control, renovate, or change the current system structures (natural or man-made systems) to align them with the needs of our civilized world. With better understanding on the

© Springer Nature Switzerland AG 2020
K. Guo and S. Liu, *Error Systems: Concepts, Theory and Applications*,
Studies in Systems, Decision and Control 275,
https://doi.org/10.1007/978-3-030-40760-5_3

system structures, we can introduce all kinds of interventions or policies to enable the systems of interest attain their optimal performance or outcomes. Moreover, by gaining better understanding on the dynamics of a system over time, decision/policy makers and practitioners can prevent policy-resistance (counter-intuitive behaviors). System theory is recognized a discipline that possesses both mathematical and logic characteristics. System theory proclaims that holism, connectedness, hierarchical structure, and dynamic equilibrium, time-dependence are common properties of all systems, which are both philosophy of system thinking and principles of using system approach. As a branch of scientific approaches, system theory helps identify the objective laws on how world is running and also offers human being a way of thinking the world. Therefore, system theory is also called system approach since it can represent concept, view, model, and mathematical methods as well. In Bertalanffy's masterpiece titled "General System Theory; Foundations, Development, Applications", he emphasized the concept of holism. System, as a organic body, is not mechanical combination or simple addition of its constituents but an organic combination of its elements working together towards a common goal. The system's features are emerging behaviors, which can not be found in its individual elements or subsystems. By quoting Aristotle's "*A whole is greater than the sum of its part,*" Bertalanffy opposed those mechanical philosophy that the wholeness (system behaviors) can be observed or inferred from the behavior of a particular element of the system. He also stated that each element of a system is in a particular location in the system hierarchy, which is also tightly coupled with other elements. The connectedness among system's elements renders system integral and holistic. The "should-be" function of a system's element will disappear once it is separated from the system structure. For example, having done the hand amputation due to traumatic injury, the removed 'limb' would never function as it should when it was an integral part of a person. The fundamental thoughts of system theory is to treat the object being investigated as a system and to analyze the structure, function, dynamic relationships between elements, system, and their environment. With better understanding on the dynamics, complexities, and uncertainties associated with the system, the ultimate goal is to find how it attain its optimal target and, consequently provide counterfactual analysis when interventions are needed to be implemented in the system. Systems are ubiquitous in the universe. From cosmos to the microscopic world, systems exist everywhere such as Milky Way, solar system, earth system, social system, transportation system, production system, human body system, bacterial system, cell system, and atom. The emergence of system theory brought profound changes on the way how people think about the world. In conventional research practices, Descartes' philosophy of 'reductionism' had been dominating the academic fields. Under such influence, the general practice in research is to divide a complicated issue or object into multiple parts and investigate each part individually. Thereafter, the characteristics of those individual parts are then used to infer the behaviors of the original issue or object. The reductionisim approach focuses on local substructures or elements and abides by the unidirectional causal-effect determinism. Although this approach had proved valid for centuries within certain confined ranges and had served as the most popular way of thinking in mainstream research communities, it can only handle simple issues or objects without being able to capture the wholeness, dynamic interactions, and circular causalities of complicated objects (i.e., systems in the language of system theory). With accelerated development in economy, technology, and society, human beings with traditional analytical thinking became incompetent in dealing with issues/objects with thousands or even millions of variables connected/networked in various ways. However, the emergence of system theory, cybernetics, and informatics paved the way for human beings to drive the rapid advancement of modern science and technologies. The widespread applications of system theory have made it become the basis for developing new theories in handling complicated system in the fields of politics, economy, military, culture, science, and society, etc. Regarding the trend of system theory, the authors think it is moving towards the formation of unified framework that summarizes the achievements obtained from the empirical and theoretical research in different fields. System thinking ensued by system theory has become a very powerful force to overturn the ingrained singular causation thinking. For ease of studying system, many ways are used to categorize system: (1) natural systems and artificial systems (whether designed by human being or not); (2) natural systems, social systems, and thinking systems (according to research subject); (3) macro systems, mesa system, micro systems, and microscopic systems(scale of the systems); (4) simple systems, complex systems (in term of structure); (5) simple small systems, simple large systems, simple giant systems, and complex giant systems, etc.(scale and structure); (6) open systems, closed systems (whether there exists interaction with environment); (7) balanced systems (systems having equilibrium), non-equilibrium systems, near-equilibrium systems, and far-from-equilibrium systems (whether there exists equilibrium).

3.1 System Structure

The totality for type and order of connections, organizations, and interactions of elements within a system is called structure. System structure can be categorized into three types: (1) temporal and spatial structures; (2) symmetric and asymmetric structures; and (3) soft and hard structures. There are typical characteristics in the system structure i.e., relative stability and absolute changeability. Relative stability means that the system tends to keep certain state (inertia of system). And absolute changeability indicates that system changes in a shape just like a parabolic helix, which is caused by the changes of the environment in which the system is embedded. System's environments are all things or objects outside the system, which have certain interactions with system. For instance, the social context and infrastructure in which a resident community resides are the community system's environment. The changes in system's environment exert impacts on the system which eventually lead to the changes in the system structure, and consequently the system behaviors. The change in weather can cause corresponding changes in people's physiological rhythm or make people sick. Of course, a system can also put impact on the environment such as the negative externalities caused by the transportation system.

System structure determines system functionality. For example, carbon atoms are the only elements of graphite and diamond and their structures determine their distinct characteristics. However, the proposition can not hold unless the following preconditions must be met:

1. System's elements are apt to achieve certain functionality;
2. System's environment have significant impact on its functionality;
3. System reacts to environment's action or impact;
4. System's structure and function are interdependent.

In the interaction between system and its environment, the favorable outputs produced by system is called functionality and the undesirable outputs are called pollution to the environment. The environment possesses the capability of providing energy and necessary elements to system, bio-accumulating and bio-remediating the outputs of system, and degrading undesirable output to harmless materials, which forms a recycle. A virtuous cycle where the undesirable outputs of system do not exceed the carrying capacity of the environment is critical to enable sustainable development of the whole ecosystem. This is why we need to gain deeper understanding on the system structure and its relationship with the system behaviors and its environment. Specifically, in order to investigate errors in a system or achieve system optimization, one must continuously examine system's structure since it is evolving over time.

3.1.1 Basic Structures of Systems

Since there are infinite number of system structures, it is impossible to investigate each of them. Therefore, we have to put emphasis on the examination of the very basic structures (protostructure) and then investigate the way how more complicated structures are formed out of those basic structures. Based on the definition of system structure in Sect. 3.1, in the following session, we mainly study the temporal and spatial dynamics of the patterns and orders that system elements are connecting, organizing, and interacting with each other.

Definition 3.1 Suppose that the system element set E is corresponding to node set N in graph theory ($E \rightarrow N$), the set C for patterns and orders on how system elements are connecting, organizing, and interacting with each other is corresponding to the link set L in graph theory ($C \rightarrow L$), the graph formed by system basic structure is called system structure graph. Different structure graphs are discussed as follows.

3.1.2 Series Connections

Figure 3.1 shows the basic series connections between system elements.

The following examples contain subsystems which have series connections between system elements (Figs. 3.2, 3.3, 3.4 and 3.5).

Due to the characteristics of connection, the functionality realization of each element in the structure of series connection is contingent on their connection. The function for describing certain functionality is denoted by $CSG(s_1, s_2, \ldots \ldots s_n)$.

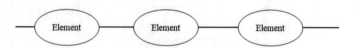

Fig. 3.1 Series connections between system elements

Fig. 3.2 Series connections
between system elements in
real application-graph 1

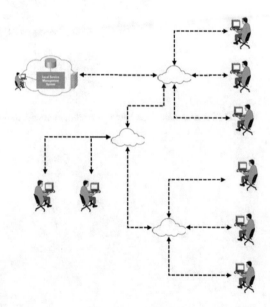

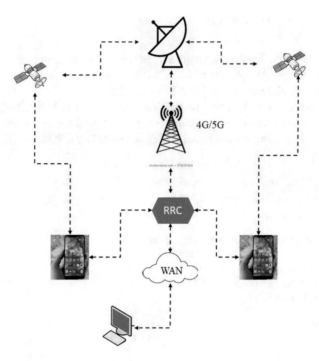

Fig. 3.3   Series connections between system elements in real application-graph 2

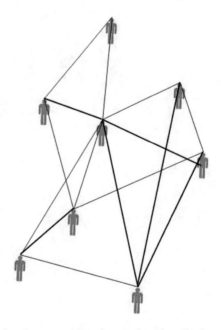

Fig. 3.4   Series connections between system elements in real application-graph 3

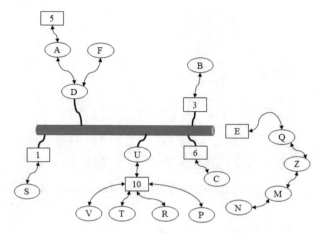

Fig. 3.5 Series connections between system elements in real application-graph 4

3.1.3 Parallel Connections

Parallel connection among system elements is demonstrated in Fig. 3.6.

The following examples include subsystem parallel connections between system elements (Fig. 3.7).

The function for describing certain functionality in system with parallel connections is denoted by $BSG(s_1, s_2, \ldots \ldots s_n)$.

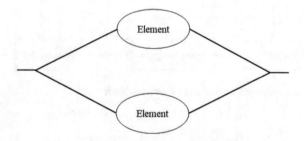

Fig. 3.6 Parallel connection among system elements

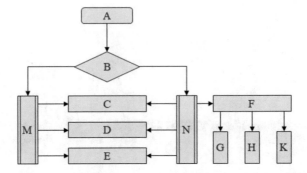

Fig. 3.7 Example demonstrating system with parallel connection among system elements

3.1.4   Connections with Feedback

Figure 3.8 shows structure graph having connections with feedback (could be either reinforcing loop or balancing loop).

The elements in system with feedback structure are interdependent on each other. The growth of element (state of an element) will be enhanced in a reinforcing loop

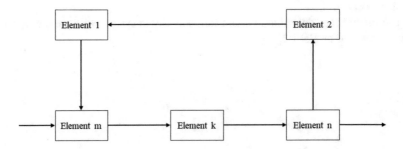

Fig. 3.8 Structure graph having connections with feedback

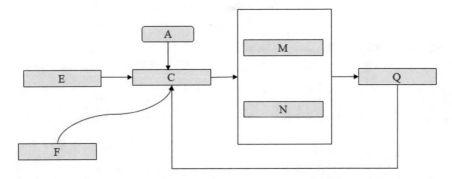

Fig. 3.9 Example demonstrating system with structure containing feedback connections

and will be weakened (dampened)in a balancing loop. The function for depicting certain functionality in system with feedback connections is denoted by $FSG(s_1, s_2, \ldots \ldots s_n)$ (Fig. 3.9).

3.1.5 Expanding and Shrinking Structure

1. Expanding structure

   Expanding structure acts like a radiating shape. In general, the functionality of leading element (mother element) in radiating structure will influence functionality realization of following elements (Fig. 3.10).

Fig. 3.10 Expanding structure

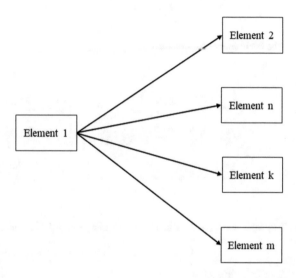

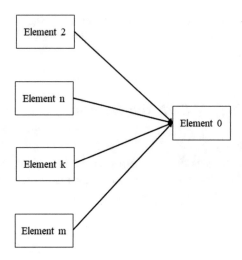

Fig. 3.11 Shrinking structure

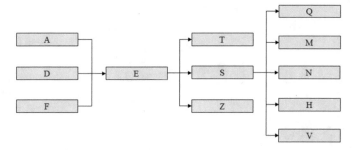

Fig. 3.12 Example demonstrating expanding and shrinking structures

2. Shrinking structure

   Shrinking structure looks like a converging shape. In general, each functionality of leading elements in converging structure will influence the functionality realization of ending element (Figs. 3.11 and 3.12).

   The function for depicting certain functionality in system with expanding and shrinking structures is denoted by $KSG(s_1, s_2, \ldots \ldots s_n)$.

3.1.6   Inclusion Structure

Basic shape of inclusion structure (Fig. 3.13 and 3.14):

   The function for depicting certain functionality in system inclusion structures is denoted by $YSG(s_1, s_2, \ldots \ldots s_n)$.

Fig. 3.13 Inclusion structure

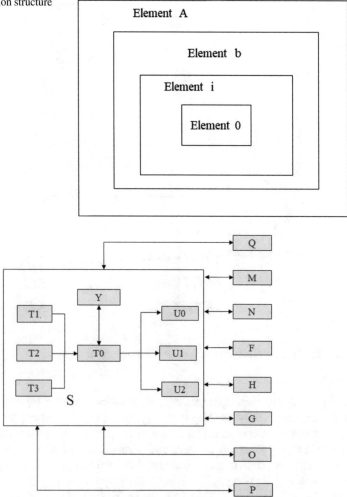

Fig. 3.14 Example demonstrating system inclusion structure

In more complicated system, the structure generally contains the combination of the above-mentioned structures or connections. Figure 3.15 exhibits a mildly complicated system having multiple aforementioned connections.

In Fig. 3.15, one can see that the included structure **S** can be decoupled into several subsystems-i.e., S_1, S_2, and S_3 where S_1 is a shrinking structure, S_2 is an expanding structure, and S_3 is an independent element.

As for any system, it is composed of subsystems in the downward hierarchies. Therefore, subsystem can be treated as an element of system/*subsystem* by the upward hierarchies. The **5** basic structures, namely (a) series structure, (b) parallel structure,

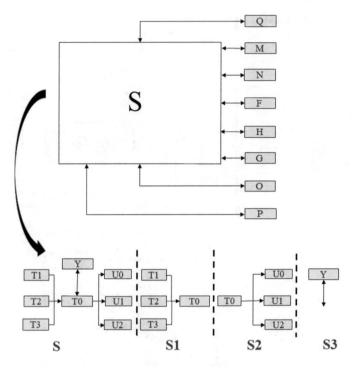

Fig. 3.15 Example demonstrating system with combined structure

(c) feedback structure, (d) expanding and shrinking structure, (e) inclusion structure, and other types can be re-categorized. Types (a), (b), and (d) can be represented by $m \times n$ format where m stands for the number of parallel series structures from starting point(observing from left to right) and n stands for the number of parallel series structures at the ending point(observing from left to right). Type (a) is $m = n = 1$ (b) is $m = n \geq 2$ (d) expanding is the case of $m = 1$ and $n \geq 2$ and shrinking is the case of $m \geq 2$ and $n = 1$. Therefore, the basic types of system are (1) $m \times n$ type, (2)feedback structure, (3) inclusion structure, and (4)other types.

Chapter 4
Structures of Error Systems

Abstract In order to gain better understanding upon the root causes, mechanism, and laws for generating errors, we must resort to the research on the system in which errors occur-i.e., error system. As system structure determines system behavior, therefore, we need to gain in-depth understanding on the study of temporal and spatial dynamics of the patterns and orders that system elements are connecting, organizing, and interacting with each other.

4.1 Basic Structures of Error System

4.1.1 Hierarchical Structure of Error Systems

1. Horizontal connections of elements in error system (Fig. 4.1)
2. Vertical connections of elements in error system
   The errors of different elements in series structure have mutual impact on each other. And the function for error of series structure system is expressed as $\mathrm{CSC}(s_1, s_2, \ldots \ldots s_n)$.
3. Hybrid vertical and horizontal connections of elements in error system
   This structure can be further decomposed into series connection, parallel connections, and expanding and shrinking connections (Figs. 4.2 and 4.3).

4.1.2 Chain Structure of Error System

1. Series structure of error system
   The errors of different elements in series structure have mutual impacts on each other. The series structure system is $\mathrm{CS}(s_1, s_1, \ldots \ldots, s_n)$ and the function for error is $\mathrm{CSC}(s_1, s_1, \ldots \ldots, s_n)$. The expression of logical proposition for error in series structure is expressed by: $X(S) = X(s_1) \vee X(s_2) \vee \ldots \ldots \vee X(s_n)$ (Figs. 4.4 and 4.5).

© Springer Nature Switzerland AG 2020

K. Guo and S. Liu, *Error Systems: Concepts, Theory and Applications*,
Studies in Systems, Decision and Control 275,
https://doi.org/10.1007/978-3-030-40760-5_4

Fig. 4.1 Horizontal connection

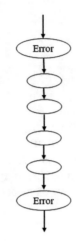

Fig. 4.2 Vertical connection of elements in error system

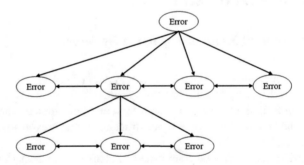

Fig. 4.3 Hybrid vertical and horizontal connections of elements in error system

2. Parallel structure of error system (Fig. 4.6)
   The parallel structure system is $BS(s_1, s_1, \ldots \ldots, s_n)$ and the function for error is $BSC(s_1, s_1, \ldots \ldots, s_n)$. The expression of logical proposition for error in basic parallel structure is expressed by: $X(S) = (X(s_{11}) \vee X(s_{12}) \vee \ldots \ldots \vee X(s_{1n}))$ $\wedge (X(s_{21}) \vee X(s_{22}) \vee \ldots \ldots \vee X(s_{2n}))$.
3. Expanding and shrinking structures

   (1) Expanding structure of error system
   (2) Shrinking structure of error system (Fig. 4.7).

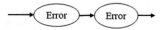

Fig. 4.4 Basic series structure of error system

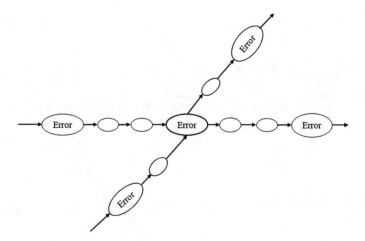

Fig. 4.5 Example for series structure of error system

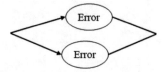

Fig. 4.6 Basic parallel structure of error system

The error at the starting points (left side of Figs. 4.8 and 4.9) in an expanding structure generally affects the errors in the ending elements. Any feature of any element in the left side (Figs. 4.10 and 4.11) in a shrinking structure generally affects the function of the ending elements. The expanding and shrinking structure of error system is $KS(s_1, s_1, \ldots \ldots, s_n)$ and the function for error is $KSC(s_1, s_1, \ldots \ldots, s_n)$. The expression of logical proposition for error in expanding and shrinking structure is expressed by: $X(S) = (X(s_1) \vee X(s_2) \vee \ldots \ldots \vee X(s_n)) \vee X(s_0)$ (Figs. 4.9 and 4.11).

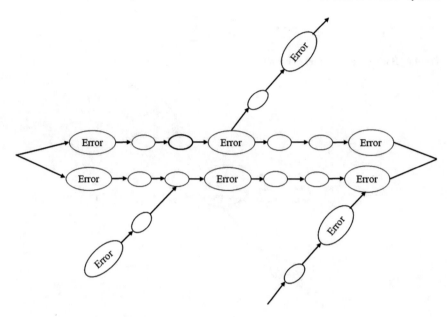

Fig. 4.7 Example for parallel structure of error system

Fig. 4.8 Basic expanding
structure of error system

Fig. 4.9 Example for
expanding structure of error
system

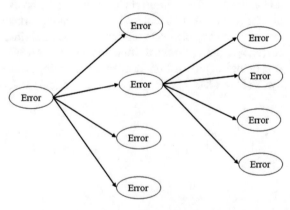

Fig. 4.10 Basic shrinking
structure of error system

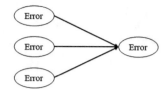

Fig. 4.11 Example for
shrinking structure of error
system

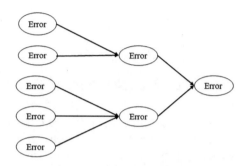

4.1.3 Inclusion Structure of Errors

In the inclusion structure of error system, although an error will definitely affect the
error of the whole system, an error in an element does not necessarily exert impact
on errors of other elements.

1. Basic inclusion structure of error system (Fig. 4.12)
2. Centered inclusion structure of error system (Fig. 4.13)
3. Multi-layered inclusion structure of error system (Fig. 4.14).

The inclusion structure of error system is $YS(s_1, s_1, \ldots \ldots, s_n)$ and the function
for error is $YSC(s_1, s_1, \ldots \ldots, s_n)$. The expression of logical proposition for error
in inclusion structure of error system is expressed by: $X(S) = X(s_1) \vee X(s_2) \vee \ldots$
$\ldots \vee X(s_n)$.

Fig. 4.12 Basic inclusion
structure of error system

Fig. 4.13 Centered
inclusion structure of error
system

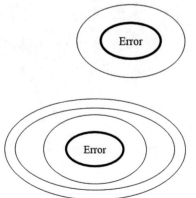

Fig. 4.14 Multi-layered
inclusion structure of error
system

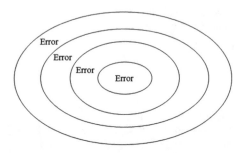

4.1.4 Feedback Structure of Errors

In feedback structure, the errors in different element of the system have mutual
impacts. Reinforcing (positive) feedback structure produces growth behavior while
balancing (negative) feedback structure generates balancing effects and tends to
decrease in size (Figs. 4.15 and 4.16).

   The feedback structure of error system is $FS(s_1, s_1, \ldots \ldots, s_n)$ and the function
for error is $FSC(s_1, s_1, \ldots \ldots, s_n)$. The expression of logical proposition for error
in feedback structure of error system is expressed by: $X(S) = X(s_1) \vee X(s_2) \vee \ldots$
$\ldots \vee X(s_n)$.

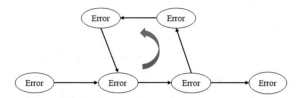

Fig. 4.15 Basic feedback structure of error system

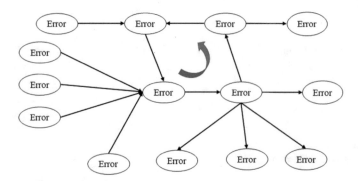

Fig. 4.16 Example for feedback structure of error system

With previous preliminary analysis, we found six types of basic structure: (1) vertical structure, horizontal structure, and part of the combined vertical and horizontal structure belong to series structure; (2) juxtaposition of multiple series structure and part of the combined vertical and horizontal structure belong to parallel structure; (3) expanding and shrinking structure; (4) inclusion structure; (5) feedback structure; (6) and other structures.

4.1.5   Changeable Structures of Error Systems

1. Transformation structure of error system (Fig. 4.17)
2. Structure with gradual change of error system (Fig. 4.18).

Transformation and gradual change of error systems are taking the forms of 6 basic transformations and their inverse transformations.

Transformation paths

(1) Universe of discourse
(2) Subsystems (or elements)
(3) Structure

    (a) Series structure
    (b) Parallel structure
    (c) Expanding and shrinking structure

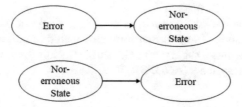

Fig. 4.17 Transformation structure of error system

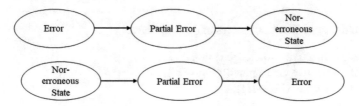

Fig. 4.18 Structure with gradual change

  (d) Inclusion structure
  (e) Feedback structure
  (f) Other structures

(4) Time.

Transformation approaches

(1) Similarity or equivalence transformation $T_x \subset \{T_{xly}, T_{xzx}, T_{xys}, T_{xjg}, T_{xsj}\}$ (similarity or equivalence); T_x^{-1} is the inverse similarity transformation. The similarity transformation includes: similarity transformations on universe of discourse, subsystems, elements, structure, and time;

(2) Displacement transformation $T_z \subset \{T_{zly}, T_{zzx}, T_{zys}, T_{zjg}, T_{zsj}\}$ (displacement); T_z^{-1} is the inverse displacement transformation. The displacement transformation includes: displacement transformations on universe of discourse, subsystems, elements, structure, and time;

(3) Addition transformation $T_{zn} \subset \{T_{znly}, T_{znzx}, T_{znys}, T_{znjg}, T_{znsj}\}$ (addition); T_{zn}^{-1} is the inverse addition transformation. The addition transformation includes: addition transformations on universe of discourse, subsystems, elements, structure, and time;

(4) Decomposition transformation $T_f \subset \{T_{fly}, T_{fzx}, T_{fys}, T_{fjg}, T_{fsj}\}$ (decomposition); T_f^{-1} is the inverse decomposition transformation. The decomposition transformation includes: decomposition transformations on universe of discourse, subsystems, elements, structure, and time;

(5) Destruction transformation $T_h \subset \{T_{hly}, T_{hzx}, T_{hys}, T_{hjg}, T_{hsj}\}$ (destruction); T_h^{-1} is the inverse destruction transformation. The destruction transformation includes: destruction transformations on universe of discourse, subsystems, elements, structure, and time;

(6) Unit transformation $T_d \subset \{T_{dly}, T_{dzx}, T_{dys}, T_{djg}, T_{dsj}\}$ (unit); T_d^{-1} is the inverse unit transformation. The unit transformation includes: unit transformations on universe of discourse, subsystems, elements, structure, and time;

(7) Transformation systems (AND, OR, INVERSE).

4.1.6 Fuzzy Structures of Error Systems

Here we list some fuzzy structures of error systems.

(1) Fuzzy subjective knowledge (Fig. 4.19);
(2) Fuzzy objective environment (Fig. 4.20);
(3) Fuzzy handling (Fig. 4.21);

Fig. 4.19 Fuzzy subjective
knowledge

Fig. 4.20 Fuzzy objective
environment

Fig. 4.21 Fuzzy handling

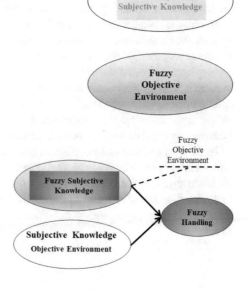

4.1.7 Stability of Error Systems

There are 3 types of stability in error systems.

1. Conditional stability: system reaches stable when some conditions are met.
2. Unconditional stability: system is stable under any condition.
3. Stable within a range: system is stable within certain range.

The stability of structure is not a basic structure but a feature of a system.

4.1.8 Discussion

In the process of investigating the structure of error system, we have discussed not
only 6 basic structures that error systems might possess but also other basic structures
of errors. The basic structures here are fundamentally the same as the static basic
structures of general systems. Therefore, the basic structures of error systems can
be categorized into: (1) static basic structures of error system which have series
structure, parallel structure, expanding and shrinking structure, inclusion structure,
feedback structure, and other basic structure; (2) dynamic basic structure of error
systems which include gradual change structure, complete transformation structure,
and fuzzy structure. Based on the analysis, we know that system structure plays
critical roles in the study and applications of systems science. As a result, in order

to unfold the mechanisms for generating errors and realizing system optimization, it is necessary to investigate both system structure and the impacts of system structure on error generation and system optimization.

In order to gain better understanding on the structures of error system and general systems, basic structures that construct them are examined. Then, more complicated systems composed of the basic structures are explored. With the basic structures for error system and general systems, the measures for preventing and eliminating errors in simple error system can be learned and exercised. Similarly, the optimization of simple error system composed of basic structures can be addressed. Given that the simple error system is well investigated, the global system optimization of complex system composed of basic structures can be easily obtained.

In most cases, it is not possible to uncover and capture all the characteristics of errors at the very early stage of studying them, 3 steps can be adopted to: (1) study the basic structures of complex system; (2) investigate the laws and approaches for understanding the errors and optimization in the basic structures; (3) examine the features, error portrayal, and optimization in the system constructed by simple system structures.

Chapter 5
Structural Change of Error Systems

Abstract A person alive can become inorganic matter after he dies. A person of role model can become an evildoer to the society due to the influences of internal factors and his surrounding environment. Therefore, a system without error becoming an error system is generally caused by change, transition, or transformation of original system structures. There was an old story in China about how an old wise farmer had divided his heirs to his three sons. This old farmer only had 17 camels. His distribution scenario was to give half of his wealth to the oldest son, $1/3$ to the second son, and $1/9$ to the youngest son. However, it was not possible for him to have the scenario implemented since he only had odd number of camels. His neighbor just passed by as he was contemplating how he should do it. Having understood the farmer's difficulty, the neighbor decided to lend the farmer one camel. Right now, the old farmer had 18 camels to be divided. The oldest son then got 9 of the 18 camels. The second son got 6 which is $1/3$ of all the camels. And the youngest son got 2 which is $1/9$ of the 18 camels. Therefore, the total number of camels allocated to three sons was 17 and the neighbor took his camel back. This story told us that addition transformation can help resolve challenging issue. Chairman Mao ever indicated in his masterpiece *On Contradiction and Practice* "Contradictions can transform to their opposite sides under certain conditions [167]." The world exhibits diverse, dynamic, vibrant states due to the mutual transformations of contradictions. System structures determine system behaviors. In a typical soccer team, a badly structured team (position of team member) dooms to fail in a match. The phenomenon that an organic molecule or a cluster transforms into different forms with the same chemical composition and configuration is called isomerism. Similar phenomenon also occurs in the social context. For instance, a same team with different arrangements in its members can achieve different functions. In Lego game, by changing the configurations and structures, one can build many different objects using same amount of Lego block pieces. Therefore, we need to investigate the mechanisms and laws on how structures of error system change over time and space in order to study the temporal and spacial characteristics of errors as well as the transitions and transformations of them. In the context of error system, we need to examine the temporal and spatial dynamics, direction, and sequence of acting system elements. The steps we will adopt is: (1) using structure diagram to illustrate related transformations; (2) employing corresponding logical

© Springer Nature Switzerland AG 2020

K. Guo and S. Liu, *Error Systems: Concepts, Theory and Applications*,
Studies in Systems, Decision and Control 275,
https://doi.org/10.1007/978-3-030-40760-5_5

connectives to investigate the laws of transformations; (3) examining the relationship between system structure and errors by understanding the relationship between
system structure and system's features(In this book, features of system indicate the
functionality of a system. We use this word "feature" instead of "function", although
it is not semantically matching the exact meaning of functionality, for preventing
generating ambiguity because function is often used to indicate the mathematical
function.) and functionality.

5.1   Addition Transformation of Error System

5.1.1   Types of Addition Transformation on Error System Structure

The expression for the logical connective of addition transformation is:

Suppose that $A((U, S(t), \vec{p}(t), T(t), L(t)), x(t) = f((u(t), \vec{p}(t)), G(t)))$ is an error
logical variable defined under judging rule G on universe of discourse U, if $T(A((U, S(t), \vec{p}(t), T(t), L(t)), x(t) = f((u(t), \vec{p}(t)), G(t)))) = \{A((U, S(t), \vec{p}(t), T(t), L(t)), x(t) = f((u(t), \vec{p}(t)), G(t))), A_1((U_1, S_1(t), \vec{p}_1(t), T_1(t), L_1(t)), x_1(t) = f_1((u_1(t), \vec{p}_1(t)), G_1(t))), A_2((U_2, S_2(t), \vec{p}_2(t), T_2(t), L_2(t)), x_2(t) = f_2((u_2(t), \vec{p}_2(t)), G_2(t))), \ldots, A_n((U_n, S_n(t), \vec{p}_n(t), T_n(t), L_n(t)), x_n(t) = f_n((u_n(t), \vec{p}_n(t)), G_n(t)))\}$, then T is called the logical connective of addition transformation regarding
$A((U, S(t), \vec{p}(t), T(t), L(t)), x(t) = f((u(t), \vec{p}(t)), G(t)))$ and judging rule G on
universe of discourse U, it is noted by T_{zj}. In $T_{zj}(A((U, S(t), \vec{p}(t), T(t), L(t)), x(t) = f((u(t), \vec{p}(t)), G(t)))) = \{A((U, S(t), \vec{p}(t), T(t), L(t)), x(t) = f((u(t), \vec{p}(t)), G(t))), A_1((U_1, S_1(t), \vec{p}_1(t), T_1(t), L_1(t)), x_1(t) = f_1((u_1(t), \vec{p}_1(t)), G_1(t))), A_2((U_2, S_2(t), \vec{p}_2(t), T_2(t), L_2(t)), x_2(t) = f_2((u_2(t), \vec{p}_2(t)), G_2(t))), \ldots, A_n((U_n, S_n(t), \vec{p}_n(t), T_n(t), L_n(t)), x_n(t) = f_n((u_n(t), \vec{p}_n(t)), G_n(t)))\}$, T_{zj} is called the
transformation connective for addition of subjects if the relationship $u(t) \to u(t)$
h $u_1(t)$ **h** $u_2(t)$ **h**, $\ldots$, $u_3(t)$ holds, which denoted by T_{zjsw}. Here, operation "**h**"
implies an approx-union operation on the elements in the right side that are obtained
from addition transformation in the left side.

   In the above case, addition transformation is conducted on the subjects in order
to achieve the expected target. For instance, suppose the universe of discourse is
$AAAA$ University and $S_i(t)$ represent colleges and other administrative departments,
Department of Artificial Intelligence is added to the School of Engineering in order
to consolidate the relevant research in the field of computer engineering.

5.1.2 Addition Transformation on Series Structure in Error Systems

In this section, addition transformations here involve adding one or the combination of the 5 basic structures (i.e., series structure, parallel structure, expanding and shrinking structure, inclusion structure, and structure with feedback) to error systems.

1. Adding series structure to existing series structure

   (1) Graphic representation of addition transformation (referring to Fig. 5.1)
   (2) Representation for logical proposition of addition transformation

   $T_{zjsw} (A((U, CS(s_1, s_2, \ldots, s_n), \vec{p}(t), T(t), L(t)), x(t) = f ((u(t), \vec{p}(t)), G(t)))) = \{A((U, CS(s_1, s_2, \ldots, s_n), \vec{p}(t), T(t), L(t)), x(t) = f((u(t), \vec{p}(t)), G(t))), A_1((U_1, CS(e_1, e_2, \ldots, e_n), \vec{p_1}(t), T_1(t), L_1(t)), x_1(t) = f_1((u_1(t), \vec{p_1}(t)), G_1(t)))\}$

   (3) Representation for the error function of addition transformation

   $CSC(s_1, s_2, \ldots, s_n, CSC(e_1, e_2, \ldots, e_n)) = CSC(s_1, s_2, \ldots, s_n, s_e, e_2, \ldots, e_n)$

2. Adding parallel structure to existing series structure

   (1) Graphic representation of addition transformation (referring to Fig. 5.2)
   (2) Representation for logical proposition of addition transformation

   $T_{zjsw} (A((U, CS(s_1, s_2, \ldots, s_n), \vec{p}(t), T(t), L(t)), x(t) = f ((u(t), \vec{p}(t)), G(t)))) = \{A((U, CS(s_1, s_2, \ldots, s_n), \vec{p}(t), T(t), L(t)), x(t) = f((u(t), \vec{p}(t)), G(t))), A_1((U_1, BS(e_1, e_2, \ldots, e_n), \vec{p_1}(t), T_1(t), L_1(t)), x_1(t) = f_1 ((u_1(t), \vec{p_1}(t)), G_1(t)))\}$

   (3) Representation for the error function of addition transformation

   $CSC(s_1, s_2, \ldots, s_n, BSC(e_1, e_2, \ldots, e_n))$

3. Adding expanding and shrinking structure to existing series structure

   (1) Graphic representation of adding expanding and shrinking structure to existing series structure (referring to Figs. 5.3 and 5.4)
   (2) Representation for logical proposition of addition transformation

   $T_{zjsw} (A((U, CS(s_1, s_2, \ldots, s_n), \vec{p}(t), T(t), L(t)), x(t) = f ((u(t), \vec{p}(t)), G(t)))) = \{A((U, CS(s_1, s_2, \ldots, s_n), \vec{p}(t), T(t), L(t)), x(t) = f((u(t), \vec{p}(t)), G(t))), A_1((U_1, KS(e_1, e_2, \ldots, e_n), \vec{p_1}(t), T_1(t), L_1(t)), x_1(t) = f_1 ((u_1(t), \vec{p_1}(t)), G_1(t)))\}$

   (3) Representation for the error function of addition transformation

   $CSC(s_1, s_2, \ldots, s_n, KSC(e_1, e_2, \ldots, e_n))$

4. Adding inclusion structure to existing series structure

   (1) Graphic representation of adding inclusion structure to existing series structure (referring to Figs. 5.5 and 5.6)

Fig. 5.1 Graphic representation of adding series structure to existing series structure

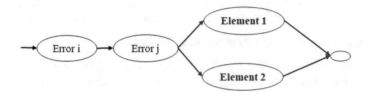

Fig. 5.2 Graphic representation of adding parallel structure to existing series structure

Fig. 5.3 Graphic representation of adding expanding structure to existing series structure

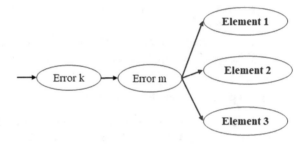

(2) Representation for logical proposition of addition transformation

$$T_{zjsw} (A((U, CS(s_1, s_2, \ldots, s_n), \vec{p}(t), T(t), L(t)), x(t) = f ((u(t), \vec{p}(t)), G(t)))) = \{A((U, CS(s_1, s_2, \ldots, s_n), \vec{p}(t), T(t), L(t)), x(t) = f((u(t), \vec{p}(t)), G(t))), A_1((U_1, YS(e_1, e_2, \ldots, e_n), \vec{p}_1(t), T_1(t), L_1(t)), x_1(t) = f_1 ((u_1(t), \vec{p}_1(t)), G_1(t)))\}.$$

(3) Representation for the error function of addition transformation

$$CSC(s_1, s_2, \ldots, s_n, YSC(e_1, e_2, \ldots, e_n)).$$

5. Adding feedback structure to existing series structure

(1) Graphic representation of adding feedback structure to existing series structure (referring to Fig. 5.7)

(2) Representation for logical proposition of addition transformation

$$T_{zjsw} (A((U, CS(s_1, s_2, \ldots, s_n), \vec{p}(t), T(t), L(t)), x(t) = f ((u(t), \vec{p}(t)), G(t)))) = \{A((U, CS(s_1, s_2, \ldots, s_n), \vec{p}(t), T(t), L(t)), x(t) = f((u(t), \vec{p}(t)), G(t))), A_1((U_1, FS(e_1, e_2, \ldots, e_n), \vec{p}_1(t), T_1(t), L_1(t)), x_1(t) = f_1 ((u_1(t), \vec{p}_1(t)), G_1(t)))\}.$$

(3) Representation for the error function of addition transformation

$$CSC(s_1, s_2, \ldots, s_n, FSC(e_1, e_2, \ldots, e_n)).$$

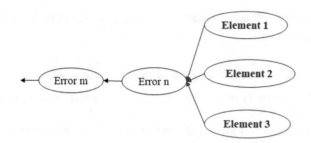

Fig. 5.4 Graphic representation of adding shrinking structure to existing series structure

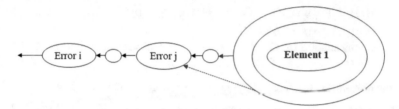

Fig. 5.5 Graphic representation of adding centered inclusion structure to existing series structure

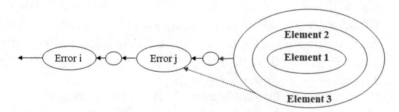

Fig. 5.6 Graphic representation of adding multi-layered inclusion structure to existing series structure

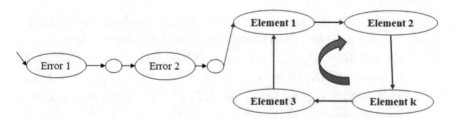

Fig. 5.7 Graphic representation of adding feedback structure to existing series structure

5.1.3 Addition Transformation on Parallel Structure in Error Systems

1. Adding series structure to one side of existing parallel structure

   (1) Graphic representation of addition transformation (referring to Figs. 5.8 and 5.9)
   (2) Representation for logical proposition of addition transformation
   $T_{zjsw}(A((U, BS(s_1, s_2, \ldots, s_n), \overrightarrow{p}(t), T(t), L(t)), x(t) = f((u(t), \overrightarrow{p}(t)),$
   $G(t)))) = \{A((U, BS(s_1, s_2, \ldots, s_n), \overrightarrow{p}(t), T(t), L(t)), x(t) = f((u(t),$
   $\overrightarrow{p}(t)), G(t))), A_1((U_1, CS(e_1, e_2, \ldots, e_n), \overrightarrow{p_1}(t), T_1(t), L_1(t)), x_1(t) =$
   $f_1((u_1(t), \overrightarrow{p_1}(t)), G_1(t)))\}.$
   (3) Representation for the error function of addition transformation
   $BSC(s_1, s_2, \ldots, s_n, CSC(e_1, e_2, \ldots, e_n)).$

2. Adding series structure to both sides of existing parallel structure

   (1) Graphic representation of addition transformation (referring to Fig. 5.10)
   (2) Representation for logical proposition of addition transformation
   $T_{zjsw}(A((U, BS(s_1, s_2, \ldots, s_n), \overrightarrow{p}(t), T(t), L(t)), x(t) = f((u(t), \overrightarrow{p}(t)),$
   $G(t)))) = \{A((U, BS(s_1, s_2, \ldots, s_n), \overrightarrow{p}(t), T(t), L(t)), x(t) = f((u(t),$
   $\overrightarrow{p}(t)), G(t))), A_1((U_1, CS_1(e_{11}, e_{12}, \ldots, e_{1n}), \overrightarrow{p_1}(t), T_1(t), L_1(t)), x_1(t) =$
   $f_1((u_1(t), \overrightarrow{p_1}(t)), G_1(t)))\}, A_2((U_2, CS_2(e_{21}, e_{22}, \ldots, e_{2n}), \overrightarrow{p_2}(t), T_2(t),$
   $L_2(t)), x_2(t) = f_2((u_2(t), \overrightarrow{p_2}(t)), G_2(t)))\}.$
   (3) Representation for the error function of addition transformation
   $BSC(s_1, s_2, \ldots, CSC_1(e_{11}, e_{12}, \ldots, e_{1n}), \ldots, CSC_2(e_{21}, e_{22}, \ldots, e_{2n}).$

3. Adding series structure to existing parallel structure to form a new series structure

   (1) Graphic representation of addition transformation (referring to Fig. 5.11)
   (2) Representation for logical proposition of addition transformation
   $T_{zjsw}(A((U, BS(s_1, s_2, \ldots, s_n), \overrightarrow{p}(t), T(t), L(t)), x(t) = f((u(t), \overrightarrow{p}(t)),$
   $G(t)))) = \{A((U, BS(s_1, s_2, \ldots, s_n), \overrightarrow{p}(t), T(t), L(t)), x(t) = f((u(t),$
   $\overrightarrow{p}(t)), G(t))), A_1((U_1, CS(e_1, e_2, \ldots, e_n), \overrightarrow{p_1}(t), T_1(t), L_1(t)), x_1(t) =$
   $f_1((u_1(t), \overrightarrow{p_1}(t)), G_1(t)))\}.$
   (3) Representation for the error function of addition transformation
   $CSC(BSC(s_1, s_2, \ldots, s_n), CSC(e_1, e_2, \ldots, e_n)).$

4. Adding parallel structure to one side of existing parallel structure

   (1) Graphic representation of addition transformation (referring to Figs. 5.12 and 5.13)

(2) Representation for logical proposition of addition transformation

$T_{zjsw}(A((U, \mathrm{BS}(s_1, s_2, \ldots, s_n), \vec{p}(t), T(t), L(t)), x(t) = f((u(t), \vec{p}(t)),$
$G(t)))) = \{A((U, \mathrm{BS}(s_1, s_2, \ldots, s_n), \vec{p}(t), T(t), L(t)), x(t) = f((u(t),$
$\vec{p}(t)), G(t))), A_1((U_1, \mathrm{BS}(e_1, e_2, \ldots, e_n), \vec{p_1}(t), T_1(t), L_1(t)), x_1(t) =$
$f_1((u_1(t), \vec{p_1}(t)), G_1(t)))\}.$

(3) Representation for the error function of addition transformation

$\mathrm{BSC}(s_1, s_2, \ldots, s_n, \mathrm{CSC}(s_1, s_2, \mathrm{BSC}(e_1, e_2, \ldots, e_n), s_n)).$

5. Adding parallel structure to both sides of existing parallel structure

(1) Graphic representation of addition transformation (referring to Fig. 5.14)
(2) Representation for logical proposition of addition transformation

$T_{zjsw}(A((U, BS(s_1, s_2, \ldots, s_n), \vec{p}(t), T(t), L(t)), x(t) = f((u(t), \vec{p}(t)),$
$G(t)))) = \{A((U, \mathrm{BS}(s_1, s_2, \ldots, s_n), \vec{p}(t), T(t), L(t)), x(t) = f((u(t),$
$\vec{p}(t)), G(t))), A_1((U_1, \mathrm{BS}_1(e_{11}, e_{12}, \ldots, e_{1n}), \vec{p_1}(t), T_1(t), L_1(t)), x_1(t) =$
$f_1((u_1(t), \vec{p_1}(t)), G_1(t)))\}, A_2((U_2, \mathrm{BS}_2(e_{21}, e_{22}, \ldots, e_{2n}), \vec{p_2}(t), T_2(t),$
$L_2(t)), x_2(t) = f_2((u_2(t), \vec{p_2}(t)), G_2(t)))\}.$

(3) Representation for the error function of addition transformation

$\mathrm{BSC}(s_1, s_2, \ldots, s_n, \mathrm{CSC}_1(s_1, s_2, \mathrm{BSC}_1(e_{11}, e_{12}, \ldots, e_{1n}), s_n), \mathrm{CSC}_2(s_1, s_2,$
$\mathrm{BSC}_2(e_{21}, e_{22}, \ldots, e_{2n})).$

6. Adding parallel structure to existing parallel structure to form a series structure

(1) Graphic representation of addition transformation (referring to Fig. 5.15)
(2) Representation for logical proposition of addition transformation

$T_{zjsw}(A((U, BS(s_1, s_2, \ldots, s_n), \vec{p}(t), T(t), L(t)), x(t) = f((u(t), \vec{p}(t)),$
$G(t)))) = \{A((U, \mathrm{BS}(s_1, s_2, \ldots, s_n), \vec{p}(t), T(t), L(t)), x(t) = f((u(t),$
$\vec{p}(t)), G(t))), A_1((U_1, \mathrm{BS}(e_1, e_2, \ldots, e_n), \vec{p_1}(t), T_1(t), L_1(t)), x_1(t) =$
$f_1((u_1(t), \vec{p_1}(t)), G_1(t)))\}.$

(3) Representation for the error function of addition transformation

$\mathrm{CSC}(\mathrm{BSC}(s_1, s_2, \ldots, s_n), \mathrm{BSC}(e_1, e_2, \ldots, e_n)).$

7. Adding expanding structure to existing parallel structure to form a series structure

(1) Graphic representation of addition transformation (referring to Fig. 5.16)
(2) Representation for logical proposition of addition transformation

$T_{zjsw}(A((U, BS(s_1, s_2, \ldots, s_n), \vec{p}(t), T(t), L(t)), x(t) = f((u(t), \vec{p}(t)),$
$G(t)))) = \{A((U, \mathrm{BS}(s_1, s_2, \ldots, s_n), \vec{p}(t), T(t), L(t)), x(t) = f((u(t),$
$\vec{p}(t)), G(t))), A_1((U_1, \mathrm{KS}(e_1, e_2, \ldots, e_n), \vec{p_1}(t), T_1(t), L_1(t)), x_1(t) =$
$f_1((u_1(t), \vec{p_1}(t)), G_1(t)))\}.$

(3) Representation for the error function of addition transformation

$\mathrm{CSC}(\mathrm{BSC}(s_1, s_2, \ldots, s_n), \mathrm{KSC}(e_1, e_2, \ldots, e_n)).$

8. Adding expanding structure to one side of existing parallel structure

(1) Graphic representation of addition transformation (referring to Figs. 5.17 and 5.18)

(2) Representation for logical proposition of addition transformation

$T_{zjsw}(A((U, BS(s_1, s_2, \ldots, s_n), \vec{p}(t), T(t), L(t)), x(t) = f((u(t), \vec{p}(t)), G(t)))) = \{A((U, BS(s_1, s_2, \ldots, s_n), \vec{p}(t), T(t), L(t)), x(t) = f((u(t), \vec{p}(t)), G(t))), A_1((U_1, KS(e_1, e_2, \ldots, e_n), \vec{p_1}(t), T_1(t), L_1(t)), x_1(t) = f_1((u_1(t), \vec{p_1}(t)), G_1(t)))\}.$

(3) Representation for the error function of addition transformation

$BSC(s_1, s_2, \ldots, s_n, CSC(s_1, s_2, KSC(e_1, e_2, \ldots, e_n), s_n)).$

9. Adding expanding structure to both sides of existing parallel structure

(1) Graphic representation of addition transformation (referring to Fig. 5.19)

(2) Representation for logical proposition of addition transformation

$T_{zjsw}(A((U, BS(s_1, s_2, \ldots, s_n), \vec{p}(t), T(t), L(t)), x(t) = f((u(t), \vec{p}(t)), G(t)))) = \{A((U, BS(s_1, s_2, \ldots, s_n), \vec{p}(t), T(t), L(t)), x(t) = f((u(t), \vec{p}(t)), G(t))), A_1((U_1, KS_1(e_{11}, e_{12}, \ldots, e_{1n}), \vec{p_1}(t), T_1(t), L_1(t)), x_1(t) = f_1((u_1(t), \vec{p_1}(t)), G_1(t)))\}, A_2((U_2, KS_2(s_{21}, s_{22}, \ldots, s_{2n}), \vec{p_2}(t), T_2(t), L_2(t)), x_2(t) = f_2((u_2(t), \vec{p_2}(t)), G_2(t)))\}.$

(3) Representation for the error function of addition transformation

$BSC(s_1, s_2, \ldots, s_n, CSC_1(s_1, s_2, KSC_1(e_{11}, e_{12}, \ldots, e_{1n}), s_n), CSC_2(s_1, s_2, KSC_2(s_{21}, s_{22}, \ldots, s_{2n}), s_n)).$

10. Adding shrinking structure to existing parallel structure to form a series structure

(1) Graphic representation of addition transformation (referring to Fig. 5.20)

(2) Representation for logical proposition of addition transformation

$T_{zjsw}(A((U, BS(s_1, s_2, \ldots, s_n), \vec{p}(t), T(t), L(t)), x(t) = f((u(t), \vec{p}(t)), G(t)))) = \{A((U, BS(s_1, s_2, \ldots, s_n), \vec{p}(t), T(t), L(t)), x(t) = f((u(t), \vec{p}(t)), G(t))), A_1((U_1, KS(e_1, e_2, \ldots, e_n), \vec{p_1}(t), T_1(t), L_1(t)), x_1(t) = f_1((u_1(t), \vec{p_1}(t)), G_1(t)))\}.$

(3) Representation for the error function of addition transformation

$CSC(BSC(s_1, s_2, \ldots, s_n), KSC(e_1, e_2, \ldots, e_n)).$

11. Adding shrinking structure to one side of existing parallel structure

(1) Graphic representation of addition transformation (referring to Figs. 5.21 and 5.22)

(2) Representation for logical proposition of addition transformation

$T_{zjsw}(A((U, BS(s_1, s_2, \ldots, s_n), \vec{p}(t), T(t), L(t)), x(t) = f((u(t), \vec{p}(t)), G(t)))) = \{A((U, BS(s_1, s_2, \ldots, s_n), \vec{p}(t), T(t), L(t)), x(t) = f((u(t), \vec{p}(t)), G(t))), A_1((U_1, KS(e_1, e_2, \ldots, e_n), \vec{p_1}(t), T_1(t), L_1(t)), x_1(t) = f_1((u_1(t), \vec{p_1}(t)), G_1(t)))\}.$

(3) Representation for the error function of addition transformation
$BSC(s_1, s_2, \ldots, s_n, CSC(s_1, s_2, KSC(e_1, e_2, \ldots, e_n), s_n))$.

12. Adding shrinking structure to both sides of existing parallel structure

(1) Graphic representation of addition transformation (referring to Fig. 5.23)
(2) Representation for logical proposition of addition transformation
$T_{zjsw}(A((U, BS(s_1, s_2, \ldots, s_n), \vec{p}(t), T(t), L(t)), x(t) = f((u(t), \vec{p}(t)),$
$G(t)))) = \{A((U, BS(s_1, s_2, \ldots, s_n), \vec{p}(t), T(t), L(t)), x(t) = f((u(t),$
$\vec{p}(t)), G(t))), A_1((U_1, KS_1(e_{11}, e_{12}, \ldots, e_{1n}), \vec{p}_1(t), T_1(t), L_1(t)), x_1(t) =$
$f_1((u_1(t), \vec{p}_1(t)), G_1(t)))\}, A_2((U_2, KS_2(e_{21}, e_{22}, \ldots, e_{2n}), \vec{p}_2(t), T_2(t),$
$L_2(t)), x_2(t) = f_2((u_2(t), \vec{p}_2(t)), G_2(t)))\}$.
(3) Representation for the error function of addition transformation
$BSC(s_1, s_2, \ldots, s_n, CSC_1(s_1, s_2, KSC_1(e_{11}, e_{12}, \ldots, e_{1n}), s_n), CSC_2(s_1, s_2,$
$KSC_2(e_{21}, e_{22}, \ldots, e_{2n}), s_n))$.

13. Adding centered inclusion structure to existing parallel structure to form a series structure

(1) Graphic representation of addition transformation (referring to Fig. 5.24)
(2) Representation for logical proposition of addition transformation
$T_{zjsw}(A((U, BS(s_1, s_2, \ldots, s_n), \vec{p}(t), T(t), L(t)), x(t) = f((u(t), \vec{p}(t)),$
$G(t)))) = \{A((U, BS(s_1, s_2, \ldots, s_n), \vec{p}(t), T(t), L(t)), x(t) = f((u(t),$
$\vec{p}(t)), G(t))), A_1((U_1, YS(e_1, e_2, \ldots, e_n), \vec{p}_1(t), T_1(t), L_1(t)), x_1(t) =$
$f_1((u_1(t), \vec{p}_1(t)), G_1(t)))\}$.
(3) Representation for the error function of addition transformation
$CSC(BSC(s_1, s_2, \ldots, s_n), YSC(e_1, e_2, \ldots, e_n))$.

14. Adding centered inclusion structure to one side of existing parallel structure

(1) Graphic representation of addition transformation (referring to Figs. 5.25 and 5.26)
(2) Representation for logical proposition of addition transformation
$T_{zjsw}(A((U, BS(s_1, s_2, \ldots, s_n), \vec{p}(t), T(t), L(t)), x(t) = f((u(t), \vec{p}(t)),$
$G(t)))) = \{A((U, BS(s_1, s_2, \ldots, s_n), \vec{p}(t), T(t), L(t)), x(t) = f((u(t),$
$\vec{p}(t)), G(t))), A_1((U_1, YS(e_1, e_2, \ldots, e_n), \vec{p}_1(t), T_1(t), L_1(t)), x_1(t) =$
$f_1((u_1(t), \vec{p}_1(t)), G_1(t)))\}$.
(3) Representation for the error function of addition transformation
$BSC(s_1, s_2, \ldots, s_n, CSC(s_1, s_2, YSC(e_1, e_2, \ldots, e_n), s_n))$.

15. Adding centered inclusion structure to both sides of existing parallel structure

(1) Graphic representation of addition transformation (referring to Fig. 5.27)

(2) Representation for logical proposition of addition transformation

$T_{zjsw}(A((U, BS(s_1, s_2, \ldots, s_n), \vec{p}(t), T(t), L(t)), x(t) = f((u(t), \vec{p}(t)), G(t)))) = \{A((U, BS(s_1, s_2, \ldots, s_n), \vec{p}(t), T(t), L(t)), x(t) = f((u(t), \vec{p}(t)), G(t))), A_1((U_1, YS_1(e_{11}, e_{12}, \ldots, e_{1n}), \vec{p_1}(t), T_1(t), L_1(t)), x_1(t) = f_1((u_1(t), \vec{p_1}(t)), G_1(t)))\}, A_2((U_2, YS_2(e_{21}, e_{22}, \ldots, e_{2n}), \vec{p_2}(t), T_2(t), L_2(t)), x_2(t) = f_2((u_2(t), \vec{p_2}(t)), G_2(t)))\}.$

(3) Representation for the error function of addition transformation

$BSC(s_1, s_2, \ldots, s_n, CSC_1(s_1, s_2, YSC_1(e_{11}, e_{12}, \ldots, e_{1n}), s_n), CSC_2(s_1, s_2, YSC_2(e_{21}, e_{22}, \ldots, e_{2n}), s_n)).$

16. Adding multi-layered inclusion structure to existing parallel structure to form a series structure

(1) Graphic representation of addition transformation (referring to Fig. 5.28)
(2) Representation for logical proposition of addition transformation

$T_{zjsw}(A((U, BS(s_1, s_2, \ldots, s_n), \vec{p}(t), T(t), L(t)), x(t) = f((u(t), \vec{p}(t)), G(t)))) = \{A((U, BS(s_1, s_2, \ldots, s_n), \vec{p}(t), T(t), L(t)), x(t) = f((u(t), \vec{p}(t)), G(t))), A_1((U_1, YS(e_1, e_2, \ldots, e_n), \vec{p_1}(t), T_1(t), L_1(t)), x_1(t) = f_1((u_1(t), \vec{p_1}(t)), G_1(t)))\}$

(3) Representation for the error function of addition transformation

$CSC(BSC(s_1, s_2, \ldots, s_n), YSC(e_1, e_2, \ldots, e_n))$

17. Adding multi-layered inclusion structure to one side of existing parallel structure

(1) Graphic representation of addition transformation (referring to Figs. 5.29 and 5.30)
(2) Representation for logical proposition of addition transformation

$T_{zjsw}(A((U, BS(s_1, s_2, \ldots, s_n), \vec{p}(t), T(t), L(t)), x(t) = f((u(t), \vec{p}(t)), G(t)))) = \{A((U, BS(s_1, s_2, \ldots, s_n), \vec{p}(t), T(t), L(t)), x(t) = f((u(t), \vec{p}(t)), G(t))), A_1((U_1, YS(e_1, e_2, \ldots, e_n), \vec{p_1}(t), T_1(t), L_1(t)), x_1(t) = f_1((u_1(t), \vec{p_1}(t)), G_1(t)))\}.$

(3) Representation for the error function of addition transformation

$BSC(s_1, s_2, \ldots, s_n, CSC(s_1, s_2, YSC(e_1, e_2, \ldots, e_n), s_n)).$

18. Adding multi-layered inclusion structure to both sides of existing parallel structure

(1) Graphic representation of addition transformation (referring to Fig. 5.31)
(2) Representation for logical proposition of addition transformation

$T_{zjsw}(A((U, BS(s_1, s_2, \ldots, s_n), \vec{p}(t), T(t), L(t)), x(t) = f((u(t), \vec{p}(t)), G(t)))) = \{A((U, BS(s_1, s_2, \ldots, s_n), \vec{p}(t), T(t), L(t)), x(t) = f((u(t), \vec{p}(t)), G(t))), A_1((U_1, YS_1(e_{11}, e_{12}, \ldots, e_{1n}), \vec{p_1}(t), T_1(t), L_1(t)), x_1(t) = f_1((u_1(t), \vec{p_1}(t)), G_1(t)))\}, A_2((U_2, YS_2(e_{21}, e_{22}, \ldots, e_{2n}), \vec{p_2}(t), T_2(t), L_2(t)), x_2(t) = f_2((u_2(t), \vec{p_2}(t)), G_2(t)))\}.$

(3) Representation for the error function of addition transformation
$BSC(s_1, s_2, \ldots, s_n, CSC_1(s_1, s_2, YSC_1(e_{11}, e_{12}, \ldots, e_{1n}), s_n), CSC_2(s_1, s_2, YSC_2(e_{21}, e_{22}, \ldots, e_{2n}), s_n))$.

19. Adding feedback structure to existing parallel structure to form a series structure

   (1) Graphic representation of addition transformation (referring to Fig. 5.32)
   (2) Representation for logical proposition of addition transformation
   $$T_{zjsw}(A((U, BS(s_1, s_2, \ldots, s_n), \vec{p}(t), T(t), L(t)), x(t) = f((u(t), \vec{p}(t)), G(t)))) = \{A((U, BS(s_1, s_2, \ldots, s_n), \vec{p}(t), T(t), L(t)), x(t) = f((u(t), \vec{p}(t)), G(t))), A_1((U_1, FS(e_1, e_2, \ldots, e_n), \vec{p_1}(t), T_1(t), L_1(t)), x_1(t) = f_1((u_1(t), \vec{p_1}(t)), G_1(t)))\}.$$
   (3) Representation for the error function of addition transformation
   $CSC(BSC(s_1, s_2, \ldots, s_n), FSC(e_1, e_2, \ldots, e_n))$.

20. Adding feedback structure to one side of existing parallel structure

   (1) Graphic representation of addition transformation (referring to Figs. 5.33 and 5.34)
   (2) Representation for logical proposition of addition transformation
   $$T_{zjsw}(A((U, BS(s_1, s_2, \ldots, s_n), \vec{p}(t), T(t), L(t)), x(t) = f((u(t), \vec{p}(t)), G(t)))) = \{A((U, BS(s_1, s_2, \ldots, s_n), \vec{p}(t), T(t), L(t)), x(t) = f((u(t), \vec{p}(t)), G(t))), A_1((U_1, FS(e_1, e_2, \ldots, e_n), \vec{p_1}(t), T_1(t), L_1(t)), x_1(t) = f_1((u_1(t), \vec{p_1}(t)), G_1(t)))\}$$
   (3) Representation for the error function of addition transformation
   $BSC(s_1, s_2, \ldots, s_n, CSC(s_1, s_2, FSC(e_1, e_2, \ldots, e_n), s_n))$

21. Adding feedback structure to both sides of existing parallel structure

   (1) Graphic representation of addition transformation (referring to Fig. 5.35)
   (2) Representation for logical proposition of addition transformation
   $$T_{zjsw}(A((U, BS(s_1, s_2, \ldots, s_n), \vec{p}(t), T(t), L(t)), x(t) = f((u(t), \vec{p}(t)), G(t)))) = \{A((U, BS(s_1, s_2, \ldots, s_n), \vec{p}(t), T(t), L(t)), x(t) = f((u(t), \vec{p}(t)), G(t))), A_1((U_1, FS_1(e_{11}, e_{12}, \ldots, e_{1n}), \vec{p_1}(t), T_1(t), L_1(t)), x_1(t) = f_1((u_1(t), \vec{p_1}(t)), G_1(t)))\}, A_2((U_2, FS_2(e_{21}, e_{22}, \ldots, e_{2n}), \vec{p_2}(t), T_2(t), L_2(t)), x_2(t) = f_2((u_2(t), \vec{p_2}(t)), G_2(t)))\}.$$
   (3) Representation for the error function of addition transformation
   $BSC(s_1, s_2, \ldots, s_n, CSC_1(s_1, s_2, FSC_1(e_{11}, e_{12}, \ldots, e_{1n}), CSC_2(s_1, s_2, FSC_2(e_{21}, e_{22}, \ldots, e_{2n})$.

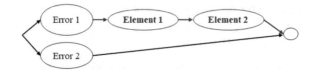

Fig. 5.8 Graphic representation of adding series structure to one side of existing parallel structure

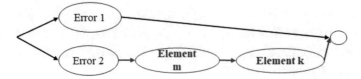

Fig. 5.9 Graphic representation of adding series structure to the other side of existing parallel structure

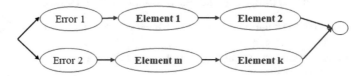

Fig. 5.10 Graphic representation of adding series structure to both sides of existing parallel structure

Fig. 5.11 Graphic representation of adding series structure to existing parallel structure to form a new series structure

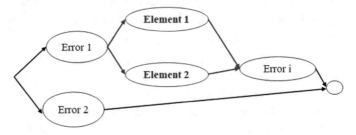

Fig. 5.12 Graphic representation of adding parallel structure to one side of existing parallel structure

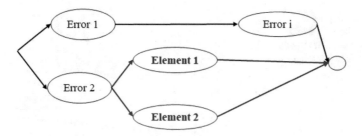

Fig. 5.13 Graphic representation of adding parallel structure to the other side of existing parallel structure

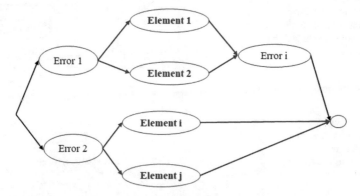

Fig. 5.14 Graphic representation of adding parallel structure to both sides of existing parallel structure

Fig. 5.15 Graphic representation of adding parallel structure to existing parallel structure to form a series structure

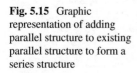

Fig. 5.16 Graphic representation of adding expanding structure to existing parallel structure to form a series structure

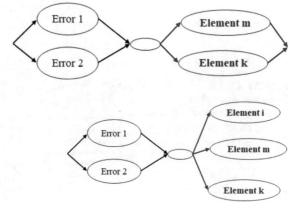

Fig. 5.17 Graphic representation of adding expanding structure to one side of existing parallel structure

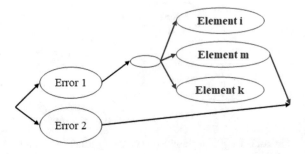

Fig. 5.18 Graphic representation of adding expanding structure to the other side of existing parallel structure

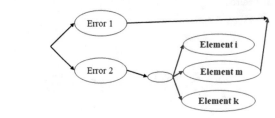

Fig. 5.19 Graphic representation of adding parallel structure to both sides of existing parallel structure

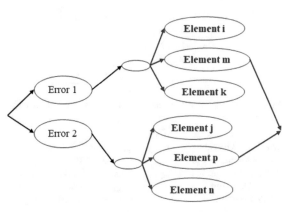

Fig. 5.20 Graphic representation of adding shrinking structure to existing parallel structure to form a series structure

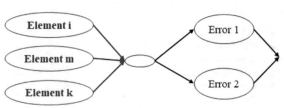

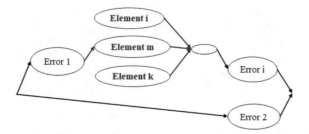

Fig. 5.21 Graphic representation of adding shrinking structure to one side of existing parallel structure

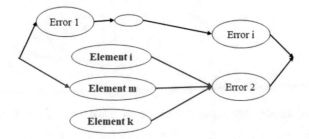

Fig. 5.22 Graphic representation of adding shrinking structure to the other side of existing parallel structure

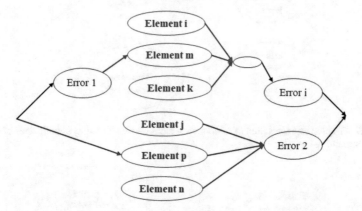

Fig. 5.23 Graphic representation of adding shrinking structure to both sides of existing parallel structure

Fig. 5.24 Graphic representation of adding centered inclusion structure to existing parallel structure to form a series structure

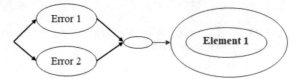

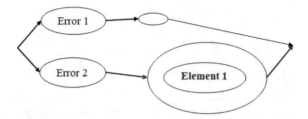

Fig. 5.25 Graphic representation of adding centered inclusion structure to one side of existing parallel structure

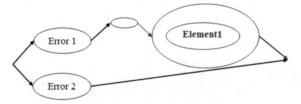

Fig. 5.26 Graphic representation of adding centered inclusion structure to the other side of existing parallel structure

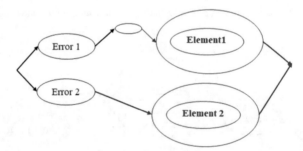

Fig. 5.27 Graphic representation of adding centered inclusion structure to both sides of existing parallel structure

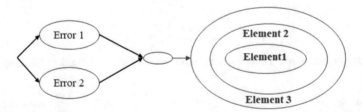

Fig. 5.28 Graphic representation of adding multi-layered inclusion structure to existing parallel structure to form a series structure

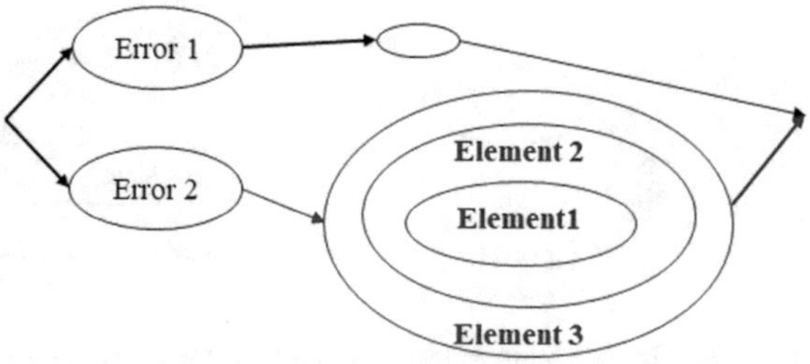

Fig. 5.29 Graphic representation of adding multi-layered inclusion structure to one side of existing parallel structure

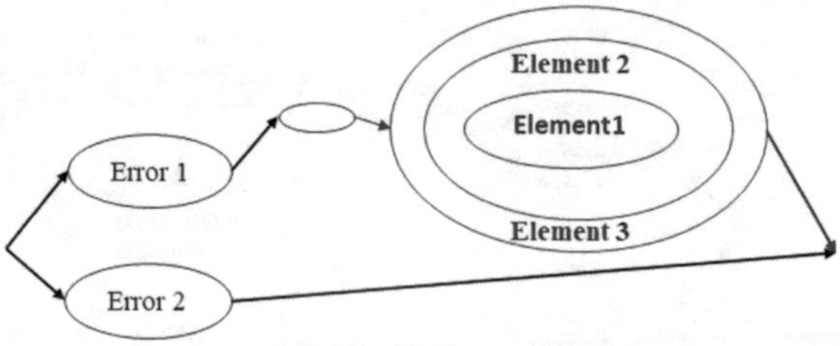

Fig. 5.30 Graphic representation of adding multi-layered inclusion structure to the other side of existing parallel structure

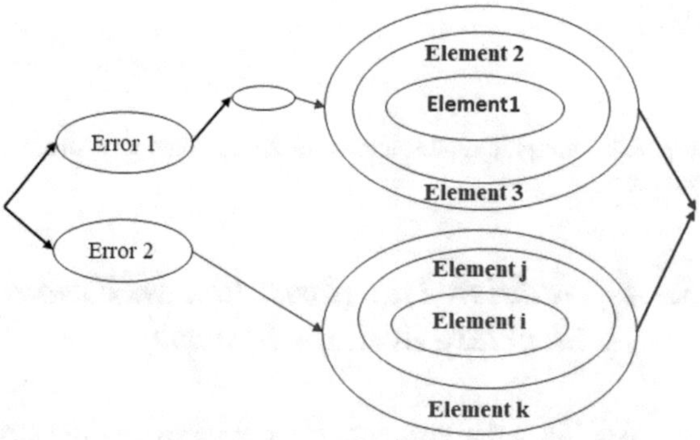

Fig. 5.31 Graphic representation of adding multi-layered inclusion structure to both sides of existing parallel structure

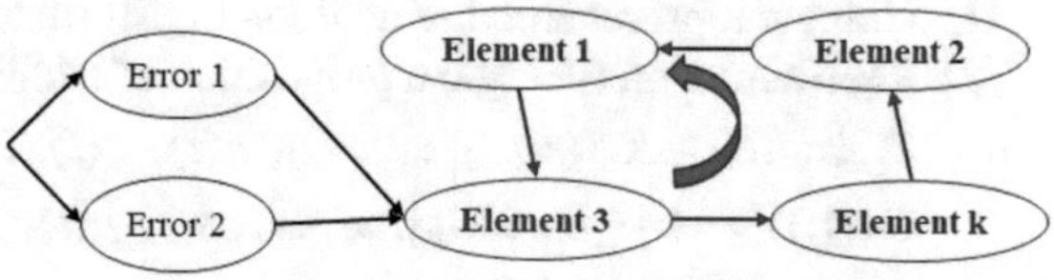

Fig. 5.32 Graphic representation of adding feedback structure to existing parallel structure to form a series structure

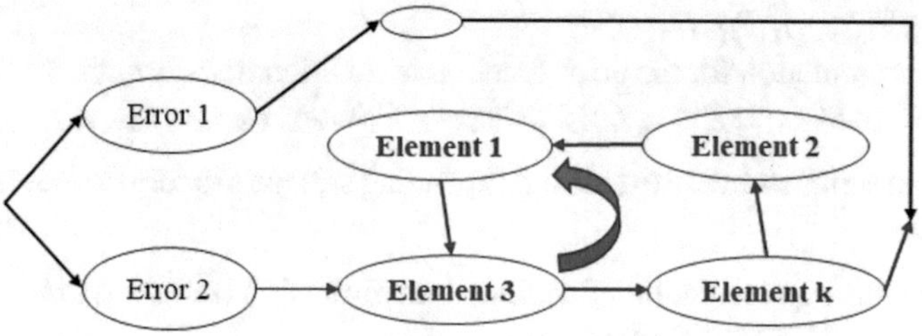

Fig. 5.33 Graphic representation of adding feedback structure to one side of existing parallel structure

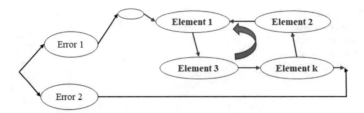

Fig. 5.34 Graphic representation of adding feedback structure to the other side of existing parallel structure

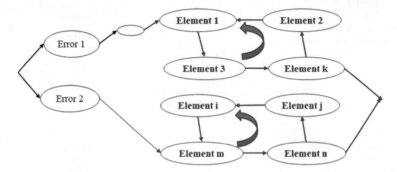

Fig. 5.35 Graphic representation of adding feedback structure to both sides of existing parallel structure

5.1.4 Addition Transformation on Expanding and Shrinking Structure in Error Systems

1. Adding series structure to existing expanding structure

   (1) Graphic representation of addition transformation (referring to Fig. 5.36)
   (2) Representation for logical proposition of addition transformation
   $T_{zjsw}(A((U, KS(s_1, s_2, \ldots, s_n), \overrightarrow{p}(t), T(t), L(t)), x(t) = f((u(t), \overrightarrow{p}(t)), G(t)))) = \{A((U, \mathrm{KS}(s_1, s_2, \ldots, s_n), \overrightarrow{p}(t), T(t), L(t)), x(t) = f((u(t), \overrightarrow{p}(t)), G(t))), A_1((U_1, \mathrm{CS}(e_1, e_2, \ldots, e_n), \overrightarrow{p_1}(t), T_1(t), L_1(t)), x_1(t) = f_1((u_1(t), \overrightarrow{p_1}(t)), G_1(t)))\}.$
   (3) Representation for the error function of addition transformation
   $\mathrm{KSC}(\mathrm{KSC}(s_1, s_2, \ldots, s_n), \mathrm{CSC}(s_1, s_2, \mathrm{CSC}(e_1, e_2, \ldots, e_n), s_n).$

2. Adding series structure to existing expanding structure to form a series structure-type 1

   (1) Graphic representation of addition transformation (referring to Fig. 5.37)

(2) Representation for logical proposition of addition transformation

$T_{zjsw}(A((U, KS(s_1, s_2, \ldots, s_n), \vec{p}(t), T(t), L(t)), x(t) = f((u(t), \vec{p}(t)),$
$G(t)))) = \{A((U, \text{KS}(s_1, s_2, \ldots, s_n), \vec{p}(t), T(t), L(t)), x(t) = f((u(t),$
$\vec{p}(t)), G(t))), A_1((U_1, \text{CS}(e_1, e_2, \ldots, e_n), \vec{p_1}(t), T_1(t), L_1(t)), x_1(t) =$
$f_1((u_1(t), \vec{p_1}(t)), G_1(t)))\}.$

(3) Representation for the error function of addition transformation
   CSC(KSC$(s_1, s_2, \ldots, s_n)$, CSC$(e_1, e_2, \ldots, e_n)$).

3. Adding series structure to existing expanding structure to form a series structure-type 2

(1) Graphic representation of addition transformation (referring to Fig. 5.38)
(2) Representation for logical proposition of addition transformation

$T_{zjsw}(A((U, KS(s_1, s_2, \ldots, s_n), \vec{p}(t), T(t), L(t)), x(t) = f((u(t), \vec{p}(t)),$
$G(t)))) = \{A((U, \text{KS}(s_1, s_2, \ldots, s_n), \vec{p}(t), T(t), L(t)), x(t) = f((u(t),$
$\vec{p}(t)), G(t))), A_1((U_1, \text{CS}(e_1, e_2, \ldots, e_n), \vec{p_1}(t), T_1(t), L_1(t)), x_1(t) =$
$f_1((u_1(t), \vec{p_1}(t)), G_1(t)))\}.$

(3) Representation for the error function of addition transformation
   CSC(KSC$(s_1, s_2, \ldots, s_n)$, CSC$(e_1, e_2, \ldots, e_n)$).

4. Adding series structure to existing shrinking structure

(1) Graphic representation of addition transformation (referring to Fig. 5.39)
(2) Representation for logical proposition of addition transformation

$T_{zjsw}(A((U, KS(s_1, s_2, \ldots, s_n), \vec{p}(t), T(t), L(t)), x(t) = f((u(t), \vec{p}(t)),$
$G(t)))) = \{A((U, \text{KS}(s_1, s_2, \ldots, s_n), \vec{p}(t), T(t), L(t)), x(t) = f((u(t),$
$\vec{p}(t)), G(t))), A_1((U_1, \text{CS}(e_1, e_2, \ldots, e_n), \vec{p_1}(t), T_1(t), L_1(t)), x_1(t) =$
$f_1((u_1(t), \vec{p_1}(t)), G_1(t)))\}.$

(3) Representation for the error function of addition transformation
   KSC(KSC$(s_1, s_2, \ldots, s_n)$, CSC$(s_1, s_2, \text{CSC}(e_1, e_2, \ldots, e_n), s_n)$).

5. Adding series structure to existing shrinking structure to form a series structure-type 1

(1) Graphic representation of addition transformation (referring to Fig. 5.40)
(2) Representation for logical proposition of addition transformation

$T_{zjsw}(A((U, KS(s_1, s_2, \ldots, s_n), \vec{p}(t), T(t), L(t)), x(t) = f((u(t), \vec{p}(t)),$
$G(t)))) = \{A((U, \text{KS}(s_1, s_2, \ldots, s_n), \vec{p}(t), T(t), L(t)), x(t) = f((u(t),$
$\vec{p}(t)), G(t))), A_1((U_1, \text{CS}(e_1, e_2, \ldots, e_n), \vec{p_1}(t), T_1(t), L_1(t)), x_1(t) =$
$f_1((u_1(t), \vec{p_1}(t)), G_1(t)))\}.$

(3) Representation for the error function of addition transformation
   CSC(KSC$(s_1, s_2, \ldots, s_n)$, CSC$(e_1, e_2, \ldots, e_n)$).

6. Adding series structure to existing shrinking structure to form a series structure-type 2

(1) Graphic representation of addition transformation (referring to Fig. 5.41)
(2) Representation for logical proposition of addition transformation

$T_{zjsw}(A((U, KS(s_1, s_2, \ldots, s_n), \vec{p}(t), T(t), L(t)), x(t) = f((u(t), \vec{p}(t)), G(t)))) = \{A((U, KS(s_1, s_2, \ldots, s_n), \vec{p}(t), T(t), L(t)), x(t) = f((u(t), \vec{p}(t)), G(t))), A_1((U_1, CS(e_1, e_2, \ldots, e_n), \vec{p_1}(t), T_1(t), L_1(t)), x_1(t) = f_1((u_1(t), \vec{p_1}(t)), G_1(t)))\}$.

(3) Representation for the error function of addition transformation
$CSC(KSC(s_1, s_2, \ldots, s_n), CSC(e_1, e_2, \ldots, e_n))$.

7. Adding parallel structure to existing expanding structure

(1) Graphic representation of addition transformation (referring to Fig. 5.42)
(2) Representation for logical proposition of addition transformation

$T_{zjsw}(A((U, KS(s_1, s_2, \ldots, s_n), \vec{p}(t), T(t), L(t)), x(t) = f((u(t), \vec{p}(t)), G(t)))) = \{A((U, KS(s_1, s_2, \ldots, s_n), \vec{p}(t), T(t), L(t)), x(t) = f((u(t), \vec{p}(t)), G(t))), A_1((U_1, BS(e_1, e_2, \ldots, e_n), \vec{p_1}(t), T_1(t), L_1(t)), x_1(t) = f_1((u_1(t), \vec{p_1}(t)), G_1(t)))\}$.

(3) Representation for the error function of addition transformation
$KSC(KSC(s_1, s_2, \ldots, s_n), CSC(s_1, s_2, BSC(e_1, e_2, \ldots, e_n), s_n)$.

8. Adding parallel structure to existing expanding structure to form a series structure-type 1

(1) Graphic representation of addition transformation (referring to Fig. 5.43)
(2) Representation for logical proposition of addition transformation

$T_{zjsw}(A((U, KS(s_1, s_2, \ldots, s_n), \vec{p}(t), T(t), L(t)), x(t) = f((u(t), \vec{p}(t)), G(t)))) = \{A((U, KS(s_1, s_2, \ldots, s_n), \vec{p}(t), T(t), L(t)), x(t) = f((u(t), \vec{p}(t)), G(t))), A_1((U_1, BS(e_1, e_2, \ldots, e_n), \vec{p_1}(t), T_1(t), L_1(t)), x_1(t) = f_1((u_1(t), \vec{p_1}(t)), G_1(t)))\}$.

(3) Representation for the error function of addition transformation
$CSC(KSC(s_1, s_2, \ldots, s_n), BSC(e_1, e_2, \ldots, e_n))$.

9. Adding parallel structure to existing expanding structure to form a series structure-type 2

(1) Graphic representation of addition transformation (referring to Fig. 5.44)
(2) Representation for logical proposition of addition transformation

$T_{zjsw}(A((U, KS(s_1, s_2, \ldots, s_n), \vec{p}(t), T(t), L(t)), x(t) = f((u(t), \vec{p}(t)), G(t)))) = \{A((U, KS(s_1, s_2, \ldots, s_n), \vec{p}(t), T(t), L(t)), x(t) = f((u(t), \vec{p}(t)), G(t))), A_1((U_1, BS(e_1, e_2, \ldots, e_n), \vec{p_1}(t), T_1(t), L_1(t)), x_1(t) = f_1((u_1(t), \vec{p_1}(t)), G_1(t)))\}$.

(3) Representation for the error function of addition transformation
    CSC(KSC($s_1, s_2, \ldots, s_n$), BSC($e_1, e_2, \ldots, e_n$)).

10. Adding parallel structure to existing shrinking structure

(1) Graphic representation of addition transformation (referring to Fig. 5.45)
(2) Representation for logical proposition of addition transformation

$T_{zjsw}(A((U, KS(s_1, s_2, \ldots, s_n), \overrightarrow{p}(t), T(t), L(t)), x(t) = f((u(t), \overrightarrow{p}(t)),$
$G(t)))) = \{A((U, \text{KS}(s_1, s_2, \ldots, s_n), \overrightarrow{p}(t), T(t), L(t)), x(t) = f((u(t),$
$\overrightarrow{p}(t)), G(t))), A_1((U_1, \text{BS}(e_1, e_2, \ldots, e_n), \overrightarrow{p_1}(t), T_1(t), L_1(t)), x_1(t) =$
$f_1((u_1(t), \overrightarrow{p-1}(t)), G_1(t)))\}.$

(3) Representation for the error function of addition transformation
    KSC(KSC($s_1, s_2, \ldots, s_n$), CSC(s_1, s_2, BSC($e_1, e_2, \ldots, e_n$), s_n).

11. Adding parallel structure to existing shrinking structure to form a series structure-type 1

(1) Graphic representation of addition transformation (referring to Fig. 5.46)
(2) Representation for logical proposition of addition transformation

$T_{zjsw}(A((U, KS(s_1, s_2, \ldots, s_n), \overrightarrow{p}(t), T(t), L(t)), x(t) = f((u(t), \overrightarrow{p}(t)),$
$G(t)))) = \{A((U, \text{KS}(s_1, s_2, \ldots, s_n), \overrightarrow{p}(t), T(t), L(t)), x(t) = f((u(t),$
$\overrightarrow{p}(t)), G(t))), A_1((U_1, \text{BS}(e_1, e_2, \ldots, e_n), \overrightarrow{p_1}(t), T_1(t), L_1(t)), x_1(t) =$
$f_1((u_1(t), \overrightarrow{p_1}(t)), G_1(t)))\}.$

(3) Representation for the error function of addition transformation
    CSC(KSC($s_1, s_2, \ldots, s_n$), BSC($e_1, e_2, \ldots, e_n$)).

12. Adding parallel structure to existing shrinking structure to form a series structure-type 2

(1) Graphic representation of addition transformation (referring to Fig. 5.47)
(2) Representation for logical proposition of addition transformation

$T_{zjsw}(A((U, KS(s_1, s_2, \ldots, s_n), \overrightarrow{p}(t), T(t), L(t)), x(t) = f((u(t), \overrightarrow{p}(t)),$
$G(t)))) = \{A((U, \text{KS}(s_1, s_2, \ldots, s_n), \overrightarrow{p}(t), T(t), L(t)), x(t) = f((u(t),$
$\overrightarrow{p}(t)), G(t))), A_1((U_1, \text{BS}(e_1, e_2, \ldots, e_n), \overrightarrow{p_1}(t), T_1(t), L_1(t)), x_1(t) =$
$f_1((u_1(t), \overrightarrow{p_1}(t)), G_1(t)))\}.$

(3) Representation for the error function of addition transformation
    CSC(KSC($s_1, s_2, \ldots, s_n$), BSC($e_1, e_2, \ldots, e_n$)).

13. Adding expanding structure to existing expanding structure-type 1

(1) Graphic representation of addition transformation (referring to Fig. 5.48)
(2) Representation for logical proposition of addition transformation

$T_{zjsw}(A((U, KS(s_1, s_2, \ldots, s_n), \overrightarrow{p}(t), T(t), L(t)), x(t) = f((u(t), \overrightarrow{p}(t)),$
$G(t)))) = \{A((U, \text{KS}(s_1, s_2, \ldots, s_n), \overrightarrow{p}(t), T(t), L(t)), x(t) = f((u(t),$

$\vec{p}(t)$), $G(t)$)), $A_1((U_1,\ KS(e_1,\ e_2,\ \ldots,\ e_n),\ \vec{p_1}(t),\ T_1(t),\ L_1(t)),\ x_1(t) =$
$f_1((u_1(t),\ \vec{p_1}(t)),\ G_1(t)))\}$.

(3) Representation for the error function of addition transformation
$KSC(s_1,\ s_2,\ \ldots,\ s_n,\ CSC(s_1,\ s_2,\ KSC(e_1,\ e_2,\ \ldots,\ e_n),\ s_n)$.

14. Adding expanding structure to existing expanding structure-type 2

(1) Graphic representation of addition transformation (referring to Fig. 5.49)
(2) Representation for logical proposition of addition transformation
$T_{zjsw}(A((U,\ KS(s_1,\ s_2,\ \ldots,\ s_n),\ \vec{p}(t),\ T(t),\ L(t)),\ x(t) = f\ ((u(t),\ \vec{p}(t)),$
$G(t)))) = \{A((U,\ KS(s_1,\ s_2,\ \ldots,\ s_n),\ \vec{p}(t),\ T(t),\ L(t)),\ x(t) = f((u(t),$
$\vec{p}(t)),\ G(t))),\ A_1((U_1,\ KS(e_1,\ e_2,\ \ldots,\ e_n),\ \vec{p_1}(t),\ T_1(t),\ L_1(t)),\ x_1(t) =$
$f_1((u_1(t),\ \vec{p_1}(t)),\ G_1(t)))\}$.
(3) Representation for the error function of addition transformation
$KSC(s_1,\ s_2,\ \ldots,\ KSC(e_1,\ e_2,\ \ldots,\ e_n),\ s_n)$.

15. Adding shrinking structure to existing shrinking structure-type 1

(1) Graphic representation of addition transformation (referring to Fig. 5.50)
(2) Representation for logical proposition of addition transformation
$T_{zjsw}(A((U,\ KS(s_1,\ s_2,\ \ldots,\ s_n),\ \vec{p}(t),\ T(t),\ L(t)),\ x(t) = f\ ((u(t),\ \vec{p}(t)),$
$G(t)))) = \{A((U,\ KS(s_1,\ s_2,\ \ldots,\ s_n),\ \vec{p}(t),\ T(t),\ L(t)),\ x(t) = f((u(t),$
$\vec{p}(t)),\ G(t))),\ A_1((U_1,\ KS(e_1,\ e_2,\ \ldots,\ e_n),\ \vec{p_1}(t),\ T_1(t),\ L_1(t)),\ x_1(t) =$
$f_1((u_1(t),\ \vec{p_1}(t)),\ G_1(t)))\}$.
(3) Representation for the error function of addition transformation
$KSC(s_1,\ s_2,\ \ldots,\ s_n,\ CSC(s_1,\ s_2,\ KSC(e_1,\ e_2,\ \ldots,\ e_n),\ s_n)$.

16. Adding shrinking structure to existing expanding structure-type 2

(1) Graphic representation of addition transformation (referring to Fig. 5.51)
(2) Representation for logical proposition of addition transformation
$T_{zjsw}(A((U,\ KS(s_1,\ s_2,\ \ldots,\ s_n),\ \vec{p}(t),\ T(t),\ L(t)),\ x(t) = f\ ((u(t),\ \vec{p}(t)),$
$G(t)))) = \{A((U,\ KS(s_1,\ s_2,\ \ldots,\ s_n),\ \vec{p}(t),\ T(t),\ L(t)),\ x(t) = f((u(t),$
$\vec{p}(t)),\ G(t))),\ A_1((U_1,\ KS(e_1,\ e_2,\ \ldots,\ e_n),\ \vec{p_1}(t),\ T_1(t),\ L_1(t)),\ x_1(t) =$
$f_1((u_1(t),\ \vec{p_1}(t)),\ G_1(t)))\}$.
(3) Representation for the error function of addition transformation
$KSC(s_1,\ s_2,\ \ldots,\ KSC(e_1,\ e_2,\ \ldots,\ e_n),\ s_n)$.

17. Adding shrinking structure to existing expanding structure to form a series structure

(1) Graphic representation of addition transformation (referring to Fig. 5.52)
(2) Representation for logical proposition of addition transformation
$T_{zjsw}(A((U,\ KS(s_1,\ s_2,\ \ldots,\ s_n),\ \vec{p}(t),\ T(t),\ L(t)),\ x(t) = f\ ((u(t),\ \vec{p}(t)),$
$G(t)))) = \{A((U,\ KS(s_1,\ s_2,\ \ldots,\ s_n),\ \vec{p}(t),\ T(t),\ L(t)),\ x(t) = f((u(t),$

$\overrightarrow{p}(t)$), $G(t)$)), $A_1((U_1,\ \mathrm{KS}_1(e_{11},\ e_{12},\ \ldots,\ e_{1n}),\ \overrightarrow{p_1}(t),\ T_1(t),\ L_1(t)),\ x_1(t) = f_1((u_1(t),\ \overrightarrow{p_1}(t)),\ G_1(t)))\}.$

(3) Representation for the error function of addition transformation
$\mathrm{CSC}(\mathrm{KSC}(s_1, s_2, \ldots, s_n),\ \mathrm{KSC}_1(e_{11}, e_{12}, \ldots, e_{1n}).$

18. Adding shrinking structure to existing expanding structure

   (1) Graphic representation of addition transformation (referring to Fig. 5.53)
   (2) Representation for logical proposition of addition transformation
   $T_{zjsw}(A((U,\ KS(s_1, s_2, \ldots, s_n),\ \overrightarrow{p}(t),\ T(t),\ L(t)),\ x(t) = f\ ((u(t),\ \overrightarrow{p}(t)),$
   $G(t)))) = \{A((U,\ \mathrm{KS}(s_1, s_2, \ldots, s_n),\ \overrightarrow{p}(t),\ T(t),\ L(t)),\ x(t) = f((u(t),$
   $\overrightarrow{p}(t)),\ G(t))),\ A_1((U_1,\ \mathrm{KS}(e_1, e_2, \ldots, e_n),\ \overrightarrow{p_1}(t),\ T_1(t),\ L_1(t)),\ x_1(t) = f_1((u_1(t),\ \overrightarrow{p_1}(t)),\ G_1(t)))\}.$
   (3) Representation for the error function of addition transformation
   $\mathrm{KSC}(s_1, s_2, \ldots, s_n),\ \mathrm{CSC}(s_1, s_2, \mathrm{KSC}(e_1, e_2, \ldots, e_n), s_n).$

19. Adding expanding structure to existing shrinking structure to form a series structure

   (1) Graphic representation of addition transformation (referring to Fig. 5.54)
   (2) Representation for logical proposition of addition transformation
   $T_{zjsw}(A((U,\ KS(s_1, s_2, \ldots, s_n),\ \overrightarrow{p}(t),\ T(t),\ L(t)),\ x(t) = f\ ((u(t),\ \overrightarrow{p}(t)),$
   $G(t)))) = \{A((U,\ \mathrm{KS}(s_1, s_2, \ldots, s_n),\ \overrightarrow{p}(t),\ T(t),\ L(t)),\ x(t) = f((u(t),$
   $\overrightarrow{p}(t)),\ G(t))),\ A_1((U_1,\ \mathrm{KS}(e_1, e_2, \ldots, e_n),\ \overrightarrow{p_1}(t),\ T_1(t),\ L_1(t)),\ x_1(t) = f_1((u_1(t),\ \overrightarrow{p_1}(t)),\ G_1(t)))\}.$
   (3) Representation for the error function of addition transformation
   $\mathrm{CSC}(\mathrm{KSC}(s_1, s_2, \ldots, s_n),\ \mathrm{KSC}_1(e_1, e_2, \ldots, e_n)).$

20. Adding expanding structure to existing shrinking structure

   (1) Graphic representation of addition transformation (referring to Fig. 5.55)
   (2) Representation for logical proposition of addition transformation
   $T_{zjsw}(A((U,\ KS(s_1, s_2, \ldots, s_n),\ \overrightarrow{p}(t),\ T(t),\ L(t)),\ x(t) = f\ ((u(t),\ \overrightarrow{p}(t)),$
   $G(t)))) = \{A((U,\ \mathrm{KS}(s_1, s_2, \ldots, s_n),\ \overrightarrow{p}(t),\ T(t),\ L(t)),\ x(t) = f((u(t),$
   $\overrightarrow{p}(t)),\ G(t))),\ A_1((U_1,\ \mathrm{KS}(e_1, e_2, \ldots, e_n),\ \overrightarrow{p_1}(t),\ T_1(t),\ L_1(t)),\ x_1(t) = f_1((u_1(t),\ \overrightarrow{p_1}(t)),\ G_1(t)))\}.$
   (3) Representation for the error function of addition transformation
   $\mathrm{KSC}(s_1, s_2, \ldots, s_n),\ \mathrm{CSC}(s_1, s_2, \mathrm{KSC}_1(e_1, e_2, \ldots, e_n), s_n).$

21. Adding inclusion structure to existing expanding structure

   (1) Graphic representation of addition transformation (referring to Figs. 5.56 and 5.57)

(2) Representation for logical proposition of addition transformation

$T_{zjsw}(A((U, KS(s_1, s_2, \ldots, s_n), \overrightarrow{p}(t), T(t), L(t)), x(t) = f((u(t), \overrightarrow{p}(t)),$
$G(t)))) = \{A((U, KS(s_1, s_2, \ldots, s_n), \overrightarrow{p}(t), T(t), L(t)), x(t) = f((u(t),$
$\overrightarrow{p}(t)), G(t))), A_1((U_1, YS(e_1, e_2, \ldots, e_n), \overrightarrow{p_1}(t), T_1(t), L_1(t)), x_1(t) =$
$f_1((u_1(t), \overrightarrow{p_1}(t)), G_1(t)))\}.$

(3) Representation for the error function of addition transformation

KSC(KSC($s_1, s_2, \ldots, s_n$), CSC(s_1, s_2, YSC($e_1, e_2, \ldots, e_n$), s_n).

22. Adding inclusion structure to existing expanding structure to form a series structure-type 1

(1) Graphic representation of addition transformation (referring to Figs. 5.58 and 5.59)

(2) Representation for logical proposition of addition transformation

$T_{zjsw}(A((U, KS(s_1, s_2, \ldots, s_n), \overrightarrow{p}(t), T(t), L(t)), x(t) = f((u(t), \overrightarrow{p}(t)),$
$G(t)))) = \{A((U, KS(s_1, s_2, \ldots, s_n), \overrightarrow{p}(t), T(t), L(t)), x(t) = f((u(t),$
$\overrightarrow{p}(t)), G(t))), A_1((U_1, YS(e_1, e_2, \ldots, e_n), \overrightarrow{p_1}(t), T_1(t), L_1(t)), x_1(t) =$
$f_1((u_1(t), \overrightarrow{p_1}(t)), G_1(t)))\}.$

(3) Representation for the error function of addition transformation

CSC(KSC($s_1, s_2, \ldots, s_n$), YSC($e_1, e_2, \ldots, e_n$)).

23. Adding inclusion structure to existing expanding structure to form a series structure-type 2

(1) Graphic representation of addition transformation (referring to Figs. 5.60 and 5.61)

(2) Representation for logical proposition of addition transformation

$T_{zjsw}(A((U, KS(s_1, s_2, \ldots, s_n), \overrightarrow{p}(t), T(t), L(t)), x(t) = f((u(t), \overrightarrow{p}(t)),$
$G(t)))) = \{A((U, KS(s_1, s_2, \ldots, s_n), \overrightarrow{p}(t), T(t), L(t)), x(t) = f((u(t),$
$\overrightarrow{p}(t)), G(t))), A_1((U_1, YS(e_1, e_2, \ldots, e_n), \overrightarrow{p_1}(t), T_1(t), L_1(t)), x_1(t) =$
$f_1((u_1(t), \overrightarrow{p_1}(t)), G_1(t)))\}.$

(3) Representation for the error function of addition transformation

CSC(KSC($s_1, s_2, \ldots, s_n$), YSC($e_1, e_2, \ldots, e_n$)).

24. Adding inclusion structure to existing shrinking structure

(1) Graphic representation of addition transformation (referring to Figs. 5.62 and 5.63)

(2) Representation for logical proposition of addition transformation

$T_{zjsw}(A((U, KS(s_1, s_2, \ldots, s_n), \overrightarrow{p}(t), T(t), L(t)), x(t) = f((u(t), \overrightarrow{p}(t)),$
$G(t)))) = \{A((U, KS(s_1, s_2, \ldots, s_n), \overrightarrow{p}(t), T(t), L(t)), x(t) = f((u(t),$
$\overrightarrow{p}(t)), G(t))), A_1((U_1, YS(e_1, e_2, \ldots, e_n), \overrightarrow{p_1}(t), T_1(t), L_1(t)), x_1(t) =$
$f_1((u_1(t), \overrightarrow{p_1}(t)), G_1(t)))\}.$

(3) Representation for the error function of addition transformation
KSC(KSC($s_1, s_2, \ldots, s_n$), CSC(s_1, s_2, YSC($e_1, e_2, \ldots, e_n$), s_n).

25. Adding inclusion structure to existing shrinking structure to form a series structure-type 1

(1) Graphic representation of addition transformation (referring to Figs. 5.64 and 5.65)

(2) Representation for logical proposition of addition transformation
$T_{zjsw}(A((U, KS(s_1, s_2, \ldots, s_n), \vec{p}(t), T(t), L(t)), x(t) = f((u(t), \vec{p}(t)),$
$G(t)))) = \{A((U, \text{KS}(s_1, s_2, \ldots, s_n), \vec{p}(t), T(t), L(t)), x(t) = f((u(t),$
$\vec{p}(t)), G(t))), A_1((U_1, \text{YS}(e_1, e_2, \ldots, e_n), \vec{p_1}(t), T_1(t), L_1(t)), x_1(t) =$
$f_1((u_1(t), \vec{p_1}(t)), G_1(t)))\}.$

(3) Representation for the error function of addition transformation
CSC(KSC($s_1, s_2, \ldots, s_n$), YSC($e_1, e_2, \ldots, e_n$)).

26. Adding inclusion structure to existing shrinking structure to form a series structure-type 2

(1) Graphic representation of addition transformation (referring to Figs. 5.66 and 5.67)

(2) Representation for logical proposition of addition transformation
$T_{zjsw}(A((U, KS(s_1, s_2, \ldots, s_n), \vec{p}(t), T(t), L(t)), x(t) = f((u(t), \vec{p}(t)),$
$G(t)))) = \{A((U, \text{KS}(s_1, s_2, \ldots, s_n), \vec{p}(t), T(t), L(t)), x(t) = f((u(t),$
$\vec{p}(t)), G(t))), A_1((U_1, \text{YS}(e_1, e_2, \ldots, e_n), \vec{p_1}(t), T_1(t), L_1(t)), x_1(t) =$
$f_1((u_1(t), \vec{p_1}(t)), G_1(t)))\}.$

(3) Representation for the error function of addition transformation
CSC(KSC($s_1, s_2, \ldots, s_n$), YSC($e_1, e_2, \ldots, e_n$)).

27. Adding feedback structure to existing expanding structure

(1) Graphic representation of addition transformation (referring to Fig. 5.68)

(2) Representation for logical proposition of addition transformation
$T_{zjsw}(A((U, KS(s_1, s_2, \ldots, s_n), \vec{p}(t), T(t), L(t)), x(t) = f((u(t), \vec{p}(t)),$
$G(t)))) = \{A((U, \text{KS}(s_1, s_2, \ldots, s_n), \vec{p}(t), T(t), L(t)), x(t) = f((u(t),$
$\vec{p}(t)), G(t))), A_1((U_1, \text{FS}(e_1, e_2, \ldots, e_n), \vec{p_1}(t), T_1(t), L_1(t)), x_1(t) =$
$f_1((u_1(t), \vec{p_1}(t)), G_1(t)))\}.$

(3) Representation for the error function of addition transformation
KSC(KSC($s_1, s_2, \ldots, s_n$), CSC(s_1, s_2, FSC($e_1, e_2, \ldots, e_n$), s_n).

28. Adding feedback structure to existing expanding structure to form a series structure-type 1

(1) Graphic representation of addition transformation (referring to Fig. 5.69)

(2) Representation for logical proposition of addition transformation

$T_{zjsw}(A((U, KS(s_1, s_2, \ldots, s_n), \overrightarrow{p}(t), T(t), L(t)), x(t) = f((u(t), \overrightarrow{p}(t)),$
$G(t)))) = \{A((U, \text{KS}(s_1, s_2, \ldots, s_n), \overrightarrow{p}(t), T(t), L(t)), x(t) = f((u(t),$
$\overrightarrow{p}(t)), G(t))), A_1((U_1, \text{FS}(e_1, e_2, \ldots, e_n), \overrightarrow{p_1}(t), T_1(t), L_1(t)), x_1(t) =$
$f_1((u_1(t), \overrightarrow{p_1}(t)), G_1(t)))\}.$

(3) Representation for the error function of addition transformation
   $\text{KSC}(s_1, s_2, \ldots, s_n, \text{FSC}(e_1, e_2, \ldots, e_n)).$

29. Adding feedback structure to existing expanding structure to form a series structure-type 2

(1) Graphic representation of addition transformation (referring to Fig. 5.70)
(2) Representation for logical proposition of addition transformation

$T_{zjsw}(A((U, KS(s_1, s_2, \ldots, s_n), \overrightarrow{p}(t), T(t), L(t)), x(t) = f((u(t), \overrightarrow{p}(t)),$
$G(t)))) = \{A((U, \text{KS}(s_1, s_2, \ldots, s_n), \overrightarrow{p}(t), T(t), L(t)), x(t) = f((u(t),$
$\overrightarrow{p}(t)), G(t))), A_1((U_1, \text{FS}(e_1, e_2, \ldots, e_n), \overrightarrow{p_1}(t), T_1(t), L_1(t)), x_1(t) =$
$f_1((u_1(t), \overrightarrow{p_1}(t)), G_1(t)))\}.$

(3) Representation for the error function of addition transformation
   $\text{CSC}(\text{KSC}(s_1, s_2, \ldots, s_n), \text{FSC}(e_1, e_2, \ldots, e_n)).$

30. Adding feedback structure to existing shrinking structure

(1) Graphic representation of addition transformation (referring to Fig. 5.71)
(2) Representation for logical proposition of addition transformation

$T_{zjsw}(A((U, KS(s_1, s_2, \ldots, s_n), \overrightarrow{p}(t), T(t), L(t)), x(t) = f((u(t), \overrightarrow{p}(t)),$
$G(t)))) = \{A((U, \text{KS}(s_1, s_2, \ldots, s_n), \overrightarrow{p}(t), T(t), L(t)), x(t) = f((u(t),$
$\overrightarrow{p}(t)), G(t))), A_1((U_1, \text{FS}(e_1, e_2, \ldots, e_n), \overrightarrow{p_1}(t), T_1(t), L_1(t)), x_1(t) =$
$f_1((u_1(t), \overrightarrow{p_1}(t)), G_1(t)))\}.$

(3) Representation for the error function of addition transformation
   $\text{KSC}(s_1, s_2, \ldots, s_n, \text{CSC}(s_i, \text{FSC}(s_1, s_2, \ldots, s_n)).$

31. Adding feedback structure to existing shrinking structure to form a series structure-type 1

(1) Graphic representation of addition transformation (referring to Fig. 5.72)
(2) Representation for logical proposition of addition transformation

$T_{zjsw}(A((U, KS(s_1, s_2, \ldots, s_n), \overrightarrow{p}(t), T(t), L(t)), x(t) = f((u(t), \overrightarrow{p}(t)),$
$G(t)))) = \{A((U, \text{KS}(s_1, s_2, \ldots, s_n), \overrightarrow{p}(t), T(t), L(t)), x(t) = f((u(t),$
$\overrightarrow{p}(t)), G(t))), A_1((U_1, \text{FS}(e_1, e_2, \ldots, e_n), \overrightarrow{p_1}(t), T_1(t), L_1(t)), x_1(t) =$
$f_1((u_1(t), \overrightarrow{p_1}(t)), G_1(t)))\}.$

(3) Representation for the error function of addition transformation
   $\text{CSC}(\text{KSC}(s_1, s_2, \ldots, s_n), \text{FSC}(e_1, e_2, \ldots, e_n)).$

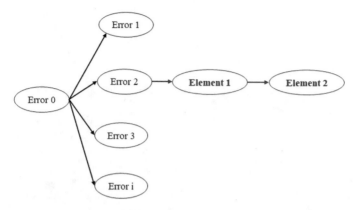

Fig. 5.36 Graphic representation of adding series structure to existing expanding structure

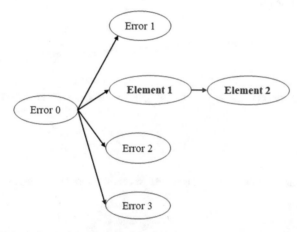

Fig. 5.37 Graphic representation of adding series structure to existing expanding structure to form a series structure-type 1

32. Adding feedback structure to existing shrinking structure to form a series structure-type 2

   (1) Graphic representation of addition transformation (referring to Fig. 5.73)
   (2) Representation for logical proposition of addition transformation
   $$T_{zjsw}(A((U, KS(s_1, s_2, \ldots, s_n), \vec{p}(t), T(t), L(t)), x(t) = f((u(t), \vec{p}(t)),$$
   $$G(t)))) = \{A((U, KS(s_1, s_2, \ldots, s_n), \vec{p}(t), T(t), L(t)), x(t) = f((u(t),$$
   $$\vec{p}(t)), G(t))), A_1((U_1, FS(e_1, e_2, \ldots, e_n), \vec{p}_1(t), T_1(t), L_1(t)), x_1(t) =$$
   $$f_1((u_1(t), \vec{p}_1(t)), G_1(t)))\}.$$
   (3) Representation for the error function of addition transformation
   $$CSC(KSC(s_1, s_2, \ldots, s_n), FSC(e_1, e_2, \ldots, e_n)).$$

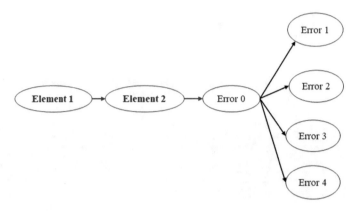

Fig. 5.38 Graphic representation of adding series structure to existing expanding structure to form a series structure-type 1

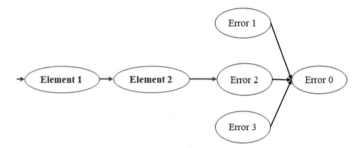

Fig. 5.39 Graphic representation of adding series structure to shrinking structure

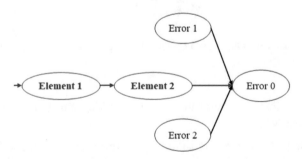

Fig. 5.40 Graphic representation of adding series structure to existing shrinking structure to form a series structure-type 1

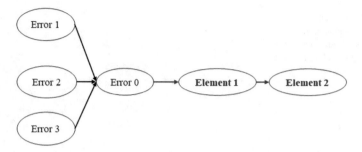

Fig. 5.41 Graphic representation of adding series structure to existing shrinking structure to form a series structure-type 2

Fig. 5.42 Graphic representation of adding parallel structure to existing expanding structure

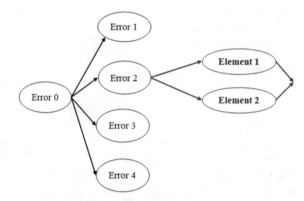

Fig. 5.43 Graphic representation of adding parallel structure to existing expanding structure to form a series structure- type 1

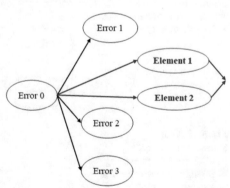

Fig. 5.44 Graphic representation of adding parallel structure to existing expanding structure to form a series structure- type 2

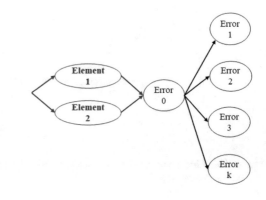

Fig. 5.45 Graphic representation of adding parallel structure to existing shrinking structure

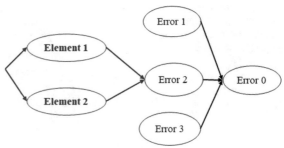

Fig. 5.46 Graphic representation of adding parallel structure to existing shrinking structure to form a series structure- type 1

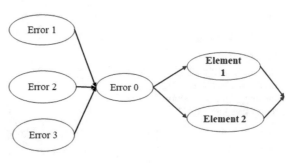

Fig. 5.47 Graphic representation of adding parallel structure to existing shrinking structure to form a series structure- type 2

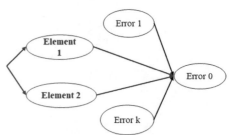

Fig. 5.48 Graphic representation of adding expanding structure to existing expanding structure-type 1

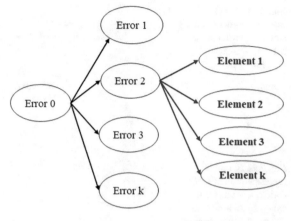

Fig. 5.49 Graphic representation of adding expanding structure to existing expanding structure-type 2

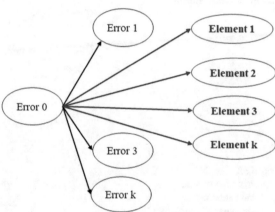

Fig. 5.50 Graphic representation of adding shrinking structure to existing shrinking structure-type 1

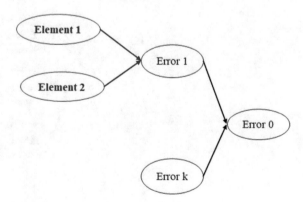

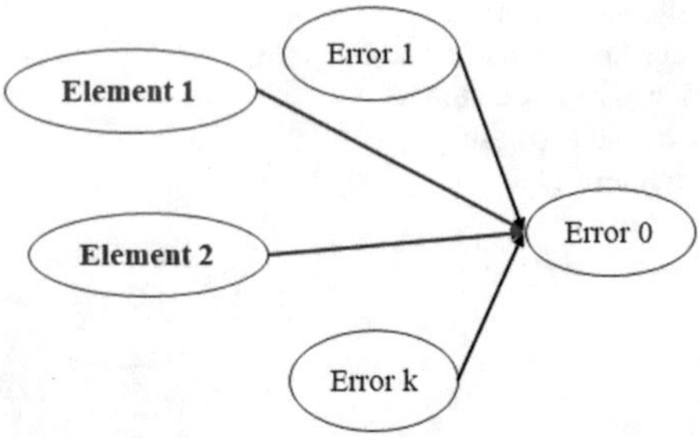

Fig. 5.51 Graphic
representation of adding
shrinking structure to
existing shrinking
structure-type 2

Fig. 5.52 Graphic
representation of adding
shrinking structure to
existing expanding structure
to form a series structure

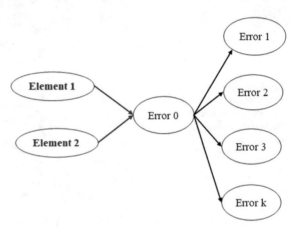

Fig. 5.53 Graphic
representation of adding
shrinking structure to
existing expanding structure

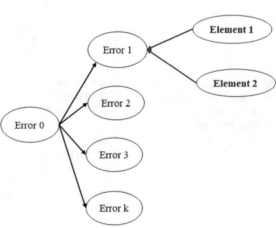

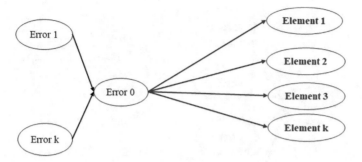

Fig. 5.54 Graphic representation of adding expanding structure to existing shrinking structure to form a series structure

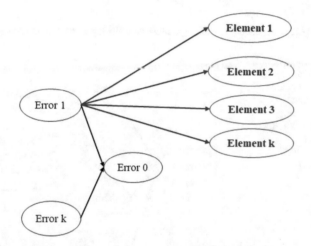

Fig. 5.55 Graphic representation of adding expanding structure to existing shrinking structure

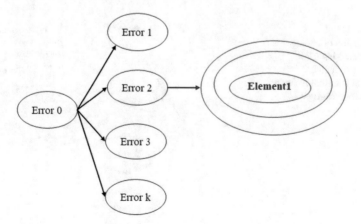

Fig. 5.56 Graphic representation of adding centered inclusion structure to existing expanding structure

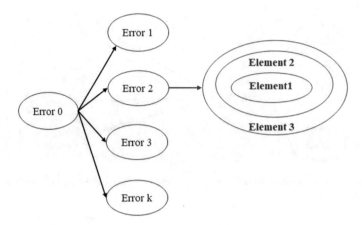

Fig. 5.57 Graphic representation of adding multi-layered inclusion structure to existing expanding structure

Fig. 5.58 Graphic representation of adding centered inclusion structure to existing expanding structure to form a series structure-type 1

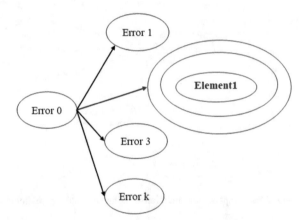

Fig. 5.59 Graphic representation of adding multi-layered inclusion structure to existing expanding structure to form a series structure-type 1

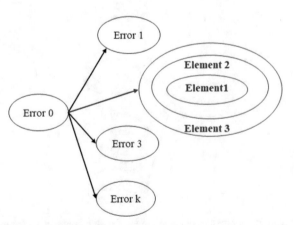

Fig. 5.60 Adding centered inclusion structure to existing expanding structure to form a series structure-type 2

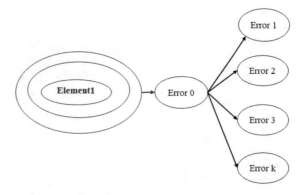

Fig. 5.61 Adding multi-layered inclusion structure to existing expanding structure to form a series structure-type 2

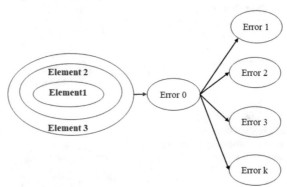

Fig. 5.62 Graphic representation of adding centered inclusion structure to existing shrinking structure

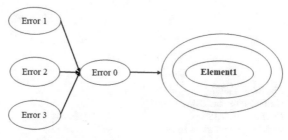

Fig. 5.63 Graphic representation of adding multi-layered inclusion structure to existing shrinking structure

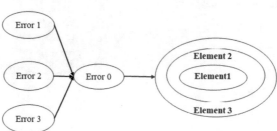

Fig. 5.64 Graphic
representation of adding
centered inclusion structure
to existing shrinking
structure to form a series
structure-type 1

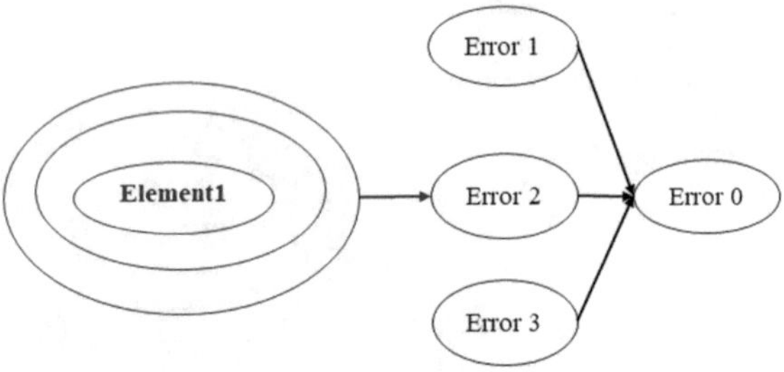

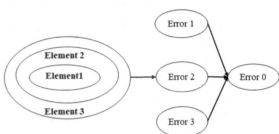

Fig. 5.65 Graphic
representation of adding
multi-layered inclusion
structure to existing
shrinking structure to form a
series structure-type 1

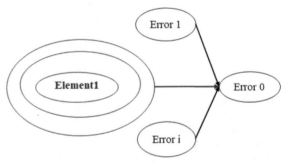

Fig. 5.66 Adding centered
inclusion structure to
existing shrinking structure
to form a series
structure-type 2

Fig. 5.67 Adding
multi-layered inclusion
structure to existing
shrinking structure to form a
series structure-type 2

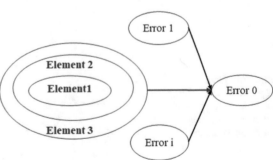

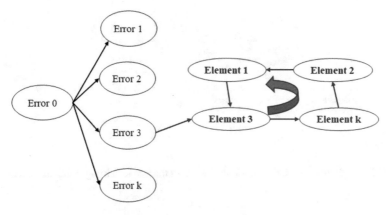

Fig. 5.68 Graphic representation of adding feedback structure to existing expanding structure

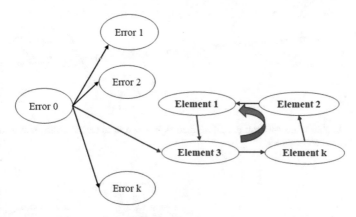

Fig. 5.69 Graphic representation of adding feedback structure to existing expanding structure to form a series structure-type 1

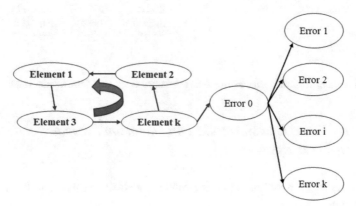

Fig. 5.70 Adding feedback structure to existing expanding structure to form a series structure-type 2

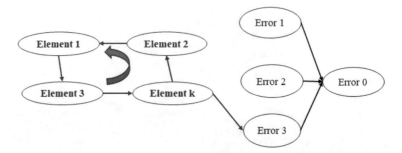

Fig. 5.71 Graphic representation of adding feedback structure to existing shrinking structure

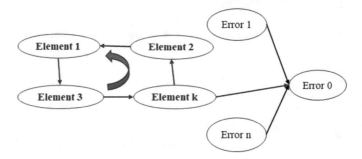

Fig. 5.72 Graphic representation of adding feedback structure to existing shrinking structure to form a series structure-type 1

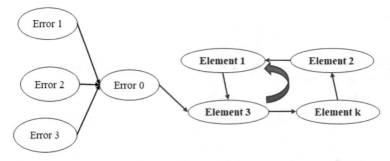

Fig. 5.73 Adding feedback structure to existing shrinking structure to form a series structure-type 2

5.1.5   *Addition Transformation on Inclusion Structure in Error Systems*

1. Adding series structure to existing centered inclusion structure to form a series structure

   (1) Graphic representation of addition transformation (referring to Fig. 5.74)

(2) Representation for logical proposition of addition transformation

$T_{zjsw}(A((U, YS(s_1, s_2, \ldots, s_n), \vec{p}(t), T(t), L(t)), x(t) = f((u(t), \vec{p}(t)), G(t)))) = \{A((U, YS(s_1, s_2, \ldots, s_n), \vec{p}(t), T(t), L(t)), x(t) = f((u(t), \vec{p}(t)), G(t))), A_1((U_1, CS(e_1, e_2, \ldots, e_n), \vec{p_1}(t), T_1(t), L_1(t)), x_1(t) = f_1((u_1(t), \vec{p_1}(t)), G_1(t)))\}.$

(3) Representation for the error function of addition transformation

$CSC(YSC(s_1, s_2, \ldots, s_n), CSC(e_1, e_2, \ldots, e_n)).$

2. Adding series structure to existing multi-layered inclusion structure to form a series structure

(1) Graphic representation of addition transformation (referring to Fig. 5.75)
(2) Representation for logical proposition of addition transformation

$T_{zjsw}(A((U, YS(s_1, s_2, \ldots, s_n), \vec{p}(t), T(t), L(t)), x(t) = f((u(t), \vec{p}(t)), G(t)))) = \{A((U, YS(s_1, s_2, \ldots, s_n), \vec{p}(t), T(t), L(t)), x(t) = f((u(t), \vec{p}(t)), G(t))), A_1((U_1, CS(e_1, e_2, \ldots, e_n), \vec{p_1}(t), T_1(t), L_1(t)), x_1(t) = f_1((u_1(t), \vec{p_1}(t)), G_1(t)))\}.$

(3) Representation for the error function of addition transformation

$CSC(YSC(s_1, s_2, \ldots, s_n), CSC(e_1, e_2, \ldots, e_n)).$

3. Adding series structure to existing centered inclusion structure to form a feedback structure

(1) Graphic representation of addition transformation (referring to Fig. 5.76)
(2) Representation for logical proposition of addition transformation

$T_{zjsw}(A((U, YS(s_1, s_2, \ldots, s_n), \vec{p}(t), T(t), L(t)), x(t) = f((u(t), \vec{p}(t)), G(t)))) = \{A((U, YS(s_1, s_2, \ldots, s_n), \vec{p}(t), T(t), L(t)), x(t) = f((u(t), \vec{p}(t)), G(t))), A_1((U_1, CS(e_1, e_2, \ldots, e_n), \vec{p_1}(t), T_1(t), L_1(t)), x_1(t) = f_1((u_1(t), \vec{p_1}(t)), G_1(t)))\}.$

(3) Representation for the error function of addition transformation

$FSC(YSC(s_1, s_2, \ldots, s_n), CSC(e_1, e_2, \ldots, e_n)).$

4. Adding series structure to existing multi-layered inclusion structure to form a feedback structure

(1) Graphic representation of addition transformation (referring to Fig. 5.77)
(2) Representation for logical proposition of addition transformation

$T_{zjsw}(A((U, YS(s_1, s_2, \ldots, s_n), \vec{p}(t), T(t), L(t)), x(t) = f((u(t), \vec{p}(t)), G(t)))) = \{A((U, YS(s_1, s_2, \ldots, s_n), \vec{p}(t), T(t), L(t)), x(t) = f((u(t), \vec{p}(t)), G(t))), A_1((U_1, CS(e_1, e_2, \ldots, e_n), \vec{p_1}(t), T_1(t), L_1(t)), x_1(t) = f_1((u_1(t), \vec{p_1}(t)), G_1(t)))\}.$

(3) Representation for the error function of addition transformation

$FSC(YSC(s_1, s_2, \ldots, s_n), CSC(e_1, e_2, \ldots, e_n)).$

5. Adding series structure to existing centered inclusion structure to form a parallel structure

(1) Graphic representation of addition transformation (referring to Fig. 5.78)
(2) Representation for logical proposition of addition transformation

$T_{zjsw}(A((U, YS(s_1, s_2, \ldots, s_n), \vec{p}(t), T(t), L(t)), x(t) = f((u(t), \vec{p}(t)),$
$G(t)))) = \{A((U, YS(s_1, s_2, \ldots, s_n), \vec{p}(t), T(t), L(t)), x(t) = f((u(t),$
$\vec{p}(t)), G(t))), A_1((U_1, CS(e_1, e_2, \ldots, e_n), \vec{p}_1(t), T_1(t), L_1(t)), x_1(t) =$
$f_1((u_1(t), \vec{p}_1(t)), G_1(t)))\}.$

(3) Representation for the error function of addition transformation
$BSC(YSC(s_1, s_2, \ldots, s_n), CSC(e_1, e_2, \ldots, e_n)).$

6. Adding series structure to existing multi-layered inclusion structure to form a parallel structure

(1) Graphic representation of addition transformation (referring to Figs. 5.79 and 5.80)
(2) Representation for logical proposition of addition transformation

$T_{zjsw}(A((U, YS(s_1, s_2, \ldots, s_n), \vec{p}(t), T(t), L(t)), x(t) = f((u(t), \vec{p}(t)),$
$G(t)))) = \{A((U, YS(s_1, s_2, \ldots, s_n), \vec{p}(t), T(t), L(t)), x(t) = f((u(t),$
$\vec{p}(t)), G(t))), A_1((U_1, CS(e_1, e_2, \ldots, e_n), \vec{p}_1(t), T_1(t), L_1(t)), x_1(t) =$
$f_1((u_1(t), \vec{p}_1(t)), G_1(t)))\}.$

(3) Representation for the error function of addition transformation
$BSC(YSC(s_1, s_2, \ldots, s_n), CSC(e_1, e_2, \ldots, e_n)).$

7. Adding parallel structure to existing inclusion structure to form a series structure

(1) Graphic representation of addition transformation (referring to Figs. 5.81 and 5.82)
(2) Representation for logical proposition of addition transformation

$T_{zjsw}(A((U, YS(s_1, s_2, \ldots, s_n), \vec{p}(t), T(t), L(t)), x(t) = f((u(t), \vec{p}(t)),$
$G(t)))) = \{A((U, YS(s_1, s_2, \ldots, s_n), \vec{p}(t), T(t), L(t)), x(t) = f((u(t),$
$\vec{p}(t)), G(t))), A_1((U_1, BS(e_1, e_2, \ldots, e_n), \vec{p}_1(t), T_1(t), L_1(t)), x_1(t) =$
$f_1((u_1(t), \vec{p}_1(t)), G_1(t)))\}.$

(3) Representation for the error function of addition transformation
$CSC(YSC(s_1, s_2, \ldots, s_n), BSC(e_1, e_2, \ldots, e_n)).$

8. Adding parallel structure to existing inclusion structure to form a parallel structure

(1) Graphic representation of addition transformation (referring to Figs. 5.83, 5.84 and 5.85)
(2) Representation for logical proposition of addition transformation

$T_{zjsw}(A((U, YS(s_1, s_2, \ldots, s_n), \vec{p}(t), T(t), L(t)), x(t) = f((u(t), \vec{p}(t)),$
$G(t)))) = \{A((U, YS(s_1, s_2, \ldots, s_n), \vec{p}(t), T(t), L(t)), x(t) = f((u(t),$

$\vec{p}(t)), G(t))), A_1((U_1, \text{BS}(e_1, e_2, \ldots, e_n), \vec{p}_1(t), T_1(t), L_1(t)), x_1(t) = f_1((u_1(t), \vec{p}_1(t)), G_1(t)))\}$.

(3) Representation for the error function of addition transformation
$\text{BSC}(\text{YSC}(s_1, s_2, \ldots, s_n), \text{BSC}(e_1, e_2, \ldots, e_n))$.

9. Adding expanding structure to existing inclusion structure to form a series structure

(1) Graphic representation of addition transformation (referring to Figs. 5.86 and 5.87)

(2) Representation for logical proposition of addition transformation
$T_{zjsw}(A((U, YS(s_1, s_2, \ldots, s_n), \vec{p}(t), T(t), L(t)), x(t) = f((u(t), \vec{p}(t)),$
$G(t)))) = \{A((U, \text{YS}(s_1, s_2, \ldots, s_n), \vec{p}(t), T(t), L(t)), x(t) = f((u(t),$
$\vec{p}(t)), G(t))), A_1((U_1, \text{KS}(e_1, e_2, \ldots, e_n), \vec{p}_1(t), T_1(t), L_1(t)), x_1(t) =$
$f_1((u_1(t), \vec{p}_1(t)), G_1(t)))\}$.

(3) Representation for the error function of addition transformation
$\text{CSC}(\text{YSC}(s_1, s_2, \ldots, s_n), \text{KSC}(e_1, e_2, \ldots, e_n))$.

10. Adding expanding structure to existing inclusion structure to form a parallel structure

(1) Graphic representation of addition transformation (referring to Figs. 5.88 and 5.89)

(2) Representation for logical proposition of addition transformation
$T_{zjsw}(A((U, YS(s_1, s_2, \ldots, s_n), \vec{p}(t), T(t), L(t)), x(t) = f((u(t), \vec{p}(t)),$
$G(t)))) = \{A((U, \text{YS}(s_1, s_2, \ldots, s_n), \vec{p}(t), T(t), L(t)), x(t) = f((u(t),$
$\vec{p}(t)), G(t))), A_1((U_1, \text{KS}(e_1, e_2, \ldots, e_n), \vec{p}_1(t), T_1(t), L_1(t)), x_1(t) =$
$f_1((u_1(t), \vec{p}_1(t)), G_1(t)))\}$.

(3) Representation for the error function of addition transformation
$\text{BSC}(\text{YSC}(s_1, s_2, \ldots, s_n), \text{KSC}(e_1, e_2, \ldots, e_n))$.

11. Adding shrinking structure to existing inclusion structure to form a series structure

(1) Graphic representation of addition transformation (referring to Figs. 5.90 and 5.91)

(2) Representation for logical proposition of addition transformation
$T_{zjsw}(A((U, YS(s_1, s_2, \ldots, s_n), \vec{p}(t), T(t), L(t)), x(t) = f((u(t), \vec{p}(t)),$
$G(t)))) = \{A((U, \text{YS}(s_1, s_2, \ldots, s_n), \vec{p}(t), T(t), L(t)), x(t) = f((u(t),$
$\vec{p}(t)), G(t))), A_1((U_1, \text{KS}(e_1, e_2, \ldots, e_n), \vec{p}_1(t), T_1(t), L_1(t)), x_1(t) =$
$f_1((u_1(t), \vec{p}_1(t)), G_1(t)))\}$.

(3) Representation for the error function of addition transformation
$\text{CSC}(\text{YSC}(s_1, s_2, \ldots, s_n), \text{KSC}(e_1, e_2, \ldots, e_n))$.

12. Adding shrinking structure to existing inclusion structure to form a parallel structure

   (1) Graphic representation of addition transformation (referring to Figs. 5.92 and 5.93)
   (2) Representation for logical proposition of addition transformation
   $T_{zjsw}(A((U, YS(s_1, s_2, \ldots, s_n), \vec{p}(t), T(t), L(t)), x(t) = f((u(t), \vec{p}(t)),$
   $G(t)))) = \{A((U, YS(s_1, s_2, \ldots, s_n), \vec{p}(t), T(t), L(t)), x(t) = f((u(t),$
   $\vec{p}(t)), G(t))), A_1((U_1, KS(e_1, e_2, \ldots, e_n), \vec{p}_1(t), T_1(t), L_1(t)), x_1(t) =$
   $f_1((u_1(t), \vec{p}_1(t)), G_1(t)))\}.$
   (3) Representation for the error function of addition transformation
   $BSC(YSC(s_1, s_2, \ldots, s_n), KSC(e_1, e_2, \ldots, e_n)).$

13. Adding inclusion structure to existing inclusion structure to form a series structure

   (1) Graphic representation of addition transformation (referring to Figs. 5.94, 5.95, 5.96, and 5.97)
   (2) Representation for logical proposition of addition transformation
   $T_{zjsw}(A((U, YS(s_1, s_2, \ldots, s_n), \vec{p}(t), T(t), L(t)), x(t) = f((u(t), \vec{p}(t)),$
   $G(t)))) = \{A((U, YS(s_1, s_2, \ldots, s_n), \vec{p}(t), T(t), L(t)), x(t) = f((u(t),$
   $\vec{p}(t)), G(t))), A_1((U_1, YS_1(e_1, e_2, \ldots, e_n), \vec{p}_1(t), T_1(t), L_1(t)), x_1(t) =$
   $f_1((u_1(t), \vec{p}_1(t)), G_1(t)))\}.$
   (3) Representation for the error function of addition transformation
   $CSC(YSC(s_1, s_2, \ldots, s_n), YSC_1(e_1, e_2, \ldots, e_n)).$

14. Adding inclusion structure to existing inclusion structure to form a parallel structure

   (1) Graphic representation of addition transformation (referring to Figs. 5.98, 5.99, 5.100 and 5.101)
   (2) Representation for logical proposition of addition transformation
   $T_{zjsw}(A((U, YS(s_1, s_2, \ldots, s_n), \vec{p}(t), T(t), L(t)), x(t) = f((u(t), \vec{p}(t)),$
   $G(t)))) = \{A((U, YS(s_1, s_2, \ldots, s_n), \vec{p}(t), T(t), L(t)), x(t) = f((u(t),$
   $\vec{p}(t)), G(t))), A_1((U_1, YS_1(e_{11}, e_{12}, \ldots, e_{1n}), \vec{p}_1(t), T_1(t), L_1(t)), x_1(t) =$
   $f_1((u_1(t), \vec{p}_1(t)), G_1(t)))\}.$
   (3) Representation for the error function of addition transformation
   $BSC(YSC(s_1, s_2, \ldots, s_n), YSC_1(e_{11}, e_{12}, \ldots, e_{1n})).$

15. Adding feedback structure to existing inclusion structure to form a series structure

   (1) Graphic representation of addition transformation (referring to Figs. 5.102 and 5.103)

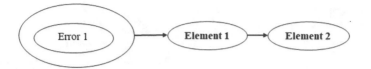

Fig. 5.74 Adding series structure to existing centered inclusion structure to form a series structure

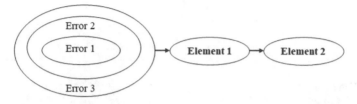

Fig. 5.75 Adding series structure to existing multi-layered inclusion structure to form a series structure

  (2) Representation for logical proposition of addition transformation
$T_{zjsw}(A((U, YS(s_1, s_2, \ldots, s_n), \overrightarrow{p}(t), T(t), L(t)), x(t) = f((u(t), \overrightarrow{p}(t)),$
$G(t)))) = \{A((U, YS(s_1, s_2, \ldots, s_n), \overrightarrow{p}(t), T(t), L(t)), x(t) = f((u(t),$
$\overrightarrow{p}(t)), G(t))), A_1((U_1, FS(e_1, e_2, \ldots, e_n), \overrightarrow{p_1}(t), T_1(t), L_1(t)), x_1(t) =$
$f_1((u_1(t), \overrightarrow{p_1}(t)), G_1(t)))\}.$
  (3) Representation for the error function of addition transformation
$CSC(YSC(s_1, s_2, \ldots, s_n), FSC(e_1, e_2, \ldots, e_n)).$

16. Adding feedback structure to existing inclusion structure to form a parallel structure

  (1) Graphic representation of addition transformation (referring to Figs. 5.104 and 5.105)
  (2) Representation for logical proposition of addition transformation
$T_{zjsw}(A((U, YS(s_1, s_2, \ldots, s_n), \overrightarrow{p}(t), T(t), L(t)), x(t) = f((u(t), \overrightarrow{p}(t)),$
$G(t)))) = \{A((U, YS(s_1, s_2, \ldots, s_n), \overrightarrow{p}(t), T(t), L(t)), x(t) = f((u(t),$
$\overrightarrow{p}(t)), G(t))), A_1((U_1, FS(e_1, e_2, \ldots, e_n), \overrightarrow{p_1}(t), T_1(t), L_1(t)), x_1(t) =$
$f_1((u_1(t), \overrightarrow{p_1}(t)), G_1(t)))\}.$
  (3) Representation for the error function of addition transformation
$BSC(YSC(s_1, s_2, \ldots, s_n), FSC(e_1, e_2, \ldots, e_n)).$

Fig. 5.76 Adding series
structure to existing centered
inclusion structure to form a
feedback structure

Fig. 5.77 Adding series
structure to existing
multi-layered inclusion
structure to form a feedback
structure

Fig. 5.78 Adding series
structure to existing centered
inclusion structure to form a
parallel structure

Fig. 5.79 Adding series
structure to existing
multi-layered inclusion
structure to form a parallel
structure

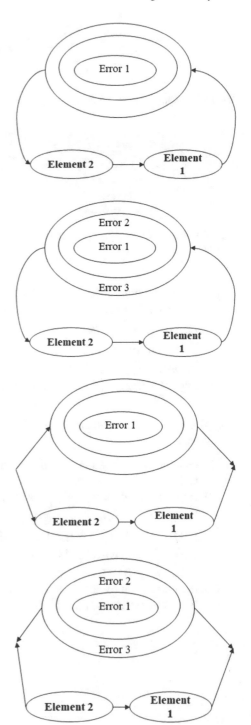

Fig. 5.80 Adding series structure to existing multi-layered inclusion structure to form a parallel structure

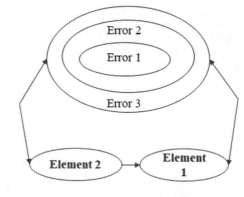

Fig. 5.81 Adding parallel structure to existing centered inclusion structure to form a series structure

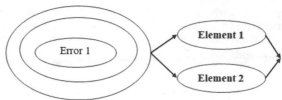

Fig. 5.82 Adding parallel structure to existing multi-layered inclusion structure to form a series structure

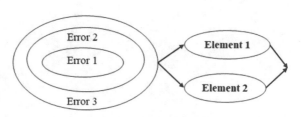

Fig. 5.83 Adding parallel structure to existing centered inclusion structure to form a parallel structure

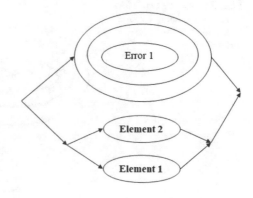

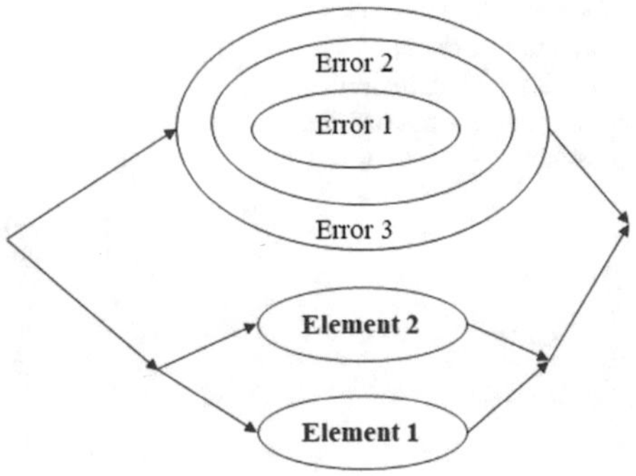

Fig. 5.84 Adding parallel
structure to existing
multi-layered inclusion
structure to form a parallel
structure-type 1

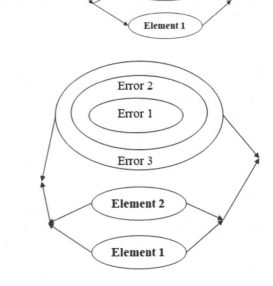

Fig. 5.85 Adding series
structure to existing
multi-layered inclusion
structure to form a parallel
structure-type 2

Fig. 5.86 Adding expanding
structure to existing centered
inclusion structure to form a
series structure

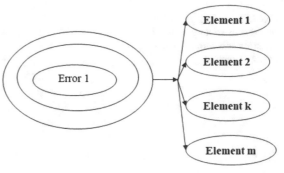

Fig. 5.87 Adding expanding
structure to existing
multi-layered inclusion
structure to form a series
structure

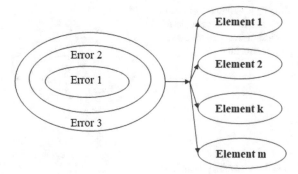

Fig. 5.88 Adding expanding
structure to existing centered
inclusion structure to form a
parallel structure

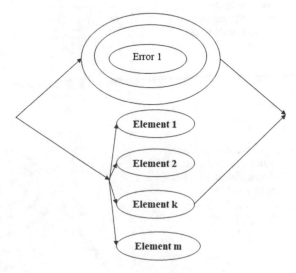

Fig. 5.89 Adding expanding
structure to existing
multi-layered inclusion
structure to form a parallel
structure

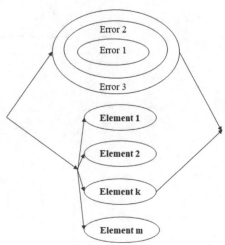

Fig. 5.90 Adding shrinking structure to existing centered inclusion structure to form a series structure

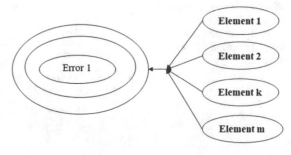

Fig. 5.91 Adding shrinking structure to existing multi-layered inclusion structure to form a series structure

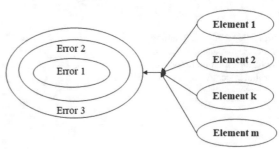

Fig. 5.92 Adding shrinking structure to existing centered inclusion structure to form a parallel structure

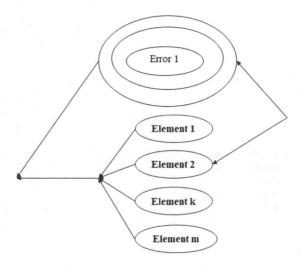

Fig. 5.93 Adding shrinking structure to existing multi-layered inclusion structure to form a parallel structure

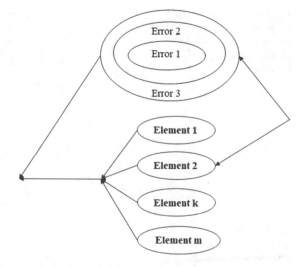

Fig. 5.94 Adding centered inclusion structure to existing centered inclusion structure to form a series structure

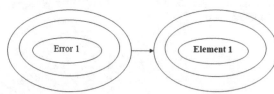

Fig. 5.95 Adding multi-layered inclusion structure to existing centered inclusion structure to form a series structure

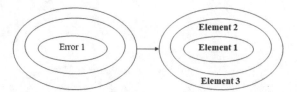

Fig. 5.96 Adding centered inclusion structure to existing multi-layered inclusion structure to form a series structure

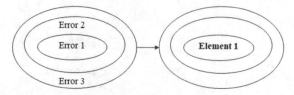

Fig. 5.97 Adding multi-layered inclusion structure to existing multi-layered inclusion structure to form a series structure

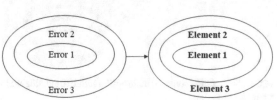

Fig. 5.98 Adding centered inclusion structure to existing centered inclusion structure to form a parallel structure

Fig. 5.99 Adding multi-layered inclusion structure to existing centered inclusion structure to form a parallel structure

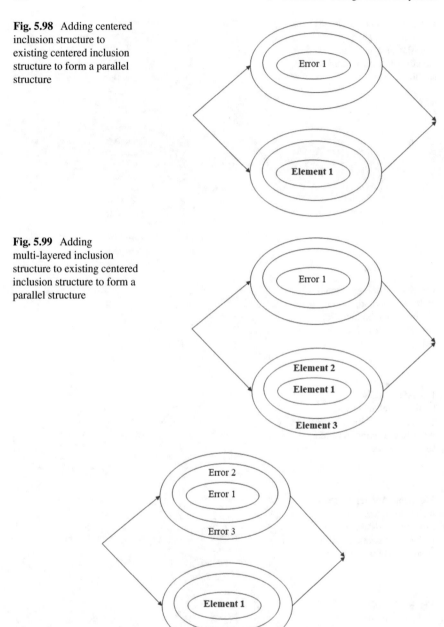

Fig. 5.100 Adding centered inclusion structure to existing multi-layered inclusion structure to form a parallel structure

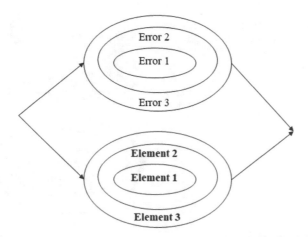

Fig. 5.101 Adding multi-layered inclusion to existing multi-layered inclusion structure to form a parallel structure

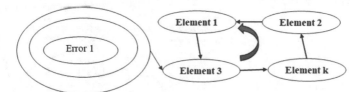

Fig. 5.102 Adding feedback structure to existing centered inclusion structure to form a series structure

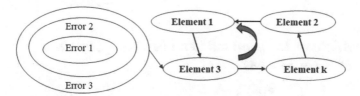

Fig. 5.103 Adding feedback inclusion structure to existing multi-layered inclusion structure to form a series structure

Fig. 5.104 Adding feedback structure to existing centered inclusion structure to form a parallel structure

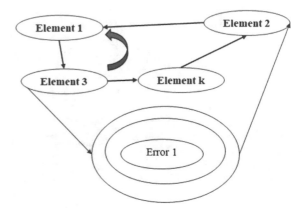

Fig. 5.105 Adding feedback structure to existing multi-layered inclusion structure to form a parallel structure

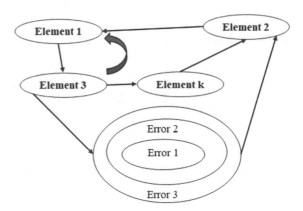

5.1.6 Addition Transformation on Feedback Structure in Error Systems

1. Adding series structure to existing feedback structure to form a series structure

   (1) Graphic representation of addition transformation (referring to Fig. 5.106)
   (2) Representation for logical proposition of addition transformation
   $$T_{zjsw}(A((U, FS(s_1, s_2, \ldots, s_n), \vec{p}(t), T(t), L(t)), x(t) = f((u(t), \vec{p}(t)),$$
   $$G(t)))) = \{A((U, \text{FS}(s_1, s_2, \ldots, s_n), \vec{p}(t), T(t), L(t)), x(t) = f((u(t),$$
   $$\vec{p}(t)), G(t))), A_1((U_1, \text{CS}(e_1, e_2, \ldots, e_n), \vec{p}_1(t), T_1(t), L_1(t)), x_1(t) =$$
   $$f_1((u_1(t), \vec{p}_1(t)), G_1(t)))\}.$$
   (3) Representation for the error function of addition transformation
   $$\text{CSC}(\text{FSC}(s_1, s_2, \ldots, s_n), \text{CSC}(e_1, e_2, \ldots, e_n)).$$

2. Adding series structure to existing feedback structure to form a new feedback structure

   (1) Graphic representation of addition transformation (referring to Fig. 5.107)

(2) Representation for logical proposition of addition transformation

$T_{zjsw}(A((U, FS(s_1, s_2, \ldots, s_n), \vec{p}(t), T(t), L(t)), x(t) = f((u(t), \vec{p}(t)),$
$G(t)))) = \{A((U, \text{FS}(s_1, s_2, \ldots, s_n), \vec{p}(t), T(t), L(t)), x(t) = f((u(t),$
$\vec{p}(t)), G(t))), A_1((U_1, \text{CS}_1(e_1, e_2, \ldots, e_n), \vec{p_1}(t), T_1(t), L_1(t)), x_1(t) =$
$f_1((u_1(t), \vec{p_1}(t)), G_1(t)))\}.$

(3) Representation for the error function of addition transformation

$\text{FSC}(s_1, s_2, \ldots, \text{CSC}(s_1, s_2, \ldots, \text{FSC}(e_1, e_2, \ldots, e_n), s_n)).$

3. Adding series structure to existing feedback structure to form a parallel structure

(1) Graphic representation of addition transformation (referring to Figs. 5.108 and 5.109)

(2) Representation for logical proposition of addition transformation

$T_{zjsw}(A((U, FS(s_1, s_2, \ldots, s_n), \vec{p}(t), T(t), L(t)), x(t) = f((u(t), \vec{p}(t)),$
$G(t)))) = \{A((U, \text{FS}(s_1, s_2, \ldots, s_n), \vec{p}(t), T(t), L(t)), x(t) = f((u(t),$
$\vec{p}(t)), G(t))), A_1((U_1, \text{CS}(e_1, e_2, \ldots, e_n), \vec{p_1}(t), T_1(t), L_1(t)), x_1(t) =$
$f_1((u_1(t), \vec{p_1}(t)), G_1(t)))\}.$

(3) Representation for the error function of addition transformation

$\text{BSC}(\text{FSC}(s_1, s_2, \ldots, s_n), \text{CSC}(e_1, e_2, \ldots, e_n)).$

4. Adding parallel structure to existing feedback structure to form a series structure

(1) Graphic representation of addition transformation (referring to Fig. 5.110)

(2) Representation for logical proposition of addition transformation

$T_{zjsw}(A((U, FS(s_1, s_2, \ldots, s_n), \vec{p}(t), T(t), L(t)), x(t) = f((u(t), \vec{p}(t)),$
$G(t)))) = \{A((U, \text{FS}(s_1, s_2, \ldots, s_n), \vec{p}(t), T(t), L(t)), x(t) = f((u(t),$
$\vec{p}(t)), G(t))), A_1((U_1, \text{BS}(e_1, e_2, \ldots, e_n), \vec{p_1}(t), T_1(t), L_1(t)), x_1(t) =$
$f_1((u_1(t), \vec{p_1}(t)), G_1(t)))\}.$

(3) Representation for the error function of addition transformation

$\text{CSC}(\text{FSC}(s_1, s_2, \ldots, s_n), \text{BSC}(e_1, e_2, \ldots, e_n)).$

5. Adding parallel structure to existing feedback structure to form a new feedback structure

(1) Graphic representation of addition transformation (referring to Fig. 5.111)

(2) Representation for logical proposition of addition transformation

$T_{zjsw}(A((U, FS(s_1, s_2, \ldots, s_n), \vec{p}(t), T(t), L(t)), x(t) = f((u(t), \vec{p}(t)),$
$G(t)))) = \{A((U, \text{FS}(s_1, s_2, \ldots, s_n), \vec{p}(t), T(t), L(t)), x(t) = f((u(t),$
$\vec{p}(t)), G(t))), A_1((U_1, \text{BS}(e_1, e_2, \ldots, e_n), \vec{p_1}(t), T_1(t), L_1(t)), x_1(t) =$
$f_1((u_1(t), \vec{p_1}(t)), G_1(t)))\}.$

(3) Representation for the error function of addition transformation

$\text{FSC}(s_1, s_2, \ldots, \text{CSC}(s_1, s_2, \text{BSC}(e_1, e_2, \ldots, e_n), s_n), s_n).$

6. Adding parallel structure to existing feedback structure to form a parallel structure-type 1

(1) Graphic representation of addition transformation (referring to Fig. 5.112)
(2) Representation for logical proposition of addition transformation

$T_{zjsw}(A((U, FS(s_1, s_2, \ldots, s_n), \vec{p}(t), T(t), L(t)), x(t) = f((u(t), \vec{p}(t)),$
$G(t)))) = \{A((U, \text{FS}(s_1, s_2, \ldots, s_n), \vec{p}(t), T(t), L(t)), x(t) = f((u(t),$
$\vec{p}(t)), G(t))), A_1((U_1, \text{BS}(e_1, e_2, \ldots, e_n), \vec{p_1}(t), T_1(t), L_1(t)), x_1(t) =$
$f_1((u_1(t), \vec{p_1}(t)), G_1(t)))\}.$

(3) Representation for the error function of addition transformation
$\text{BSC}(\text{FSC}(s_1, s_2, \ldots, s_n), \text{BSC}(e_1, e_2, \ldots, e_n)).$

7. Adding parallel structure to existing feedback structure to form a parallel structure-type 2

(1) Graphic representation of addition transformation (referring to Fig. 5.113)
(2) Representation for logical proposition of addition transformation

$T_{zjsw}(A((U, FS(s_1, s_2, \ldots, s_n), \vec{p}(t), T(t), L(t)), x(t) = f((u(t), \vec{p}(t)),$
$G(t)))) = \{A((U, \text{FS}(s_1, s_2, \ldots, s_n), \vec{p}(t), T(t), L(t)), x(t) = f((u(t),$
$\vec{p}(t)), G(t))), A_1((U_1, \text{BS}(e_1, e_2, \ldots, e_n), \vec{p_1}(t), T_1(t), L_1(t)), x_1(t) =$
$f_1((u_1(t), \vec{p_1}(t)), G_1(t)))\}.$

(3) Representation for the error function of addition transformation
$\text{BSC}(\text{CSC}(e_i, e_j, \text{FSC}(s_1, s_2, \ldots, s_n), e_{n-m}), \text{BSC}(e_1, e_2, \ldots, e_k)).$

8. Adding expanding structure to existing feedback structure to form a series structure

(1) Graphic representation of addition transformation (referring to Fig. 5.114)
(2) Representation for logical proposition of addition transformation

$T_{zjsw}(A((U, FS(s_1, s_2, \ldots, s_n), \vec{p}(t), T(t), L(t)), x(t) = f((u(t), \vec{p}(t)),$
$G(t)))) = \{A((U, \text{FS}(s_1, s_2, \ldots, s_n), \vec{p}(t), T(t), L(t)), x(t) = f((u(t),$
$\vec{p}(t)), G(t))), A_1((U_1, \text{KS}(e_1, e_2, \ldots, e_n), \vec{p_1}(t), T_1(t), L_1(t)), x_1(t) =$
$f_1((u_1(t), \vec{p_1}(t)), G_1(t)))\}.$

(3) Representation for the error function of addition transformation
$\text{CSC}(\text{FSC}(s_1, s_2, \ldots, s_n), \text{KSC}(e_1, e_2, \ldots, e_n)).$

9. Adding expanding structure to existing feedback structure to form a new feedback structure

(1) Graphic representation of addition transformation (referring to Fig. 5.115)
(2) Representation for logical proposition of addition transformation

$T_{zjsw}(A((U, FS(s_1, s_2, \ldots, s_n), \vec{p}(t), T(t), L(t)), x(t) = f((u(t), \vec{p}(t)),$
$G(t)))) = \{A((U, \text{FS}(s_1, s_2, \ldots, s_n), \vec{p}(t), T(t), L(t)), x(t) = f((u(t),$
$\vec{p}(t)), G(t))), A_1((U_1, \text{KS}(e_1, e_2, \ldots, e_n), \vec{p_1}(t), T_1(t), L_1(t)), x_1(t) =$
$f_1((u_1(t), \vec{p_1}(t)), G_1(t)))\}.$

(3) Representation for the error function of addition transformation
$\text{FSC}(s_1, s_2, \ldots, s_n, \text{CSC}(s_k, \text{KSC}(s_1, s_2, \ldots, s_n))).$

10. Adding expanding structure to existing feedback structure to form a parallel structure

(1) Graphic representation of addition transformation (referring to Fig. 5.116)
(2) Representation for logical proposition of addition transformation

$T_{zjsw}(A((U, FS(s_1, s_2, \ldots, s_n), \vec{p}(t), T(t), L(t)), x(t) = f((u(t), \vec{p}(t)), G(t)))) = \{A((U, \mathrm{FS}(s_1, s_2, \ldots, s_n), \vec{p}(t), T(t), L(t)), x(t) = f((u(t), \vec{p}(t)), G(t))), A_1((U_1, \mathrm{KS}(e_1, e_2, \ldots, e_n), \vec{p_1}(t), T_1(t), L_1(t)), x_1(t) = f_1((u_1(t), \vec{p_1}(t)), G_1(t)))\}.$

(3) Representation for the error function of addition transformation

$\mathrm{BSC}(\mathrm{FSC}(s_1, s_2, \ldots, s_n), \mathrm{KSC}(e_1, e_2, \ldots, e_k, e_m)).$

11. Adding shrinking structure to existing feedback structure to form a series structure

(1) Graphic representation of addition transformation (referring to Fig. 5.117)
(2) Representation for logical proposition of addition transformation

$T_{zjsw}(A((U, FS(s_1, s_2, \ldots, s_n), \vec{p}(t), T(t), L(t)), x(t) = f((u(t), \vec{p}(t)), G(t)))) = \{A((U, \mathrm{FS}(s_1, s_2, \ldots, s_n), \vec{p}(t), T(t), L(t)), x(t) = f((u(t), \vec{p}(t)), G(t))), A_1((U_1, \mathrm{KS}(e_1, e_2, \ldots, e_n), \vec{p_1}(t), T_1(t), L_1(t)), x_1(t) = f_1((u_1(t), \vec{p_1}(t)), G_1(t)))\}.$

(3) Representation for the error function of addition transformation

$\mathrm{CSC}(\mathrm{FSC}(s_1, s_2, \ldots, s_n), \mathrm{KSC}(e_1, e_2, \ldots, e_n)).$

12. Adding shrinking structure to existing feedback structure to form a new feedback structure

(1) Graphic representation of addition transformation (referring to Fig. 5.118)
(2) Representation for logical proposition of addition transformation

$T_{zjsw}(A((U, FS(s_1, s_2, \ldots, s_n), \vec{p}(t), T(t), L(t)), x(t) = f((u(t), \vec{p}(t)), G(t)))) = \{A((U, \mathrm{FS}(s_1, s_2, \ldots, s_n), \vec{p}(t), T(t), L(t)), x(t) = f((u(t), \vec{p}(t)), G(t))), A_1((U_1, \mathrm{KS}(e_1, e_2, \ldots, e_n), \vec{p_1}(t), T_1(t), L_1(t)), x_1(t) = f_1((u_1(t), \vec{p_1}(t)), G_1(t)))\}.$

(3) Representation for the error function of addition transformation

$\mathrm{FSC}(s_1, s_2, \ldots, s_n, \mathrm{CSC}(s_k, \mathrm{KSC}(e_1, e_2, \ldots, e_n))).$

13. Adding shrinking structure to existing feedback structure to form a parallel structure

(1) Graphic representation of addition transformation (referring to Fig. 5.119)
(2) Representation for logical proposition of addition transformation

$T_{zjsw}(A((U, FS(s_1, s_2, \ldots, s_n), \vec{p}(t), T(t), L(t)), x(t) = f((u(t), \vec{p}(t)), G(t)))) = \{A((U, \mathrm{FS}(s_1, s_2, \ldots, s_n), \vec{p}(t), T(t), L(t)), x(t) = f((u(t), \vec{p}(t)), G(t))), A_1((U_1, \mathrm{KS}(e_1, e_2, \ldots, e_n), \vec{p_1}(t), T_1(t), L_1(t)), x_1(t) = f_1((u_1(t), \vec{p_1}(t)), G_1(t)))\}.$

(3) Representation for the error function of addition transformation
BSC(FSC($s_1, s_2, \ldots, s_n$), KSC($e_1, e_2, \ldots, e_k, e_m$)).

14. Adding inclusion structure to existing feedback structure to form a series structure

(1) Graphic representation of addition transformation (referring to Figs. 5.120 and 5.121)

(2) Representation for logical proposition of addition transformation
$T_{zjsw}(A((U, FS(s_1, s_2, \ldots, s_n), \overrightarrow{p}(t), T(t), L(t)), x(t) = f((u(t), \overrightarrow{p}(t)),$
$G(t)))) = \{A((U, FS(s_1, s_2, \ldots, s_n), \overrightarrow{p}(t), T(t), L(t)), x(t) = f((u(t),$
$\overrightarrow{p}(t)), G(t))), A_1((U_1, YS(e_1, e_2, \ldots, e_n), \overrightarrow{p_1}(t), T_1(t), L_1(t)), x_1(t) =$
$f_1((u_1(t), \overrightarrow{p_1}(t)), G_1(t)))\}.$

(3) Representation for the error function of transformation
CSC(FSC($s_1, s_2, \ldots, s_n$), YSC($e_1, e_2, \ldots, e_n$)).

15. Adding inclusion structure to existing feedback structure to form a new feedback structure

(1) Graphic representation of addition transformation (referring to Figs. 5.122 and 5.123)

(2) Representation for logical proposition of addition transformation
$T_{zjsw}(A((U, FS(s_1, s_2, \ldots, s_n), \overrightarrow{p}(t), T(t), L(t)), x(t) = f((u(t), \overrightarrow{p}(t)),$
$G(t)))) = \{A((U, FS(s_1, s_2, \ldots, s_n), \overrightarrow{p}(t), T(t), L(t)), x(t) = f((u(t),$
$\overrightarrow{p}(t)), G(t))), A_1((U_1, YS(e_1, e_2, \ldots, e_n), \overrightarrow{p_1}(t), T_1(t), L_1(t)), x_1(t) =$
$f_1((u_1(t), \overrightarrow{p_1}(t)), G_1(t)))\}.$

(3) Representation for the error function of addition transformation
FSC($s_1, s_2, \ldots, s_n$, CSC(YSC($e_1, e_2, \ldots, e_m$), $s_1, s_2, \ldots, s_n$)).

16. Adding inclusion structure to existing feedback structure to form a parallel structure

(1) Graphic representation of addition transformation (referring to Figs. 5.124 and 5.125)

(2) Representation for logical proposition of addition transformation
$T_{zjsw}(A((U, FS(s_1, s_2, \ldots, s_n), \overrightarrow{p}(t), T(t), L(t)), x(t) = f((u(t), \overrightarrow{p}(t)),$
$G(t)))) = \{A((U, FS(s_1, s_2, \ldots, s_n), \overrightarrow{p}(t), T(t), L(t)), x(t) = f((u(t),$
$\overrightarrow{p}(t)), G(t))), A_1((U_1, YS(e_1, e_2, \ldots, e_n), \overrightarrow{p_1}(t), T_1(t), L_1(t)), x_1(t) =$
$f_1((u_1(t), \overrightarrow{p_1}(t)), G_1(t)))\}.$

(3) Representation for the error function of addition transformation
BSC(FSC($s_1, s_2, \ldots, s_n$), YSC($e_1, e_2, \ldots, e_k, e_m$)).

17. Adding feedback structure to existing feedback structure to form a series structure

(1) Graphic representation of addition transformation (referring to Fig. 5.126)

(2) Representation for logical proposition of addition transformation

$T_{zjsw}(A((U, FS(s_1, s_2, \ldots, s_n), \vec{p}(t), T(t), L(t)), x(t) = f((u(t), \vec{p}(t)),$
$G(t)))) = \{A((U, \mathrm{FS}(s_1, s_2, \ldots, s_n), \vec{p}(t), T(t), L(t)), x(t) = f((u(t),$
$\vec{p}(t)), G(t))), A_1((U_1, \mathrm{FS}_1(s_{11}, s_{12}, \ldots, s_{1n}, \vec{p_1}(t), T_1(t), L_1(t)), x_1(t) =$
$f_1((u_1(t), \vec{p_1}(t)), G_1(t)))\}.$

(3) Representation for the error function of addition transformation
$\mathrm{CSC}(\mathrm{FSC}(s_1, s_2, \ldots, s_n), \mathrm{FSC}_1(s_{11}, s_{12}, \ldots, s_{1n})).$

18. Adding feedback structure to existing feedback structure to form a new feedback structure

(1) Graphic representation of addition transformation (referring to Fig. 5.127)
(2) Representation for logical proposition of addition transformation

$T_{zjsw}(A((U, FS(s_1, s_2, \ldots, s_n), \vec{p}(t), T(t), L(t)), x(t) = f((u(t), \vec{p}(t)),$
$G(t)))) = \{A((U, \mathrm{FS}(s_1, s_2, \ldots, s_n), \vec{p}(t), T(t), L(t)), x(t) = f((u(t),$
$\vec{p}(t)), G(t))), A_1((U_1, \mathrm{FS}(e_1, e_2, \ldots, e_n), \vec{p_1}(t), T_1(t), L_1(t)), x_1(t) =$
$f_1((u_1(t), \vec{p_1}(t)), G_1(t)))\}.$

(3) Representation for the error function of addition transformation
$\mathrm{FSC}(s_1, s_2, \ldots, s_n), \mathrm{CSC}(\mathrm{FSC}(e_1, e_2, \ldots, e_m), s_1, s_2, \ldots, s_k)).$

19. Adding feedback structure to existing feedback structure to form a parallel structure

(1) Graphic representation of addition transformation (referring to Fig. 5.128)
(2) Representation for logical proposition of addition transformation

$T_{zjsw}(A((U, FS(s_1, s_2, \ldots, s_n), \vec{p}(t), T(t), L(t)), x(t) = f((u(t), \vec{p}(t)),$
$G(t)))) = \{A((U, \mathrm{FS}(s_1, s_2, \ldots, s_n), \vec{p}(t), T(t), L(t)), x(t) = f((u(t),$
$\vec{p}(t)), G(t))), A_1((U_1, FS_1(e_{11}, e_{12}, \ldots, e_{1n}), \vec{p_1}(t), T_1(t), L_1(t)), x_1(t) =$
$f_1((u_1(t), \vec{p_1}(t)), G_1(t)))\}.$

(3) Representation for the error function of addition transformation
$\mathrm{BSC}(\mathrm{FSC}(s_1, s_2, \ldots, s_n), FSC_1(e_{11}, e_{12}, \ldots, e_{1n})).$

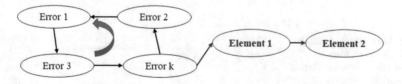

Fig. 5.106 Adding series structure to existing feedback structure to form a series structure

Fig. 5.107 Adding series structure to existing feedback structure to form a new feedback structure

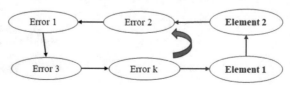

Fig. 5.108 Adding series
structure to existing
feedback structure to form a
parallel structure-type 1

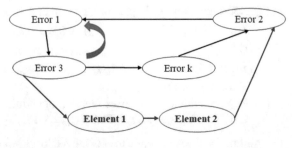

Fig. 5.109 Adding series
structure to existing
feedback structure to form a
parallel structure-type 2

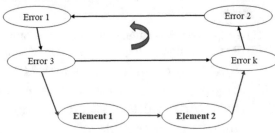

Fig. 5.110 Adding parallel
structure to existing
feedback structure to form a
series structure

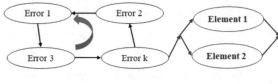

Fig. 5.111 Adding parallel
structure to existing
feedback structure to form a
new feedback structure

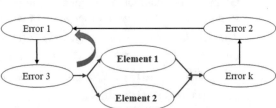

Fig. 5.112 Adding parallel
structure to existing
feedback structure to form a
parallel structure-type 1

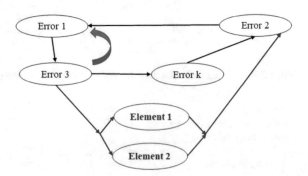

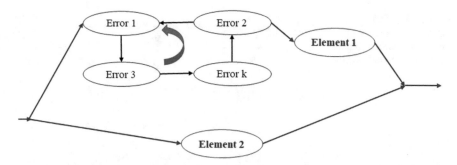

Fig. 5.113 Adding parallel structure to existing feedback structure to form a parallel structure-type 2

Fig. 5.114 Adding expanding structure to existing feedback structure to form a series structure

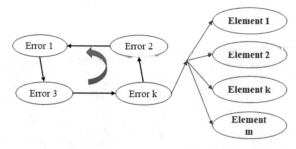

Fig. 5.115 Adding expanding structure to existing feedback structure to form a new feedback structure

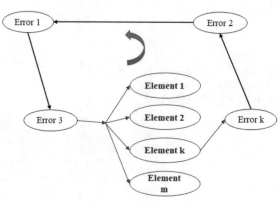

Fig. 5.116 Adding
expanding structure to
existing feedback structure
to form a parallel structure

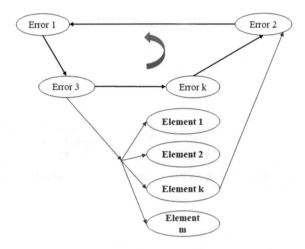

Fig. 5.117 Adding
shrinking structure to
existing feedback structure
to form a series structure

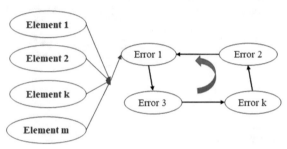

Fig. 5.118 Adding
shrinking structure to
existing feedback structure
to form a new feedback
structure

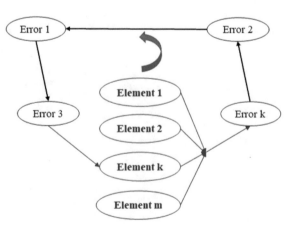

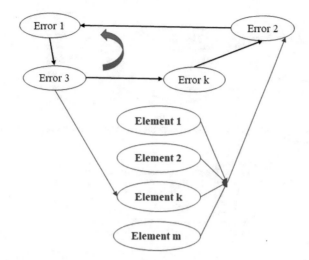

Fig. 5.119 Adding shrinking structure to existing feedback structure to form a parallel structure

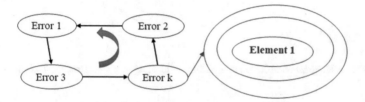

Fig. 5.120 Adding centered inclusion structure to existing feedback structure to form a series structure

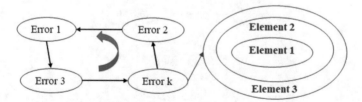

Fig. 5.121 Adding multi-layered inclusion structure to existing feedback structure to form a series structure

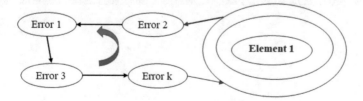

Fig. 5.122 Adding centered inclusion structure to existing feedback structure to form a new feedback structure

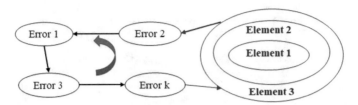

Fig. 5.123 Adding multi-layered inclusion structure to existing feedback structure to form a new feedback structure

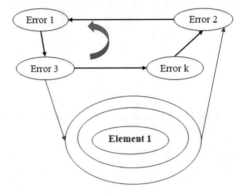

Fig. 5.124 Adding centered inclusion structure to existing feedback structure to form a parallel structure

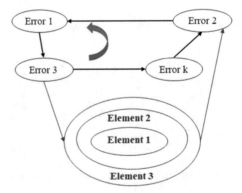

Fig. 5.125 Adding multi-layered inclusion structure to existing feedback structure to form a parallel structure

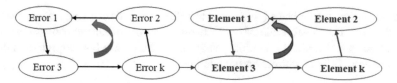

Fig. 5.126 Adding feedback structure to existing feedback structure to form a series structure

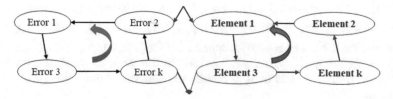

Fig. 5.127 Adding feedback structure to existing feedback structure to form a new feedback structure

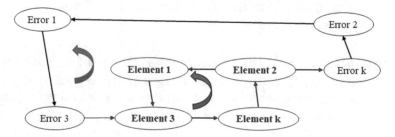

Fig. 5.128 Adding feedback structure to existing feedback structure to form a parallel structure

5.2 Displacement Transformation on Error System Structures

Displacement transformation is a process of substituting A with B for the purpose of achieving desired outcome or objective. In reality, broken electronic appliances and machines got back to normal by replacing malfunctioning part with new spare part. In the the process of examining the reliability or safety of a very complicated system, we need to understand the impact on system reliability or safety when one or multiple subsystems or elements fail(s). Therefore, it is necessary to study the patterns and laws of transition and transformation of errors in a system. Having done that, one can prevent errors from happening and successfully eliminate them once they occurred. In order to eliminate the error occurred in a subsystem of system **S**, following steps are employed: (1) system **S** is decomposed into a set $\{s_1, s_2, \ldots, s_i, s_k, s_n\}$; (2) identify the subsystem with error-i.e., s_i; (3) replace s_i with a subsystem (s_τ) having the same functionality. For example, in the management of chain stores, it is imperative to study the decomposition transformation, displacement transformation, equivalence (similarity) transformation, destruction transformation, and their inverse transformations on the mature enterprise management system used in the successful chain store before integrating it into a newly-built one in another country with different culture and regulations.

5.2.1   Types of Displacement Transformation on Error System Structures

The expression for the logical proposition of displacement transformation is:

Suppose that $\{A_i((V_i(t), S_{iu}(t), \overrightarrow{p_{iu}}(t), T_{iu}(t), L_{iu}(t)), x_i(t) = f_i((u_i(t), G_{iA}(t))),$ $i = 1, 2, ..., m\}$ represents the set of error logical variables defined under judging rule G on universe of discourse V, if $T_z\{A_i((V_i(t), S_{iu}(t), \overrightarrow{p_{iu}}(t), T_{iu}(t), L_{iu}(t)),$ $x_i(t) = f_i((u_i(t), G_{iA}(t))), i = 1, 2, ..., m\} = \{B_j((V_j(t), S_{jv}(t), \overrightarrow{p_{jv}}(t), T_{jv}(t),$ $L_{jv}(t)), y_j(t) = f_j((v_j(t), G_{jB}(t))), j = 1, 2, ..., n\}$, then T_z is called the logical connective of displacement transformation regarding $\{A_i((V_i(t), S_{iu}(t), \overrightarrow{p_{iu}}(t), T_{iu}(t),$ $L_{iu}(t)), x_i(t) = f_i((u_i(t), G_{iA}(t))), i = 1, 2, ..., m\}$ and judging rule G on universe of discourse $U(t)$. Here, $\{B_j((V_j(t), S_{jv}(t), \overrightarrow{p_{jv}}(t), T_{jv}(t), L_{jv}(t)), y_j(t) = f_j((v_j(t),$ $G_{jB}(t))), j = 1, 2, ..., n\}$ is used to replace $\{A_i((V_i(t), S_{iu}(t), \overrightarrow{p_{iu}}(t), T_{iu}(t), L_{iu}(t)),$ $x_i(t) = f_i((u_i(t), G_{iA}(t))), i = 1, 2, ..., m\}$. In this case, the desired error value $\{y_j(t), j = 1, 2, ..., n\}$ is used to replace the undesired error value $\{x_i(t), i = 1, 2, ..., m\}$.

There are more displacement transformations listed as below.

(1) In the case of $\{x_i(t), i = 1, 2, ..., m\} \rightarrow \{y_j(t), j = 1, 2, ..., n\}$, T_z has conducted opposite displacement transformation on error value if $x_i(t) = -y_j(t)(i = 1, 2, ..., m, j = 1, 2, ..., n)$ and it is noted by T_{zfcz}

(2) Suppose that $a = max\{x_i(t), i = 1, 2, ..., m\}, b = min\{x_i(t), i = 1, 2, ..., m\}, c = max\{y_j(t), j = 1, 2, ..., n\}, d = min\{y_j(t), j = 1, 2, ..., n\}$. If $c < a$ and $d < b$, then T_z has conducted soothing displacement transformation on error value noted by T_{zycz};

(3) Suppose that $a = max\{x_i(t), i = 1, 2, ..., m\}, b = min\{x_i(t), i = 1, 2, ..., m\}, c = max\{y_j(t), j = 1, 2, ..., n\}, d = min\{y_j(t), j = 1, 2, ..., n\}$. If $c > a$ and $d > b$, then T_z has conducted worsening displacement transformation on error value noted by T_{zhcz};

(4) In the displacement transformation $\{x_i(t), i = 1, 2, ..., m\} \rightarrow \{y_j(t), j = 1, 2, ..., n\}$, if $y_j(t) = kx_i(t)$ $i = 1, 2, ..., m$ and $j = 1, 2, ..., n$, then T_z has conducted amplification displacement transformation on error value noted by T_{zkcz}.

(a) If $k \geq 1$, then T_{zkcz} has conducted positive amplification displacement transformation on error value noted by T_{zkzcz};

(b) If $k \leq -1$, then T_{zkcz} has conducted negative amplification displacement transformation on error value noted by T_{zkfcz};

(c) If $0 < k < 1$, then T_{zkcz} has conducted positive decreasing displacement transformation on error value noted by T_{zkzscz};

(d) If $-1 < k < 0$, then T_{zkcz} has conducted negative decreasing displacement transformation on error value noted by T_{zkfscz};

(e) If $k = 0$, then T_{zkcz} has conducted error-elimination displacement transformation noted by T_{zkhlcz}.

1. Principles for displancement transformation

Three principles must be held when considering displacement transformation

(1) Objective needs
(2) Determined by actual conditions
(3) Minimum costs

2. Types of displacement transformations

In general, object u(t) not only has horizontal and vertical structure but also has many hierarchical and complex relationships (referring to Fig. 5.129). Therefore, it is feasible to conduct displacement transformation on objects at different hierarchies. In $\{A_i((V_i(t), S_{iu}(t), \vec{p}_{iu}(t), T_{iu}(t), L_{iu}(t)), x_i(t) = f_i((u_i(t), G_{iA}(t))) = A_i(u_i(t), x_i(t)), i = 1, 2, \ldots, m\}$ and $\{B_j((V_j(t), S_{jv}(t), \vec{p}_{jv}(t), T_{jv}(t), L_{jv}(t)), y_j(t) = f_j((v_j(t), G_{jB}(t))) = B_j(v_j(t), y_j(t)), j = 1, 2, \ldots, n\}$, if $\{S_{iu}(t), i = 1, 2, \ldots, m\} \rightarrow \{S_{jv}(t), j = 1, 2, \ldots, n\}$, then T_z is called subject displacement transformation regarding $\{A_i((V_i(t), S_{iu}(t), \vec{p}_{iu}(t), T_{iu}(t), L_{iu}(t)), x_i(t) = f_i((u_i(t), G_{iA}(t)))), i = 1, 2, \ldots, m\}$ and judging rule $G_A(t)$ on universe of discourse $U(t)$ denoted by T_{zsw}. The transformation on other parameters can follow the same steps here.

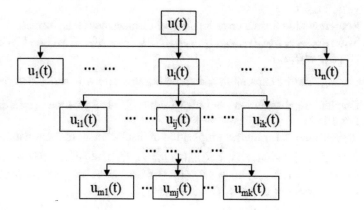

Fig. 5.129 Structure of object system

5.2.2   Displacement Transformation on Error System Structures

Suppose that the structure of system **S** is denoted by Fig. 5.130.

1. Replacing existing series structure with new series structure

   (1) Graphic representation of displacement transformation (referring to Fig. 5.131)

   (2) Representation for logical proposition of displacement transformation

   $T_z\{A_i((U_i(t), \mathrm{CS}(s_1, s_2, \ldots, s_n)_{iu}(t), \vec{p_{iu}}(t), T_{iu}(t), L_{iu}(t)), x_i(t) = f_i(u_i(t),$
   $G_{iA}(t))), i = 1, 2, \ldots, m\} = \{B_j((V_j(t), \mathrm{CS}(s_1, s_2, \ldots, s_n)_{jv}t, \vec{p_{jv}}(t), T_{jv}(t),$
   $L_{jv}(t)), y_j(t) = g_j(v_j(t), G_{jB}(t))), j = 1, 2, \ldots, n\}$

   (3) Symbolic representation for displacement transformation

   $S = S(s_1, s_2, \ldots, \mathrm{CS}(s_1, s_2, \ldots, s_m)_{iu}(t), s_n) \longrightarrow S(s_1, s_2, \ldots, \mathrm{CS}(s_1, s_2, \ldots,$
   $s_m)_{jv}(t), s_n)$

   (4) Representation for the error function of displacement transformation

   $\mathrm{SC}(s_1, s_2, \ldots, \mathrm{CSC}(s_1, s_2, \ldots, s_m)_{iu}(t), s_n) \longrightarrow \mathrm{SC}(s_1, s_2, \ldots, \mathrm{CSC}(s_1, s_2,$
   $\ldots, s_m)_{jv}(t), s_n)$

2. Replacing existing parallel structure with series structure

   (1) Graphic representation of displacement transformation (referring to Fig. 5.132)

   (2) Representation for logical proposition of displacement transformation

   $T_z\{A_i((U_i(t), \mathrm{BS}(s_1, s_2, \ldots, s_n)_{iu}(t), \vec{p_{iu}}(t), T_{iu}(t), L_{iu}(t)), x_i(t) = f_i(u_i(t),$
   $G_{iA}(t))), i = 1, 2, \ldots, m\} = \{B_j((V_j(t), \mathrm{CS}(s_1, s_2, \ldots, s_n)_{jv}t, \vec{p_{jv}}(t), T_{jv}(t),$
   $L_{jv}(t)), y_j(t) = g_j(v_j(t), G_{jB}(t))), j = 1, 2, \ldots, n\}$

   (3) Symbolic representation for displacement transformation

   $S = S(s_1, s_2, \ldots, \mathrm{BS}(s_1, s_2, \ldots, s_m)_{iu}(t), s_n) \longrightarrow S(s_1, s_2, \ldots, \mathrm{CS}(s_1, s_2, \ldots,$
   $s_m)_{jv}(t), s_n)$

   (4) Representation for the error function of displacement transformation

   $\mathrm{SC}(s_1, s_2, \ldots, \mathrm{BSC}(s_1, s_2, \ldots, s_m)_{iu}(t), s_n) \longrightarrow \mathrm{SC}(s_1, s_2, \ldots, \mathrm{CSC}(s_1, s_2,$
   $\ldots, s_m)_{jv}(t), s_n)$

3. Replacing existing expanding and shrinking structure with series structure

   (1) Graphic representation of displacement transformation (referring to Fig. 5.133)

   (2) Representation for logical proposition of displacement transformation

   $T_z\{A_i((U_i(t), \mathrm{KS}(s_1, s_2, \ldots, s_n)_{iu}(t), \vec{p_{iu}}(t), T_{iu}(t), L_{iu}(t)), x_i(t) = f_i(u_i(t),$
   $G_{iA}(t))), i = 1, 2, \ldots, m\} = \{B_j((V_j(t), \mathrm{CS}(s_1, s_2, \ldots, s_n)_{jv}t, \vec{p_{jv}}(t), T_{jv}(t),$
   $L_{jv}(t)), y_j(t) = g_j(v_j(t), G_{jB}(t))), j = 1, 2, \ldots, n\}$

(3) Symbolic representation for displacement transformation

$S = S(s_1, s_2, \ldots, KS(s_1, s_2, \ldots, s_m)_{iu}(t), s_n) \longrightarrow S(s_1, s_2, \ldots, CS(s_1, s_2, \ldots, s_m)_{jv}(t), s_n)$

(4) Representation for the error function of displacement transformation

$SC(s_1, s_2, \ldots, KSC(s_1, s_2, \ldots, s_m)_{iu}(t), s_n) \longrightarrow SC(s_1, s_2, \ldots, CSC(s_1, s_2, \ldots, s_m)_{jv}(t), s_n)$

4. Replacing existing inclusion structure with series structure

(1) Graphic representation of displacement transformation (referring to Fig. 5.134)

(2) Representation for logical proposition of displacement transformation

$T_z\{A_i((U_i(t), YS(s_1, s_2, \ldots, s_n)_{iu}(t), \overrightarrow{p_{iu}}(t), T_{iu}(t), L_{iu}(t)), x_i(t) = f_i(u_i(t), G_{iA}(t))), i = 1, 2, \ldots, m\} = \{B_j((V_j(t), CS(s_1, s_2, \ldots, s_n)_{jv}t, \overrightarrow{p_{jv}}(t), T_{jv}(t), L_{jv}(t)), y_j(t) = g_j(v_j(t), G_{jB}(t))), j = 1, 2, \ldots, n\}$

(3) Symbolic representation for displacement transformation

$S = S(s_1, s_2, \ldots, YS(s_1, s_2, \ldots, s_m)_{iu}(t), s_n) \longrightarrow S(s_1, s_2, \ldots, CS(s_1, s_2, \ldots, s_m)_{jv}(t), s_n)$

(4) Representation for the error function of displacement transformation

$SC(s_1, s_2, \ldots, YSC(s_1, s_2, \ldots, s_m)_{iu}(t), s_n) \longrightarrow SC(s_1, s_2, \ldots, CSC(s_1, s_2, \ldots, s_m)_{jv}(t), s_n)$

5. Replacing existing feedback structure with series structure

(1) Graphic representation of displacement transformation (referring to Fig. 5.135)

(2) Representation for logical proposition of displacement transformation

$T_z\{A_i((U_i(t), FS(s_1, s_2, \ldots, s_n)_{iu}(t), \overrightarrow{p_{iu}}(t), T_{iu}(t), L_{iu}(t)), x_i(t) = f_i(u_i(t), G_{iA}(t))), i = 1, 2, \ldots, m\} = \{B_j((V_j(t), CS(s_1, s_2, \ldots, s_n)_{jv}t, \overrightarrow{p_{jv}}(t), T_{jv}(t), L_{jv}(t)), y_j(t) = g_j(v_j(t), G_{jB}(t))), j = 1, 2, \ldots, n\}$

(3) Symbolic representation for displacement transformation

$S = S(s_1, s_2, \ldots, FS(s_1, s_2, \ldots, s_m)_{iu}(t), s_n) \longrightarrow S(s_1, s_2, \ldots, CS(s_1, s_2, \ldots, s_m)_{jv}(t), s_n)$

(4) Representation for the error function of displacement transformation

$SC(s_1, s_2, \ldots, FSC(s_1, s_2, \ldots, s_m)_{iu}(t), s_n) \longrightarrow SC(s_1, s_2, \ldots, CSC(s_1, s_2, \ldots, s_m)_{jv}(t), s_n)$

6. Replacing existing sereies structure with parallel structure

(1) Graphic representation of displacement transformation (referring to Fig. 5.136)

(2) Representation for logical proposition of displacement transformation

$T_z\{A_i((U_i(t), CS(s_1, s_2, \ldots, s_n)_{iu}(t), \overrightarrow{p_{iu}}(t), T_{iu}(t), L_{iu}(t)), x_i(t) = f_i(u_i(t), G_{iA}(t))), i = 1, 2, \ldots, m\} = \{B_j((V_j(t), BS(s_1, s_2, \ldots, s_n)_{jv}t, \overrightarrow{p_{jv}}(t), T_{jv}(t), L_{jv}(t)), y_j(t) = g_j(v_j(t), G_{jB}(t))), j = 1, 2, \ldots, n\}$

(3) Symbolic representation for displacement transformation

$S = S(s_1, s_2, \ldots, CS(s_1, s_2, \ldots, s_m)_{iu}(t), s_n) \longrightarrow S(s_1, s_2, \ldots, BS(s_1, s_2, \ldots, s_m)_{jv}(t), s_n)$

(4) Representation for the error function of displacement transformation

$SC(s_1, s_2, \ldots, CS(s_1, s_2, \ldots, s_m)_{iu}(t), s_n) \longrightarrow SC(s_1, s_2, \ldots, BS(s_1, s_2, \ldots, s_m)_{jv}(t), s_n)$

7. Replacing existing parallel structure with new parallel structure

(1) Graphic representation of displacement transformation (referring to Fig. 5.137)

(2) Representation for logical proposition of displacement transformation

$T_z\{A_i((U_i(t), BS(s_1, s_2, \ldots, s_n)_{iu}(t), \vec{p_{iu}}(t), T_{iu}(t), L_{iu}(t)), x_i(t) = f_i(u_i(t), G_{iA}(t))), i = 1, 2, \ldots, m\} = \{B_j((V_j(t), BS(s_1, s_2, \ldots, s_n)_{jv}t, \vec{p_{jv}}(t), T_{jv}(t), L_{jv}(t)), y_j(t) = g_j(v_j(t), G_{jB}(t))), j = 1, 2, \ldots, n\}$

(3) Symbolic representation for displacement transformation

$S = S(s_1, s_2, \ldots, BS(s_1, s_2, \ldots, s_m)_{iu}(t), s_n) \longrightarrow S(s_1, s_2, \ldots, BS(s_1, s_2, \ldots, s_m)_{jv}(t), s_n)$

(4) Representation for the error function of displacement transformation

$SC(s_1, s_2, \ldots, BSC(s_1, s_2, \ldots, s_m)_{iu}(t), s_n) \longrightarrow SC(s_1, s_2, \ldots, BSC(s_1, s_2, \ldots, s_m)_{iv}(t), s_n)$

8. Replacing existing expanding and shrinking structure with parallel structure

(1) Graphic representation of displacement transformation (referring to Fig. 5.138)

(2) Representation for logical proposition of displacement transformation

$T_z\{A_i((U_i(t), KS(s_1, s_2, \ldots, s_n)_{iu}(t), \vec{p_{iu}}(t), T_{iu}(t), L_{iu}(t)), x_i(t) = f_i(u_i(t), G_{iA}(t))), i = 1, 2, \ldots, m\} = \{B_j((V_j(t), BS(s_1, s_2, \ldots, s_n)_{jv}t, \vec{p_{jv}}(t), T_{jv}(t), L_{jv}(t)), y_j(t) = g_j(v_j(t), G_{jB}(t))), j = 1, 2, \ldots, n\}$

(3) Symbolic representation for displacement transformation

$S = S(s_1, s_2, \ldots, KS(s_1, s_2, \ldots, s_m)_{iu}(t), s_n) \longrightarrow S(s_1, s_2, \ldots, BS(s_1, s_2, \ldots, s_m)_{jv}(t), s_n)$

(4) Representation for the error function of displacement transformation

$SC(s_1, s_2, \ldots, KSC(s_1, s_2, \ldots, s_m)_{iu}(t), s_n) \longrightarrow SC(s_1, s_2, \ldots, BSC(s_1, s_2, \ldots, s_m)_{jv}(t), s_n)$

9. Replacing existing inclusion structure with parallel structure

(1) Graphic representation of displacement transformation (referring to Fig. 5.139)

(2) Representation for logical proposition of displacement transformation

$T_z\{A_i((U_i(t), YS(s_1, s_2, \ldots, s_n)_{iu}(t), \vec{p_{iu}}(t), T_{iu}(t), L_{iu}(t)), x_i(t) = f_i(u_i(t), G_{iA}(t))), i = 1, 2, \ldots, m\} = \{B_j((V_j(t), BS(s_1, s_2, \ldots, s_n)_{jv}t, \vec{p_{jv}}(t), T_{jv}(t), L_{jv}(t)), y_j(t) = g_j(v_j(t), G_{jB}(t))), j = 1, 2, \ldots, n\}$

(3) Symbolic representation for displacement transformation
$$S = S(s_1, s_2, \ldots, YS(s_1, s_2, \ldots, s_m)_{iu}(t), s_n) \longrightarrow S(s_1, s_2, \ldots, BS(s_1, s_2, \ldots, s_m)_{jv}(t), s_n)$$

(4) Representation for the error function of displacement transformation
$$SC(s_1, s_2, \ldots, YSC(s_1, s_2, \ldots, s_m)_{iu}(t), s_n) \longrightarrow SC(s_1, s_2, \ldots, BSC(s_1, s_2, \ldots, s_m)_{jv}(t), s_n)$$

10. Replacing existing feedback structure with parallel structure

(1) Graphic representation of displacement transformation (referring to Fig. 5.140)

(2) Representation for logical proposition of displacement transformation
$$T_z\{A_i((U_i(t), FS(s_1, s_2, \ldots, s_n)_{iu}(t), \overrightarrow{p_{iu}}(t), T_{iu}(t), L_{iu}(t)), x_i(t) = f_i(u_i(t), G_{iA}(t))), i = 1, 2, \ldots, m\} = \{B_j((V_j(t), BS(s_1, s_2, \ldots, s_n)_{jv}t, \overrightarrow{p_{jv}}(t), T_{jv}(t), L_{jv}(t)), y_j(t) = g_j(v_j(t), G_{jB}(t))), j = 1, 2, \ldots, n\}$$

(3) Symbolic representation for displacement transformation
$$S = S(s_1, s_2, \ldots, FS(s_1, s_2, \ldots, s_m)_{iu}(t), s_n) \longrightarrow S(s_1, s_2, \ldots, BS(s_1, s_2, \ldots, s_m)_{jv}(t), s_n)$$

(4) Representation for the error function of displacement transformation
$$SC(s_1, s_2, \ldots, FSC(s_1, s_2, \ldots, s_m)_{iu}(t), s_n) \longrightarrow SC(s_1, s_2, \ldots, BSC(s_1, s_2, \ldots, s_m)_{jv}(t), s_n)$$

11. Replacing existing series structure with expanding and shrinking structure

(1) Graphic representation of displacement transformation (referring to Fig. 5.141)

(2) Representation for logical proposition of displacement transformation
$$T_z\{A_i((U_i(t), CS(s_1, s_2, \ldots, s_n)_{iu}(t), \overrightarrow{p_{iu}}(t), T_{iu}(t), L_{iu}(t)), x_i(t) = f_i(u_i(t), G_{iA}(t))), i = 1, 2, \ldots, m\} = \{B_j((V_j(t), KS(s_1, s_2, \ldots, s_n)_{jv}t, \overrightarrow{p_{jv}}(t), T_{jv}(t), L_{jv}(t)), y_j(t) = g_j(v_j(t), G_{jB}(t))), j = 1, 2, \ldots, n\}$$

(3) Symbolic representation for displacement transformation
$$S = S(s_1, s_2, \ldots, CS(s_1, s_2, \ldots, s_m)_{iu}(t), s_n) \longrightarrow S(s_1, s_2, \ldots, KS(s_1, s_2, \ldots, s_m)_{jv}(t), s_n)$$

(4) Representation for the error function of displacement transformation
$$SC(s_1, s_2, \ldots, CSC(s_1, s_2, \ldots, s_m)_{iu}(t), s_n) \longrightarrow SC(s_1, s_2, \ldots, KSC(s_1, s_2, \ldots, s_m)_{jv}(t), s_n)$$

12. Replacing existing parallel structure with expanding and shrinking structure

(1) Graphic representation of displacement transformation (referring to Fig. 5.142)

(2) Representation for logical proposition of displacement transformation
$$T_z\{A_i((U_i(t), BS(s_1, s_2, \ldots, s_n)_{iu}(t), \overrightarrow{p_{iu}}(t), T_{iu}(t), L_{iu}(t)), x_i(t) = f_i(u_i(t), G_{iA}(t))), i = 1, 2, \ldots, m\} = \{B_j((V_j(t), KS(s_1, s_2, \ldots, s_n)_{jv}t, \overrightarrow{p_{jv}}(t), T_{jv}(t), L_{jv}(t)), y_j(t) = g_j(v_j(t), G_{jB}(t))), j = 1, 2, \ldots, n\}$$

(3) Symbolic representation for displacement transformation

$S = S(s_1, s_2, \ldots, BS(s_1, s_2, \ldots, s_m)_{iu}(t), s_n) \longrightarrow S(s_1, s_2, \ldots, KS(s_1, s_2, \ldots, s_m)_{jv}(t), s_n)$

(4) Representation for the error function of displacement transformation

$SC(s_1, s_2, \ldots, BSC(s_1, s_2, \ldots, s_m)_{iu}(t), s_n) \longrightarrow SC(s_1, s_2, \ldots, KSC(s_1, s_2, \ldots, s_m)_{jv}(t), s_n)$

13. Replacing existing expanding and shrinking structure with new expanding and shrinking structure

(1) Graphic representation of displacement transformation (referring to Fig. 5.143)

(2) Representation for logical proposition of displacement transformation

$T_z\{A_i((U_i(t), KS(s_1, s_2, \ldots, s_n)_{iu}(t), \vec{p}_{iu}(t), T_{iu}(t), L_{iu}(t)), x_i(t) = f_i(u_i(t), G_{iA}(t))), i = 1, 2, \ldots, m\} = \{B_j((V_j(t), KS(s_1, s_2, \ldots, s_n)_{jv}t, \vec{p}_{jv}(t), T_{jv}(t), L_{jv}(t)), y_j(t) = g_j(v_j(t), G_{jB}(t))), j = 1, 2, \ldots, n\}$

(3) Symbolic representation for displacement transformation

$S = S(s_1, s_2, \ldots, KS(s_1, s_2, \ldots, s_m)_{iu}(t), s_n) \longrightarrow S(s_1, s_2, \ldots, KS(s_1, s_2, \ldots, s_m)_{jv}(t), s_n)$

(4) Representation for the error function of displacement transformation

$SC(s_1, s_2, \ldots, KSC(s_1, s_2, \ldots, s_m)_{iu}(t), s_n) \longrightarrow SC(s_1, s_2, \ldots, KSC(s_1, s_2, \ldots, s_m)_{jv}(t), s_n)$

14. Replacing existing inclusion structure with expanding and shrinking structure

(1) Graphic representation of displacement transformation (referring to Fig. 5.144)

(2) Representation for logical proposition of displacement transformation

$T_z\{A_i((U_i(t), YS(s_1, s_2, \ldots, s_n)_{iu}(t), \vec{p}_{iu}(t), T_{iu}(t), L_{iu}(t)), x_i(t) = f_i(u_i(t), G_{iA}(t))), i = 1, 2, \ldots, m\} = \{B_j((V_j(t), KS(s_1, s_2, \ldots, s_n)_{jv}t, \vec{p}_{jv}(t), T_{jv}(t), L_{jv}(t)), y_j(t) = g_j(v_j(t), G_{jB}(t))), j = 1, 2, \ldots, n\}$

(3) Symbolic representation for displacement transformation

$S = S(s_1, s_2, \ldots, YS(s_1, s_2, \ldots, s_m)_{iu}(t), s_n) \longrightarrow S(s_1, s_2, \ldots, KS(s_1, s_2, \ldots, s_m)_{jv}(t), s_n)$

(4) Representation for the error function of displacement transformation

$SC(s_1, s_2, \ldots, YSC(s_1, s_2, \ldots, s_m)_{iu}(t), s_n) \longrightarrow SC(s_1, s_2, \ldots, KSC(s_1, s_2, \ldots, s_m)_{jv}(t), s_n)$

15. Replacing existing feedback structure with expanding and shrinking structure

(1) Graphic representation of displacement transformation (referring to Fig. 5.145)

(2) Representation for logical proposition of displacement transformation

$T_z\{A_i((U_i(t), FS(s_1, s_2, \ldots, s_n)_{iu}(t), \vec{p}_{iu}(t), T_{iu}(t), L_{iu}(t)), x_i(t) = f_i(u_i(t), G_{iA}(t))), i = 1, 2, \ldots, m\} = \{B_j((V_j(t), KS(s_1, s_2, \ldots, s_n)_{jv}t, \vec{p}_{jv}(t), T_{jv}(t), L_{jv}(t)), y_j(t) = g_j(v_j(t), G_{jB}(t))), j = 1, 2, \ldots, n\}$

(3) Symbolic representation for displacement transformation

$S = S(s_1, s_2, \ldots, FS(s_1, s_2, \ldots, s_m)_{iu}(t), s_n) \longrightarrow S(s_1, s_2, \ldots, KS(s_1, s_2, \ldots, s_m)_{jv}(t), s_n)$

(4) Representation for the error function of displacement transformation

$SC(s_1, s_2, \ldots, FSC(s_1, s_2, \ldots, s_m)_{iu}(t), s_n) \longrightarrow SC(s_1, s_2, \ldots, KSC(s_1, s_2, \ldots, s_m)_{jv}(t), s_n)$

16. Replacing existing series structure with inclusion structure

(1) Graphic representation of displacement transformation (referring to Fig. 5.146)

(2) Representation for logical proposition of displacement transformation

$T_z\{A_i((U_i(t), CS(s_1, s_2, \ldots, s_n)_{iu}(t), \vec{p_{iu}}(t), T_{iu}(t), L_{iu}(t)), x_i(t) = f_i(u_i(t), G_{iA}(t))), i = 1, 2, \ldots, m\} = \{B_j((V_j(t), YS(s_1, s_2, \ldots, s_n)_{jv}t, \vec{p_{jv}}(t), T_{jv}(t), L_{jv}(t)), y_j(t) = g_j(v_j(t), G_{jB}(t))), j = 1, 2, \ldots, n\}$

(3) Symbolic representation for displacement transformation

$S = S(s_1, s_2, \ldots, CS(s_1, s_2, \ldots, s_m)_{iu}(t), s_n) \longrightarrow S(s_1, s_2, \ldots, YS(s_1, s_2, \ldots, s_m)_{jv}(t), s_n)$

(4) Representation for the error function of displacement transformation

$SC(s_1, s_2, \ldots, CSC(s_1, s_2, \ldots, s_m)_{iu}(t), s_n) \longrightarrow SC(s_1, s_2, \ldots, YSC(s_1, s_2, \ldots, s_m)_{jv}(t), s_n)$

17. Replacing existing parallel structure with inclusion structure

(1) Graphic representation of displacement transformation (referring to Fig. 5.147)

(2) Representation for logical proposition of displacement transformation

$T_z\{A_i((U_i(t), BS(s_1, s_2, \ldots, s_n)_{iu}(t), \vec{p_{iu}}(t), T_{iu}(t), L_{iu}(t)), x_i(t) = f_i(u_i(t), G_{iA}(t))), i = 1, 2, \ldots, m\} = \{B_j((V_j(t), YS(s_1, s_2, \ldots, s_n)_{jv}t, \vec{p_{jv}}(t), T_{jv}(t), L_{jv}(t)), y_j(t) = g_j(v_j(t), G_{jB}(t))), j = 1, 2, \ldots, n\}$

(3) Symbolic representation for displacement transformation

$S = S(s_1, s_2, \ldots, BS(s_1, s_2, \ldots, s_m)_{iu}(t), s_n) \longrightarrow S(s_1, s_2, \ldots, YS(s_1, s_2, \ldots, s_m)_{jv}(t), s_n)$

(4) Representation for the error function of displacement transformation

$SC(s_1, s_2, \ldots, BSC(s_1, s_2, \ldots, s_m)_{iu}(t), s_n) \longrightarrow SC(s_1, s_2, \ldots, YSC(s_1, s_2, \ldots, s_m)_{jv}(t), s_n)$

18. Replacing existing expanding and shrinking structure with inclusion structure

(1) Graphic representation of displacement transformation (referring to Fig. 5.148)

(2) Representation for logical proposition of displacement transformation

$T_z\{A_i((U_i(t), KS(s_1, s_2, \ldots, s_n)_{iu}(t), \vec{p_{iu}}(t), T_{iu}(t), L_{iu}(t)), x_i(t) = f_i(u_i(t), G_{iA}(t))), i = 1, 2, \ldots, m\} = \{B_j((V_j(t), YS(s_1, s_2, \ldots, s_n)_{jv}t, \vec{p_{jv}}(t), T_{jv}(t), L_{jv}(t)), y_j(t) = g_j(v_j(t), G_{jB}(t))), j = 1, 2, \ldots, n\}$

(3) Symbolic representation for displacement transformation

$S = S(s_1, s_2, \ldots, KS(s_1, s_2, \ldots, s_m)_{iu}(t), s_n) \longrightarrow S(s_1, s_2, \ldots, YS(s_1, s_2, \ldots, s_m)_{jv}(t), s_n)$

(4) Representation for the error function of displacement transformation

$SC(s_1, s_2, \ldots, KSC(s_1, s_2, \ldots, s_m)_{iu}(t), s_n) \longrightarrow SC(s_1, s_2, \ldots, YSC(s_1, s_2, \ldots, s_m)_{jv}(t), s_n)$

19. Replacing existing inclusion structure with new inclusion structure

(1) Graphic representation of displacement transformation (referring to Fig. 5.149)

(2) Representation for logical proposition of displacement transformation

$T_z\{A_i((U_i(t), YS(s_1, s_2, \ldots, s_n)_{iu}(t), \vec{p_{iu}}(t), T_{iu}(t), L_{iu}(t)), x_i(t) = f_i(u_i(t),$
$G_{iA}(t))), i = 1, 2, \ldots, m\} = \{B_j((V_j(t), YS(s_1, s_2, \ldots, s_n)_{jv}t, \vec{p_{jv}}(t), T_{jv}(t),$
$L_{jv}(t)), y_j(t) = g_j(v_j(t), G_{jB}(t))), j = 1, 2, \ldots, n\}$

(3) Symbolic representation for displacement transformation

$S = S(s_1, s_2, \ldots, YS(s_1, s_2, \ldots, s_m)_{iu}(t), s_n) \longrightarrow S(s_1, s_2, \ldots, YS(s_1, s_2, \ldots, s_m)_{jv}(t), s_n)$

(4) Representation for the error function of displacement transformation

$SC(s_1, s_2, \ldots, YSC(s_1, s_2, \ldots, s_m)_{iu}(t), s_n) \longrightarrow SC(s_1, s_2, \ldots, YSC(s_1, s_2, \ldots, s_m)_{jv}(t), s_n)$

20. Replacing existing feedback structure with inclusion structure

(1) Graphic representation of displacement transformation (referring to Fig. 5.150)

(2) Representation for logical proposition of displacement transformation

$T_z\{A_i((U_i(t), FS(s_1, s_2, \ldots, s_n)_{iu}(t), \vec{p_{iu}}(t), T_{iu}(t), L_{iu}(t)), x_i(t) = f_i(u_i(t),$
$G_{iA}(t))), i = 1, 2, \ldots, m\} = \{B_j((V_j(t), YS(s_1, s_2, \ldots, s_n)_{jv}t, \vec{p_{jv}}(t), T_{jv}(t),$
$L_{jv}(t)), y_j(t) = g_j(v_j(t), G_{jB}(t))), j = 1, 2, \ldots, n\}$

(3) Symbolic representation for displacement transformation

$S = S(s_1, s_2, \ldots, FS(s_1, s_2, \ldots, s_m)_{iu}(t), s_n) \longrightarrow S(s_1, s_2, \ldots, YS(s_1, s_2, \ldots, s_m)_{jv}(t), s_n)$

(4) Representation for the error function of displacement transformation

$SC(s_1, s_2, \ldots, FSC(s_1, s_2, \ldots, s_m)_{iu}(t), s_n) \longrightarrow SC(s_1, s_2, \ldots, YSC(s_1, s_2, \ldots, s_m)_{jv}(t), s_n)$

21. Replacing existing series structure with feedback structure

(1) Graphic representation of displacement transformation (referring to Fig. 5.151)

(2) Representation for logical proposition of displacement transformation

$T_z\{A_i((U_i(t), CS(s_1, s_2, \ldots, s_n)_{iu}(t), \vec{p_{iu}}(t), T_{iu}(t), L_{iu}(t)), x_i(t) = f_i(u_i(t),$
$G_{iA}(t))), i = 1, 2, \ldots, m\} = \{B_j((V_j(t), FS(s_1, s_2, \ldots, s_n)_{jv}t, \vec{p_{jv}}(t), T_{jv}(t),$
$L_{jv}(t)), y_j(t) = g_j(v_j(t), G_{jB}(t))), j = 1, 2, \ldots, n\}$

(3) Symbolic representation for displacement transformation
$$S = S(s_1, s_2, \ldots, CS(s_1, s_2, \ldots, s_m)_{iu}(t), s_n) \longrightarrow S(s_1, s_2, \ldots, FS(s_1, s_2, \ldots, s_m)_{jv}(t), s_n)$$

(4) Representation for the error function of displacement transformation
$$SC(s_1, s_2, \ldots, CSC(s_1, s_2, \ldots, s_m)_{iu}(t), s_n) \longrightarrow SC(s_1, s_2, \ldots, FSC(s_1, s_2, \ldots, s_m)_{jv}(t), s_n)$$

22. Replacing existing parallel structure with feedback structure

(1) Graphic representation of displacement transformation (referring to Fig. 5.152)

(2) Representation for logical proposition of displacement transformation
$$T_z\{A_i((U_i(t), BS(s_1, s_2, \ldots, s_n)_{iu}(t), \overrightarrow{p_{iu}}(t), T_{iu}(t), L_{iu}(t)), x_i(t) = f_i(u_i(t),$$
$$G_{iA}(t))), i = 1, 2, \ldots, m\} = \{B_j((V_j(t), FS(s_1, s_2, \ldots, s_n)_{jv}t, \overrightarrow{p_{jv}}(t), T_{jv}(t),$$
$$L_{jv}(t)), y_j(t) = g_j(v_j(t), G_{jB}(t))), j = 1, 2, \ldots, n\}$$

(3) Symbolic representation for displacement transformation
$$S = S(s_1, s_2, \ldots, BS(s_1, s_2, \ldots, s_m)_{iu}(t), s_n) \longrightarrow S(s_1, s_2, \ldots, FS(s_1, s_2, \ldots, s_m)_{jv}(t), s_n)$$

(4) Representation for the error function of displacement transformation
$$SC(s_1, s_2, \ldots, BSC(s_1, s_2, \ldots, s_m)_{iu}(t), s_n) \longrightarrow SC(s_1, s_2, \ldots, FSC(s_1, s_2, \ldots, s_m)_{jv}(t), s_n)$$

23. Replacing existing expanding and shrinking structure with feedback structure

(1) Graphic representation of displacement transformation (referring to Fig. 5.153)

(2) Representation for logical proposition of displacement transformation
$$T_z\{A_i((U_i(t), KS(s_1, s_2, \ldots, s_n)_{iu}(t), \overrightarrow{p_{iu}}(t), T_{iu}(t), L_{iu}(t)), x_i(t) = f_i(u_i(t),$$
$$G_{iA}(t))), i = 1, 2, \ldots, m\} = \{B_j((V_j(t), FS(s_1, s_2, \ldots, s_n)_{jv}t, \overrightarrow{p_{jv}}(t), T_{jv}(t),$$
$$L_{jv}(t)), y_j(t) = g_j(v_j(t), G_{jB}(t))), j = 1, 2, \ldots, n\}$$

(3) Symbolic representation for displacement transformation
$$S = S(s_1, s_2, \ldots, KS(s_1, s_2, \ldots, s_m)_{iu}(t), s_n) \longrightarrow S(s_1, s_2, \ldots, FS(s_1, s_2, \ldots, s_m)_{jv}(t), s_n)$$

(4) Representation for the error function of displacement transformation
$$SC(s_1, s_2, \ldots, KSC(s_1, s_2, \ldots, s_m)_{iu}(t), s_n) \longrightarrow SC(s_1, s_2, \ldots, FSC(s_1, s_2, \ldots, s_m)_{jv}(t), s_n)$$

24. Replacing existing inclusion structure with feedback structure

(1) Graphic representation of displacement transformation (referring to Fig. 5.154)

(2) Representation for logical proposition of displacement transformation

$T_z\{A_i((U_i(t), \text{YS}(s_1, s_2, \ldots, s_n)_{iu}(t), \overrightarrow{p_{iu}}(t), T_{iu}(t), L_{iu}(t)), x_i(t) = f_i(u_i(t),$
$G_{iA}(t))), i = 1, 2, \ldots, m\} = \{B_j((V_j(t), \text{FS}(s_1, s_2, \ldots, s_n)_{jv}t, \overrightarrow{p_{jv}}(t), T_{jv}(t),$
$L_{jv}(t)), y_j(t) = g_j(v_j(t), G_{jB}(t))), j = 1, 2, \ldots, n\}$

(3) Symbolic representation for displacement transformation

$S = S(s_1, s_2, \ldots, \text{YS}(s_1, s_2, \ldots, s_m)_{iu}(t), s_n) \longrightarrow S(s_1, s_2, \ldots, \text{FS}(s_1, s_2, \ldots, s_m)_{jv}(t), s_n)$

(4) Representation for the error function of displacement transformation

$\text{SC}(s_1, s_2, \ldots, \text{YSC}(s_1, s_2, \ldots, s_m)_{iu}(t), s_n) \longrightarrow \text{SC}(s_1, s_2, \ldots, \text{FSC}(s_1, s_2, \ldots, s_m)_{jv}(t), s_n)$

25. Replacing existing feedback structure with new feedback structure

(1) Graphic representation of displacement transformation (referring to Fig. 5.155)

(2) Representation for logical proposition of displacement transformation

$T_z\{A_i((U_i(t), \text{FS}(s_1, s_2, \ldots, s_n)_{iu}(t), \overrightarrow{p_{iu}}(t), T_{iu}(t), L_{iu}(t)), x_i(t) = f_i(u_i(t),$
$G_{iA}(t))), i = 1, 2, \ldots, m\} = \{B_j((V_j(t), \text{FS}(s_1, s_2, \ldots, s_n)_{jv}t, \overrightarrow{p_{jv}}(t), T_{jv}(t),$
$L_{jv}(t)), y_j(t) = g_j(v_j(t), G_{jB}(t))), j = 1, 2, \ldots, n\}$

(3) Symbolic representation for displacement transformation

$S = S(s_1, s_2, \ldots, \text{FS}(s_1, s_2, \ldots, s_m)_{iu}(t), s_n) \longrightarrow S(s_1, s_2, \ldots, \text{FS}(s_1, s_2, \ldots, s_m)_{jv}(t), s_n)$

(4) Representation for the error function of displacement transformation

$\text{SC}(s_1, s_2, \ldots, \text{FSC}(s_1, s_2, \ldots, s_m)_{iu}(t), s_n) \longrightarrow \text{SC}(s_1, s_2, \ldots, \text{FSC}(s_1, s_2, \ldots, s_m)_{jv}(t), s_n)$

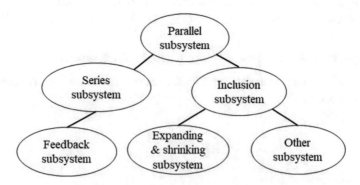

Fig. 5.130 Structural illustration of system S

Fig. 5.131 Replacing existing series structure with new series structure

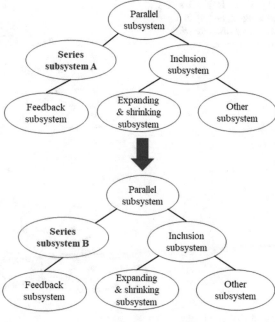

Fig. 5.132 Replacing existing parallel structure with series structure

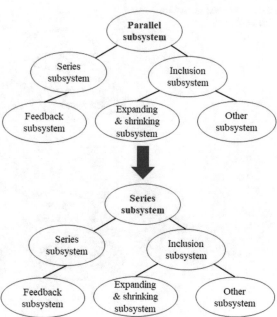

Fig. 5.133 Replacing
existing expanding and
shrinking structure with
series structure

Fig. 5.134 Replacing
existing inclusion structure
with series structure

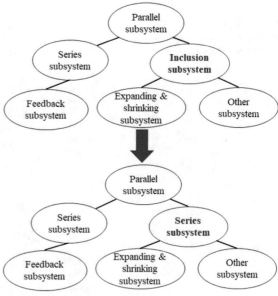

Fig. 5.135 Replacing existing feedback structure with series structure

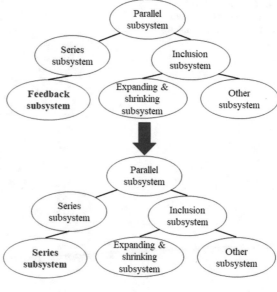

Fig. 5.136 Replacing existing sereies structure with parallel structure

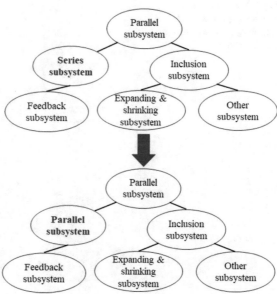

Fig. 5.137 Replacing
existing parallel structure
with new parallel structure

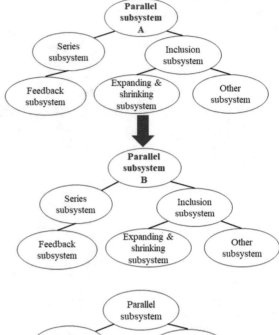

Fig. 5.138 Replacing
existing expanding and
shrinking structure with
parallel structure

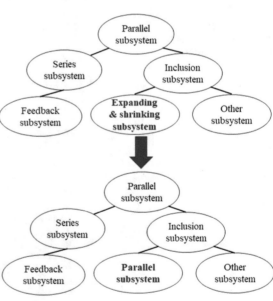

Fig. 5.139 Replacing existing inclusion structure with parallel structure

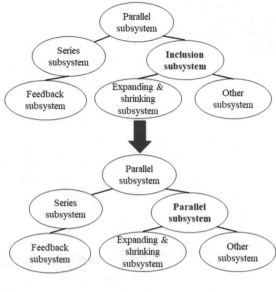

Fig. 5.140 Replacing existing feedback structure with parallel structure

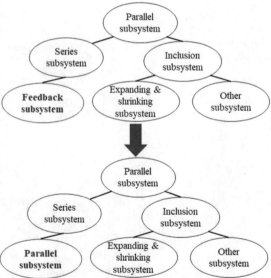

Fig. 5.141 Replacing
existing series structure with
expanding and shrinking
structure

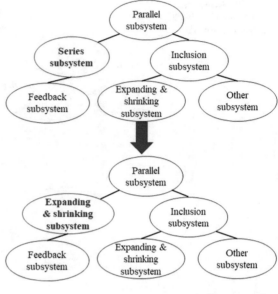

Fig. 5.142 Replacing
existing parallel structure
with expanding and
shrinking structure

Fig. 5.143 Replacing existing expanding and shrinking structure with new expanding and shrinking structure

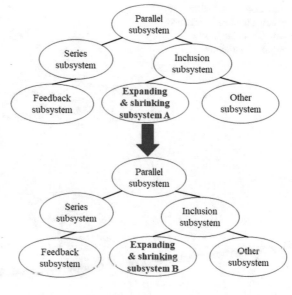

Fig. 5.144 Replacing existing inclusion structure with expanding and shrinking structure

Fig. 5.145 Replacing
existing feedback structure
with expanding and
shrinking structure

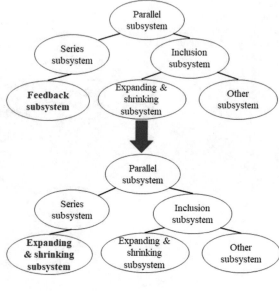

Fig. 5.146 Replacing
existing series structure with
inclusion structure

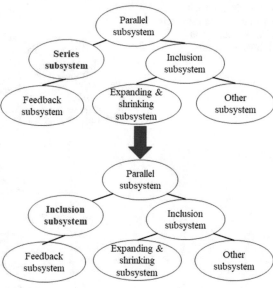

Fig. 5.147 Replacing existing parallel structure with inclusion structure

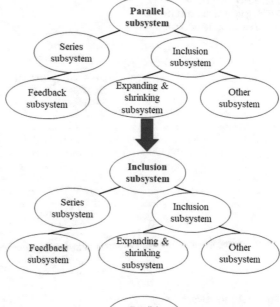

Fig. 5.148 Replacing existing expanding and shrinking structure with inclusion structure

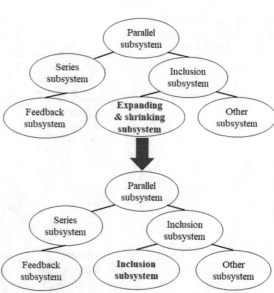

Fig. 5.149 Replacing
existing inclusion structure
with new inclusion structure

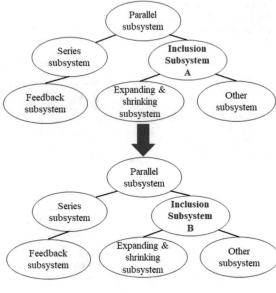

Fig. 5.150 Replacing
existing feedback structure
with inclusion structure

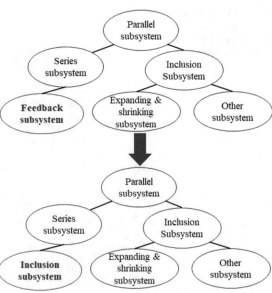

Fig. 5.151 Replacing
existing series structure with
feedback structure

Fig. 5.152 Replacing
existing parallel structure
with feedback structure

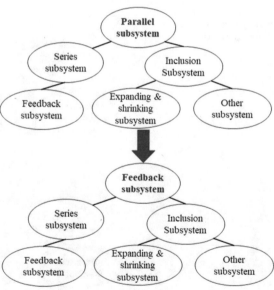

Fig. 5.153 Replacing
existing expanding and
shrinking structure with
feedback structure

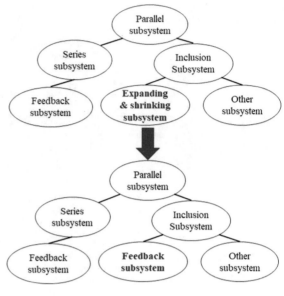

Fig. 5.154 Replacing
existing inclusion structure
with feedback structure

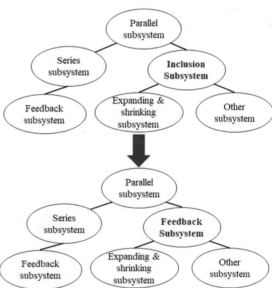

Fig. 5.155 Replacing existing feedback structure with new feedback structure

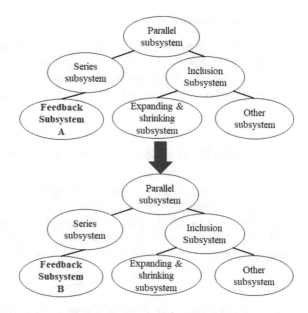

5.2.3 Inverse Displacement Transformation on Error System Structures

The expression for error function of inverse displacement transformation is:
$SC(s_1, s_2, \ldots, FSC(s_1, s_2, \ldots, s_m)_{iv}(t), s_n) \longrightarrow SC(s_1, s_2, \ldots, FSC(s_1, s_2, \ldots, s_m)_{iu}(t), s_n)$.

Theorem 5.1 *Suppose that:(1) the features GY_j of system S have additivity with respect to the features GY_{ji} ($i=1, 2, \ldots, n$) of its subsystem S_i, and the features GY_{ji} ($i=1, 2, \ldots, n$) of subsystem S_i are independent, (2) subsystems other than subsystem S_i are non-erroneous; errors in system are eliminated if a non-erroneous subsystem S_i' with equivalent features GY_{ji} ($i=1, 2, \ldots, n$) is used to replace the erroneous subsystem S_i.*

Proof Based on the hypotheses (1) and (2), system S becomes non-erroneous once a non-erroneous subsystem S_i' with equivalent features GY_{ji} ($i=1, 2, \ldots, n$) is used to replace the erroneous subsystem S_i. Proof is finished.

Theorem 5.2 *Suppose that the features GY_j of system S have no additivity with respect to the features GY_{ji} ($i=1, 2, \ldots, n$) of its subsystem S_i-i.e., there exists the following relationship:$S(GY_j)=S(S_1(GY_{j1} (a_1, b_1)), S_2(GY_{j2} (a_2, b_2)), \ldots, S_i(GY_{ji} (a_i, b_i)), \ldots, S_n(GY_{jn} (a_n, b_n)))$, where (a_i, b_i) stands for the value range of the features of GY_{ji} ($i=1, 2, \ldots, n$) of its ith subsystem S_i and there is no error in other subsystems of system S but subsystem S_i, then errors in system S are eliminated if a non-erroneous subsystem S_i' with equivalent features GY_{ji} ($i=1, 2, \ldots, n$) is used to replace the erroneous subsystem S_i.*

Proof Proof is omitted here.

Based on Theorems 5.1 and 5.2, errors in system are eliminated if a non-erroneous subsystem S_i' with equivalent features GY_{ji} ($i=1, 2, \ldots$, n) is used to replace the erroneous subsystem S_i whether the features GY_j of system **S** have additivity with respect to the features GY_{ji} ($i=1, 2, \ldots$, n) of its subsystem S_i or not.

5.3   Similarity Transformation on Error System Structures

In enterprise management, transnational companies, by employing the concept of similarity transformation, built new plants in differnt countries and regions by refer-ring to the manuals, operation procedures, equipment, process design, product design, and facility design ever used in other operating plants. Design engineers devise a new product by using similar modules in another products.

5.3.1   *Expression for Logical Proposition of Similarity Transformation on Error System Structure*

Definition 5.1   Suppose that $A(u(t), x(t)) = A((U(t), S(t), \vec{p}(t), T(t), L(t)), x(t) = f(u(t), G(t)))$ is an error logical variable defined under judging rule $G(t)$ on universe of discourse $U(t)$, if $T(A((U(t), S(t), \vec{p}(t), T(t), L(t)), x(t) = f(u(t), G(t)))) = A((U(t), S(t), \vec{p}(t), T(t), L(t))', x'(t) = f(u'(t), G'(t)))$, then T has conducted similarity transformation on $A((U(t), S(t), \vec{p}(t), T(t), L(t)), x(t) = f(u(t), G(t)))$ within judging rule $G(t)$ defined on domain $U(t)$, it is noted by T_x.

Here, we provide the definition of similarity transformations on universe of discourse, subject, property, quantifier, error function, error value, judging rules, and their certain combination thereof.

1. Similarity transformation on universe of discourse
   In $A((U(t), S(t), \vec{p}(t), T(t), L(t))', x'(t) = f(u'(t), G'(t)))$, if $T_x(A((U(t), S(t), \vec{p}(t), T(t), L(t))', x'(t) = f(u'(t), G'(t)))) = A((U'(t), S(t), \vec{p}(t), T(t), L(t)), x'(t) = f(u'(t), G'(t)))$, then T_x has done similarity transformation on universe of discourse noted by T_{xly}. In this case, if $U_1(t) = kU_2(t), k > 0, U_2(t)$ is used to replace $U_1(t)$, or $U_1(t)$ is used to replace $U_2(t)$. For example, when talking about labor resources in China, the domain of Guagndong province is $U_2(t)$ and the domain of whole China is $U_1(t)$, there exists a factor k between $U_1(t)$ and $U_2(t)$ that makes $U_1(t) = kU_2(t)$ hold.

2. Similarity transformation on subject

In $A((U(t), S(t), \vec{p}(t), T(t), L(t))', x'(t) = f\ (u'(t), G'(t)))$, if $T_x(A((U(t), S(t), \vec{p}(t), T(t), L(t))', x'(t) = f\ (u'(t), G'(t)))) = A((U(t), S'(t), \vec{p}(t), T(t), L(t)), x'(t) = f\ (u'(t), G'(t)))$, then T_x has conducted similarity transformation on subject noted by T_{xsw}. In most cases, T_{xsw} generally makes geometric similarity transformation.

3. Spatial similarity transformation

In $A((U(t), S(t), \vec{p}(t), T(t), L(t))', x'(t) = f\ (u'(t), G'(t)))$, if $T_x(A((U(t), S(t), \vec{p}(t), T(t), L(t))', x'(t) = f\ (u'(t), G'(t)))) = A((U(t), S(t), \overrightarrow{p'(t)}, T(t), L(t)), x'(t) = f\ (u'(t), G'(t)))$, then T_x has done spatial similarity transformation noted by T_{xkj}.

4. Property similarity transformation

In $A((U(t), S(t), \vec{p}(t), T(t), L(t))', x'(t) = f\ (u'(t), G'(t)))$, if $T_x(A((U(t), S(t), \vec{p}(t), T(t), L(t))', x'(t) = f\ (u'(t), G'(t)))) = A((U(t), S(t), \vec{p}(t), T'(t), L(t)), x'(t) = f\ (u'(t), G'(t)))$, then T_x has done similarity transformation on property noted by T_{xtx}. Considering advantages of aerodynamic design of rocket and airplane, the design can provide high-speed train with more advanced features, which is an example of property similarity transformation.

5. Quantifier similarity transformation

In $A((U(t), S(t), \vec{p}(t), T(t), L(t))', x'(t) = f\ (u'(t), G'(t)))$, if $T_x(A((U(t), S(t), \vec{p}(t), T(t), L(t))', x'(t) = f\ (u'(t), G'(t)))) = A((U(t), S(t), \vec{p}(t), T(t), L'(t)), x'(t) = f\ (u'(t), G'(t)))$, then T_x has conducted quantifier similarity transformation noted by T_{xlz}. For example, product A has silver color and similarity transformation can be made to have more variations, e.g., Silver Chalice, Silver Sand, and Silver Tree, etc.

6. Similarity transformation on error value

In $A((U(t), S(t), \vec{p}(t), T(t), L(t))', x'(t) = f\ (u'(t), G'(t)))$, if $T_x(A((U(t), S(t), \vec{p}(t), T(t), L(t))', x'(t) = f\ (u'(t), G'(t)))) = A((U(t), S(t), \vec{p}(t), T(t), L(t))', x'(t) = f\ (u(t), G(t)))$, where $x'(t) \in [x - \varepsilon, x + \varepsilon]$, then T_x has conducted similarity transformation on error value noted by T_{xcz}.

7. Similarity transformation on judging rule

In $A((U(t), S(t), \vec{p}(t), T(t), L(t))', x'(t) = f\ (u'(t), G'(t)))$, if $T_x(A((U(t), S(t), \vec{p}(t), T(t), L(t))', x(t) = f\ (u(t), G'(t)))) = A((U(t), S(t), \vec{p}(t), T(t), L(t))', x'(t) = f\ (u(t), G(t)))$, then T_x has done similarity transformation on judging rule noted by T_{xgz}. For instance, $G_1(t)$ represents the regulations in university AAAA in 2003 and $G_2(t)$ stands for regulations in the same university in 2010 where $G_1(t)$ was replaced with $G_2(t)$.

8. Similarity transformation on error function f

In $A((U(t), S(t), \vec{p}(t), T(t), L(t))', x'(t) = f\ (u'(t), G'(t)))$, if $T_x(A((U(t), S(t), \vec{p}(t), T(t), L(t))', x(t) = f\ (u(t), G'(t)))) = A((U(t), S(t), \vec{p}(t), T(t), L(t))', x(t) = f'\ (u(t), G(t)))$, then T_x has conducted similarity transformation on on error function f noted by T_{xhs}.

9. Temporal similarity transformation

   In $A((U(t), S(t), \vec{p}(t), T(t), L(t))', x'(t) = f(u'(t), G'(t)))$, if $T_x(A((U(t), S(t), \vec{p}(t), T(t), L(t))', x(t) = f(u(t), G'(t)))) = A((U(t'), S(t'), \vec{p}(t'), T(t'), L(t')), x(t') = f(u(t'), G(t')))$, then T_x has done temporal similarity transformation on object of interests noted by T_{xsj}.

10. Comprehensive similarity transformation

   In $A((U(t), S(t), \vec{p}(t), T(t), L(t))', x'(t) = f(u'(t), G'(t)))$, if $T_x(A((U(t), S(t), \vec{p}(t), T(t), L(t))', x(t) = f(u(t), G'(t)))) = A((U'(t'), S'(t'), \vec{p}'(t'), T'(t'), L'(t')), x'(t') = f'(u'(t'), G'(t')))$, then T_x has conducted comprehensive similarity transformation on object of interests noted by T_{xq}. This transformation T_{xq} has simultaneously made similarity transformation on universe of discourse, subject, property, quantifier, error function, error value, judging rules, and their certain combination thereof.

   Therefore, connectives of similarity transformations include $T_x \subseteq \{T_{xly}, T_{xsw}, T_{xkj}, T_{xtx}, T_{xlz}, T_{xgz}, T_{xhs}, T_{xq}\}$. T_x^{-1} represents corresponding inverse similarity transformation.

5.3.2 Types of Similarity Transformation on Error System Structures

1. Similarity transformation on series structure

   (1) Graphic representation of similarity transformation (referring to Figs. 5.156 and 5.157)
   (2) Representation for logical proposition of similarity transformation
   $$T_z(A((U(t), CS(s_1, s_2, \ldots, s_n), \vec{p}(t), T(t), L(t)), x(t) = f(u(t), G(t)))) =$$
   $$A((U'(t), CS'(s_1, s_2, \ldots, s_n), \vec{p}'(t), T'(t), L'(t)), x'(t) = f(u'(t), G'(t))).$$
   (3) Symbolic representation for similarity transformation
   $$S = CS(s_1, s_2, \ldots, s_i) \longrightarrow CS(s_1, s_2, \ldots, s_n)$$
   $$S = CS(s_1, s_2, \ldots, s_n) \longrightarrow CS(s_1, s_2, \ldots, s_i), \text{ where } n > i.$$
   (4) Representation for the error function of similarity transformation
   $$CSC(s_1, s_2, \ldots, s_i) \longrightarrow CSC(s_1, s_2, \ldots, s_n)$$
   $$CSC(s_1, s_2, \ldots, s_n) \longrightarrow CSC(s_1, s_2, \ldots, s_i), \text{ where } n > i.$$

2. Similarity transformation on parallel structure: Type 1

   (1) Graphic representation of similarity transformation (referring to Figs. 5.158 and 5.159)

(2) Representation for logical proposition of similarity transformation

$T_z(A((U(t), BS(s_1, s_2, \ldots, s_n), \overrightarrow{p}(t), T(t), L(t)), x(t) = f(u(t), G(t)))) =$
$A((U'(t), BS'(s_1, s_2, \ldots, s_n), \overrightarrow{p'}(t), T'(t), L'(t)), x'(t) = f(u'(t), G'(t))).$

(3) Symbolic representation for similarity transformation

$S = BS(CS_1(s_1, s_2, \ldots, s_{i1}), CS_2(s_1, s_2, \ldots, s_{i2})) \longrightarrow BS(CS_1(s_1, s_2, \ldots, s_{n1}), CS_2(s_1, s_2, \ldots, s_{n2}))$
$S = BS(CS_1(s_1, s_2, \ldots, s_{n1}), CS_2(s_1, s_2, \ldots, s_{n2})) \longrightarrow BS(CS_1(s_1, s_2, \ldots, s_{i1}), CS_2(s_1, s_2, \ldots, s_{i2})),$ where $n > i$.

(4) Representation for the error function of similarity transformation

$BSC(CSC_1(s_1, s_2, \ldots, s_{i1}), CSC_2(s_1, s_2, \ldots, s_{i2})) \longrightarrow BSC(CSC_1(s_1, s_2, \ldots, s_{n1}), CSC_2(s_1, s_2, \ldots, s_{n2}))$
$BSC(CSC_1(s_1, s_2, \ldots, s_{n1}), CSC_2(s_1, s_2, \ldots, s_{n2})) \longrightarrow BSC(CSC_1(s_1, s_2, \ldots, s_{i1}), CSC_2(s_1, s_2, \ldots, s_{i2})),$ where $n > i$.

3. Similarity transformation on parallel structure: Type 2

(1) Graphic representation of similarity transformation (referring to Figs. 5.160 and 5.161)

(2) Representation for logical proposition of similarity transformation

$T_z(A((U(t), BS(s_1, s_2, \ldots, s_n), \overrightarrow{p}(t), T(t), L(t)), x(t) = f(u(t), G(t)))) =$
$A((U'(t), BS'(s_1, s_2, \ldots, s_n), \overrightarrow{p'}(t), T'(t), L'(t)), x'(t) = f(u'(t), G'(t))).$

(3) Symbolic representation for similarity transformation

$S = BS(CS_j(s_1, s_2, \ldots, s_{ij}), CS_k(s_1, s_2, \ldots, s_{ik})) \longrightarrow BS(CS_j(s_1, s_2, \ldots, s_{nj}), CS_k(s_1, s_2, \ldots, s_{nk}), CS_m(s_1, s_2, \ldots, s_{nm}))$
$S = BS(CS_j(s_1, s_2, \ldots, s_{nj}), CS_k(s_1, s_2, \ldots, s_{nk}), CS_m(s_1, s_2, \ldots, s_{nm})) \longrightarrow BS(CS_j(s_1, s_2, \ldots, s_{ij}), CS_k(s_1, s_2, \ldots, s_{ik})),$ where $n > i$.

(4) Representation for the error function of similarity transformation

$BSC(CSC_j(s_1, s_2, \ldots, s_{ij}), CSC_k(s_1, s_2, \ldots, s_{ik})) \longrightarrow BSC(CSC_j(s_1, s_2, \ldots, s_{nj}), CSC_k(s_1, s_2, \ldots, s_{nk}), CSC_m(s_1, s_2, \ldots, s_{nm}))$
$BSC(CSC_j(s_1, s_2, \ldots, s_{nj}), CSC_k(s_1, s_2, \ldots, s_{nk}), CSC_m(s_1, s_2, \ldots, s_{nm})) \longrightarrow BSC(CSC_j(s_1, s_2, \ldots, s_{ij}), CSC_k(s_1, s_2, \ldots, s_{ik})),$ where $n > i$.

4. Similarity transformation on expanding structure

(1) Graphic representation of similarity transformation (referring to Figs. 5.162 and 5.163)

(2) Representation for logical proposition of similarity transformation

$T_z(A((U(t), KS(s_1, s_2, \ldots, s_n), \overrightarrow{p}(t), T(t), L(t)), x(t) = f(u(t), G(t)))) =$
$A((U'(t), KS'(s_1, s_2, \ldots, s_n), \overrightarrow{p'}(t), T'(t), L'(t)), x'(t) = f(u'(t), G'(t))).$

(3) Symbolic representation for similarity transformation

$S = KS(s_1, s_2, \ldots, s_i) \longrightarrow KS(s_1, s_2, \ldots, s_n)$
$S = KS(s_1, s_2, \ldots, s_n) \longrightarrow KS(s_1, s_2, \ldots, s_i),$ where $n > i$.

(4) Representation for the error function of similarity transformation

$KSC(s_1, s_2, \ldots, s_i) \longrightarrow KSC(s_1, s_2, \ldots, s_n)$
$KSC(s_1, s_2, \ldots, s_n) \longrightarrow KSC(s_1, s_2, \ldots, s_i),$ where $n > i$.

5. Similarity transformation on inclusion structure

   (1) Graphic representation of similarity transformation (referring to Figs. 5.164 and 5.165)

   (2) Representation for logical proposition of similarity transformation

   $T_z(A((U(t), \mathrm{YS}(s_1, s_2, \ldots, s_n), \overrightarrow{p}(t), T(t), L(t)), x(t) = f(u(t), G(t)))) =$

   $A((U'(t), \mathrm{YS}'(s_1, s_2, \ldots, s_n), \overrightarrow{p'}(t), T'(t), L'(t)), x'(t) = f(u'(t), G'(t))).$

   (3) Symbolic representation for similarity transformation

   $S = \mathrm{YS}(s_1, s_2, \ldots, s_i) \longrightarrow \mathrm{YS}(s_1, s_2, \ldots, s_n)$

   $S = \mathrm{YS}(s_1, s_2, \ldots, s_n) \longrightarrow \mathrm{YS}(s_1, s_2, \ldots, s_i),$ where $n > i.$

   (4) Representation for the error function of similarity transformation

   $\mathrm{YSC}(s_1, s_2, \ldots, s_i) \longrightarrow \mathrm{YSC}(s_1, s_2, \ldots, s_n)$

   $\mathrm{YSC}(s_1, s_2, \ldots, s_n) \longrightarrow \mathrm{YSC}(s_1, s_2, \ldots, s_i),$ where $n > i.$

6. Similarity transformation on feedback structure

   (1) Graphic representation of similarity transformation (referring to Figs. 5.166 and 5.167)

   (2) Representation for logical proposition of similarity transformation

   $T_z(A((U(t), \mathrm{FS}(s_1, s_2, \ldots, s_n), \overrightarrow{p}(t), T(t), L(t)), x(t) = f(u(t), G(t)))) =$

   $A((U'(t), \mathrm{FS}'(s_1, s_2, \ldots, s_n), \overrightarrow{p'}(t), T'(t), L'(t)), x'(t) = f(u'(t), G'(t))).$

   (3) Symbolic representation for similarity transformation

   $S = \mathrm{FS}(s_1, s_2, \ldots, s_i) \longrightarrow \mathrm{FS}(s_1, s_2, \ldots, s_n)$

   $S = \mathrm{FS}(s_1, s_2, \ldots, s_n) \longrightarrow \mathrm{FS}(s_1, s_2, \ldots, s_i),$ where $n > i.$

   (4) Representation for the error function of similarity transformation

   $\mathrm{FSC}(s_1, s_2, \ldots, s_i) \longrightarrow \mathrm{FSC}(s_1, s_2, \ldots, s_n)$

   $\mathrm{FSC}(s_1, s_2, \ldots, s_n) \longrightarrow \mathrm{FSC}(s_1, s_2, \ldots, s_i),$ where $n > i.$

Theorem 5.3 *Suppose that the featurens GY_j of system S have additivity with respect to the features GY_{ji} (i = 1, 2, ..., n) of its subsystem S_i, and the features GY_{ji} (i = 1, 2, ..., n) of subsystem S_i are independent, errors with respect to GY_j in subsystems other than subsystem S_i in system S will not be affected if similarity transformation is conducted on S_i.*

Proof Based on the hypothesis, the errors in other subsystems will not be affected either because any change in the feature GY_{ji} (i = 1, 2, ..., n) of subsystem S_i will not influence the features of other subsystems in the system **S**. Proof is finished.

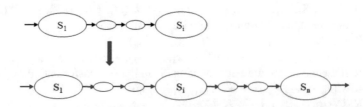

Fig. 5.156 Graphic representation of similarity transformation on series structure-expanding

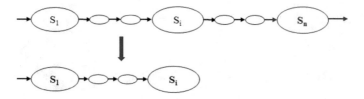

Fig. 5.157 Graphic representation of similarity transformation on series structure-shrinking

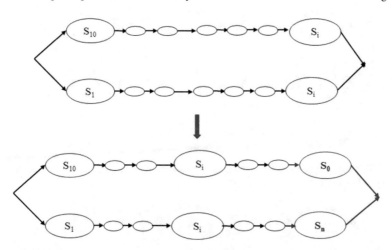

Fig. 5.158 Graphic representation of similarity transformation on parallel structure-expanding one of parallel structure's sides

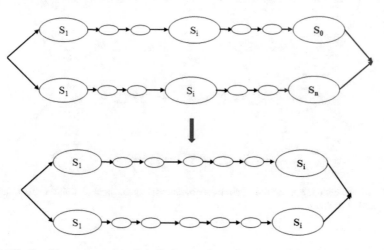

Fig. 5.159 Graphic representation of similarity transformation on parallel structure-shrinking one of parallel structure's sides

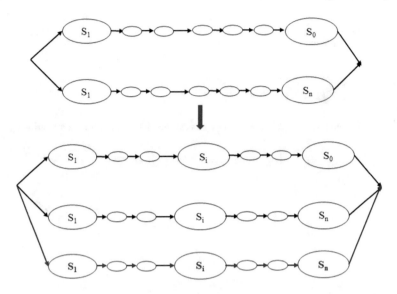

Fig. 5.160 Graphic representation of similarity transformation on parallel structure-adding more side to parallel structure

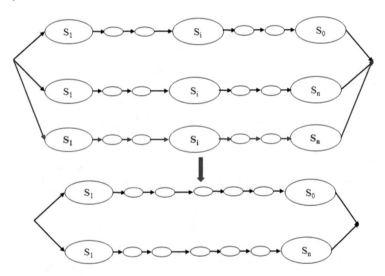

Fig. 5.161 Graphic representation of similarity transformation on parallel structure–removing side from parallel structure

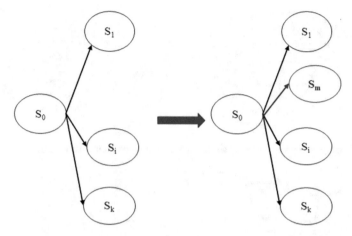

Fig. 5.162 Graphic representation of similarity transformation on expanding structure-adding elements

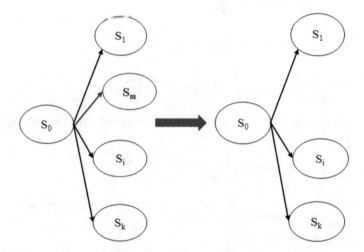

Fig. 5.163 Graphic representation of similarity transformation on expanding structure-removing elements

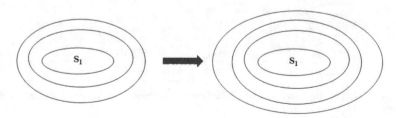

Fig. 5.164 Graphic representation of similarity transformation on inclusion structure-expanding

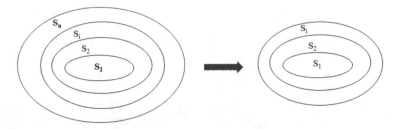

Fig. 5.165 Graphic representation of similarity transformation on inclusion structure-shrinking

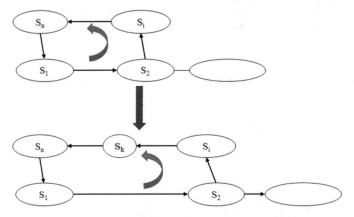

Fig. 5.166 Graphic representation of similarity transformation on feedback structure-adding element

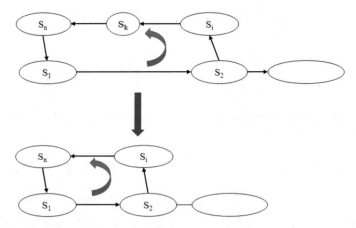

Fig. 5.167 Graphic representation of similarity transformation on feedback structure-removing element

Theorem 5.4 *Suppose that the features GY_j of system S have additivity with respect to the features GY_{ji} (i = 1, 2, ..., n) of its subsystem S_i, and the features GY_{ji} (i = 1, 2, ..., n) of subsystem S_i are independent, the scales of change in GY_j of system S and features GY_{ji} (i = 1, 2, ..., n) of its subsystem S_i are same if similarity transformation is conducted on S_i.*

Proof Based on the hypothesis, the features GY_j of system S have additivity with respect to the features GY_{ji} (i = 1, 2, ..., n) of its subsystem S_i, therefore, the features GY_{ji} (i = 1, 2, ..., n) of its subsystem S_i will not cast effect on features of other subsystems, the conclusion holds. Proof is finished.

Theorem 5.5 *Suppose that the features GY_j of system S have no additivity with respect to the features GY_{ji} (i = 1, 2, ..., n) of its subsystem S_i-i.e., there exists the following relationship: $S(GY_j)=S(S_1(GY_{j1}\ (a_1, b_1)), S_2(GY_{j2}\ (a_2, b_2)), ..., S_i(GY_{ji}\ (a_i, b_i)), ..., S_n(GY_{jn}\ (a_n, b_n)))$, where (a_i, b_i) stands for the value range of the features of GY_{ji} (i = 1, 2, ..., n) of its ith subsystem S_i, errors with respect to GY_j in subsystems other than subsystem S_i in system S will not be affected as long as the feature of new subsystem S'_k is still within the range of $GY_{jk}\ (a_k, b_k)$) if similarity transformation is conducted on S_k.*

Proof Based on the hypothesis, the features GY_{ji} (i = 1, 2, ..., n) of other subsystem will not be affected since the feature of new subsystem S'_k does not exceed the range of $GY_{jk}\ (a_k, b_k)$), and the errors in other subsystem will not be changed either. Proof is finished.

Based on Theorems 5.1 and 5.2, errors in system are eliminated if a non-erroneous subsystem S'_i with equivalent features GY_{ji} (i = 1, 2, ..., n) is used to replace the erroneous subsystem S_i whether the features GY_j of system S have additivity with respect to the features GY_{ji} (i = 1, 2, ..., n) of its subsystem S_i or not.

5.4 Decomposition Transformation on Error System Structures

Before 2004, one of the many reasons causing traffic congestion was the design of bus system where too many bus lines were using a single bus stop in Guangzhou's transportation management system. Transportation administration employed the concept of decomposition transformation to build multiple adjacent bus stops to resolve this problem. In another example, a factory wants to move a giant new machine into its production building. One can choose to disassemble the big machine into modules and reassemble it after those modules are moved into factory building.

5.4.1   Expression for Logical Proposition of Decomposition Transformation on Error System Structure

The expression for the logical proposition of decomposition transformation is:

Suppose that $A((U(t), S(t), \vec{p}(t), T(t), L(t)), x(t) = f(u(t), G(t)))$ is an error logical variable defined under judging rule $G(t)$ on universe of discourse $U(t)$, if
$T(A((U(t), S(t), \vec{p}(t), T(t), L(t)), x(t) = f(u(t), G(t)))) = \{A_1((V_1(t), S_1(t), \vec{p}_1(t), T_1(t), L_1(t)), x_1(t) = f_1((u_1(t), G_1(t)))), A_2((V_2(t), S_2(t), \vec{p}_2(t), T_2(t), L_2(t)), x_2(t) = f_2((u_2(t), G_2(t)))), \ldots, A_n((V_n(t), S_n(t), \vec{p}_n(t), T_n(t), L_n(t)), x_n(t) = f_n((u_n(t), G_n(t))))\}$, where $u(t) = u_1(t)\,\mathbf{h}\,u_2(t)\mathbf{h}, \ldots, \mathbf{h}\,u_n(t)$, then T has done decomposition transformation on $A((U(t), S(t), \vec{p}(t), T(t), L(t)), x(t) = f(u(t), G(t)))$ within judging rule $G(t)$ defined on universe of discourse $U(t)$ noted by T_f.

1. There are three cases of decomposition transformations listed as below:

    (1) In $\{x_i(t), i = 1, 2, \ldots, n\}$, T_f is called connective of error-generating decomposition transformation if $x_i(t) \geq x(t)$ $(i = 1, 2, \ldots, n)$ noted by T_{fz}
    (2) In $\{x_i(t), i = 1, 2, \ldots, n\}$, T_f is called connective of error-eliminating decomposition transformation if $x_i(t) \leq 0$ $(i = 1, 2, \ldots, n)$ noted by T_{fx};
    (3) In $\{x_i(t), i = 1, 2, \ldots, n\}$, T_f is called connective of error-increasing decomposition transformation if $x_i(t) = kx(t)$ $(i = 1, 2, \ldots, n)$ noted by T_{fk}. The value of k determines the characteristics of relevant connectives of decomposition transformation.
        (a) If $k \geq 1$, then T_{fk} is called connective of decomposition transformation with positive amplification on error value noted by T_{fzk};
        (b) If $k \leq -1$, then T_{fk} is called connective of decomposition transformation with negative amplification on error value noted by T_{ffk};
        (c) If $0 < k < 1$, then T_{fk} is called connective of decomposition transformation with positive reduction on error value noted by T_{fzs};
        (d) If $-1 < k < 0$, then T_{fk} is called connective of decomposition transformation with negative reduction on error value noted by T_{ffs};
        (e) If $k = 0$, then T_{fk} is called connective of decomposition transformation with effect of error removal noted by T_{fhl}.

2. Principles of using decomposition transformation connectives

    (1) Objective needs
    (2) Determined by actual conditions
    (3) Minimum costs

3. Types of decomposition transformations

(1) Physical decomposition

Suppose that **u** represents a car with diesel engine, **u** can be decomposed into cooling subsystem, transmission subsystem, fuel supply subsystem, braking subsystem, air management subsystem, and emission management subsystem, etc. In general, object **u(t)**'s structure extends in both vertical and horizontal direction and it also changes over time.

(2) Mathematical decomposition

If the object of interests **u(t)** is represented by a mathematical formulation (e.g. differential equations, difference equations, generic algebraic equation, or other mathematical models), for example

(a) u(t): $\frac{du}{dx} = u^2 + 6u + 25$

(b) u(t): $x_t = a_1 x_{t-1} + \cdots + a_n x_{t-n}$

(c) ... ...

(d) u(t): $x^5 - 3x + 1 = 0$

We can employ Lyapunov methods to conduct decomposition on differential equations.

(3) Decomposition types determined by actual needs;

(4) Comprehensive decomposition

Based on the definition on T_f and the elements of the error logical variable $A((U(t), S(t), \overrightarrow{p}(t), T(t), L(t)), x(t) = f (u(t), G(t)))$, T_f can conduct transformation on object u(t), error value x(t), time t, and judging rules for error G(t) of $A((U(t), S(t), \overrightarrow{p}(t), T(t), L(t)), x(t) = f (u(t), G(t)))$, therefore $T_f \subseteq \{T_{fly}, T_{fsw}, T_{fgz}, T_{fsj}\}$. The type of error logical variable $B((U(t), S(t), \overrightarrow{p}(t), T(t), L(t)), x(t) = g (u(t), G(t)))$ will not be changed as long as T_f does not conduct transformation on error function **g**.

Regarding the decomposition transformation on error system structure, we can decompose the whole system structure into basic structures (subsystem structures) which we have understood the mechanisms and rules of error transition and transformations in them if we have already known the profile of system structure. Hereby, if system **S** is constructed by (or composed of) **6** basic structures, then **S** has the following decomposition transformations:

(1) $S = CS(S_{c1}, S_{c2}, S_{ci}, S_{cnc})$; or

(2) $S = BS(S_{b1}, S_{b2}, S_{bi}, S_{bnb})$; or

(3) $S = KS(S_{k1}, S_{k2}, S_{ki}, S_{knk})$; or

(4) $S = YS(S_{y1}, S_{y2}, S_{yi}, S_{yny})$; or

(5) $S = FS(S_{f1}, S_{f2}, S_{fi}, S_{fnf})$; or

(6) $S = QS(S_{q1}, S_{q2}, S_{qi}, S_{qnq})$.

5.4.2   Decomposition Transformation on Error System Structures

1. Decomposition transformation on series structure

   (1) Graphic representation of decomposition transformations (referring to Figs. 5.168, 5.169, 5.170, and 5.171)
   (2) Representation for logical proposition of decomposition transformation

   $T(A((U(t), S(t), \vec{p}(t), T(t), L(t)), x(t) = f(u(t), G(t)))) = \{A_1((V_1(t), S_1(t), \vec{p_1}(t), T_1(t), L_1(t)), x_1(t) = f_1((u_1(t), G_1(t))), A_2((V_2(t), S_2(t), \vec{p_2}(t), T_2(t), L_2(t)), x_2(t) = f_2((u_2(t), G_2(t)))), \dots, A_n((V_n(t), S_n(t), \vec{p_n}(t), T_n(t), L_n(t)), x_n(t) = f_n((u_n(t), G_n(t))))\}.$

   (3) Symbolic representation for decomposition transformation

   $S = CS(s_1, s_2, \dots, s_i, s_n) \longrightarrow \{CS_1(s_1), CS_2(s_2), \dots, CS_i(s_i), CS_n(s_n)\}.$

   (4) Representation for the error function of decomposition transformation

   $CSC(s_1, s_2, \dots, s_i, s_n) \longrightarrow \{CSC_1(s_1), CSC_2(s_2), \dots, CSC_i(s_i), CSC_i(s_n)\}.$

2. Decomposition transformation on parallel structure

   (1) Graphic representation of decomposition transformation (referring to Figs. 5.172, 5.173, 5.173, 5.174, 5.175, and 5.176)
   (2) Representation for logical proposition of decomposition transformation

   $T(A((U(t), S(t), \vec{p}(t), T(t), L(t)), x(t) = f(u(t), G(t)))) = \{A_1((V_1(t), S_1(t), \vec{p_1}(t), T_1(t), L_1(t)), x_1(t) = f_1((u_1(t), G_1(t))), A_2((V_2(t), S_2(t), \vec{p_2}(t), T_2(t), L_2(t)), x_2(t) = f_2((u_2(t), G_2(t)))), \dots, A_n((V_n(t), S_n(t), \vec{p_n}(t), T_n(t), L_n(t)), x_n(t) = f_n((u_n(t), G_n(t))))\}.$

   (3) Symbolic representation for decomposition transformation

   $S = BS(s_1, s_2, \dots, s_i, s_n) \longrightarrow \{S_1(s_1), S_2(s_2), \dots, S_i(s_i), S_n(s_n)\}.$

   (4) Representation for the error function of decomposition transformation

   $BSC(s_1, s_2, \dots, s_i, s_n) \longrightarrow \{SC_1(s_1), SC_2(s_2), \dots, SC_i(s_i), SC_i(s_n)\}.$

3. Decomposition transformation on expanding and shrinking structure

   (1) Graphic representation of decomposition transformation (referring to Figs. 5.177, 5.178, 5.179, 5.180, 5.181, 5.182, 5.183, and 5.184)
   (2) Representation for logical proposition of decomposition transformation

   $T(A((U(t), S(t), \vec{p}(t), T(t), L(t)), x(t) = f(u(t), G(t)))) = \{A_1((V_1(t), S_1(t), \vec{p_1}(t), T_1(t), L_1(t)), x_1(t) = f_1((u_1(t), G_1(t))), A_2((V_2(t), S_2(t), \vec{p_2}(t), T_2(t), L_2(t)), x_2(t) = f_2((u_2(t), G_2(t)))), \dots, A_n((V_n(t), S_n(t), \vec{p_n}(t), T_n(t), L_n(t)), x_n(t) = f_n((u_n(t), G_n(t))))\}.$

   (3) Symbolic representation for decomposition transformation

   $S = KS(s_1, s_2, \dots, s_i, s_n) \longrightarrow \{KS_1(s_1), KS_2(s_2), \dots, KS_i(s_i), KS_n(s_n)\}.$

   (4) Representation for the error function of decomposition transformation

   $KSC(s_1, s_2, \dots, s_i, s_n) \longrightarrow \{KSC_1(s_1), KSC_2(s_2), \dots, KSC_i(s_i), KSC_i(s_n)\}.$

4. Decomposition transformation on inclusion structure

   (1) Graphic representation of decomposition transformation(referring to Figs. 5.185, 5.186, and 5.187)
   (2) Representation for logical proposition of decomposition transformation

   $T(A((U(t), S(t), \vec{p}(t), T(t), L(t)), x(t) = f(u(t), G(t))))= \{A_1((V_1(t),$
   $S_1(t), \vec{p}_1(t), T_1(t), L_1(t)), x_1(t) = f_1((u_1(t), G_1(t))), A_2((V_2(t), S_2(t),$
   $\vec{p}_2(t), T_2(t), L_2(t)), x_2(t) = f_2((u_2(t), G_2(t))), \ldots, A_n((V_n(t), S_n(t), \vec{p}_n(t),$
   $T_n(t), L_n(t)), x_n(t)= f_n((u_n(t), G_n(t))))\}.$

   (3) Symbolic representation for decomposition transformation
   $S = YS(s_1, s_2, \ldots, s_i, s_n) \longrightarrow \{YS_1(s_1), YS_2(s_2), \ldots, BS_i(s_i), CS_n(s_n)\}.$
   (4) Representation for the error function of decomposition transformation
   $YSC(s_1, s_2, \ldots, s_i, s_n) \longrightarrow \{YSC_1(s_1), YSC_2(s_2), \ldots, BSC_i(s_i), BSC_i(s_n)\}.$

5. Decomposition transformation on feedback structure

   (1) Graphic representation of decomposition transformation (referring to Figs. 5.188, 5.189, 5.190, and 5.191)
   (2) Representation for logical proposition of decomposition transformation

   $T(A((U(t), S(t), \vec{p}(t), T(t), L(t)), x(t) = f(u(t), G(t))))= \{A_1((V_1(t),$
   $S_1(t), \vec{p}_1(t), T_1(t), L_1(t)), x_1(t) = f_1((u_1(t), G_1(t))), A_2((V_2(t), S_2(t),$
   $\vec{p}_2(t), T_2(t), L_2(t)), x_2(t) = f_2((u_2(t), G_2(t))), \ldots, A_n((V_n(t), S_n(t), \vec{p}_n(t),$
   $T_n(t), L_n(t)), x_n(t)= f_n((u_n(t), G_n(t))))\}.$

   (3) Symbolic representation for decomposition transformation
   $S = YS(s_1, s_2, \ldots, s_i, s_n) \longrightarrow \{CS_1(s_1), CS_2(s_2), \ldots, BS_i(s_i), KS_n(s_n)\}.$
   (4) Representation for the error function of decomposition transformation
   $YSC(s_1, s_2, \ldots, s_i, s_n) \longrightarrow \{CSC_1(s_1), CSC_2(s_2), \ldots, BSC_i(s_i), KSC_i(s_n)\}.$

Theorem 5.6 *Suppose that the features GY_j of system S have additivity with respect to the features GY_{ji} ($i=1, 2, \ldots, n$) of its subsystem S_i, and the features GY_{ji} ($i=1, 2, \ldots, n$) of subsystem S_i are independent, system S and its features GY_j can be decomposed into subsystems S_i and corresponding features GY_{ji} ($i=1, 2, \ldots, n$), errors with respect to GY_j in subsystems other than subsystem S_i will not be affected if decomposition transformation is conducted on S_i.*

Proof Based on the hypothesis, the errors in other subsystems will not be affected because any change in the feature GY_{ji} ($i=1, 2, \ldots, n$) of subsystem S_i will not influence the features of other subsystems in the system **S**.
Proof is finished.

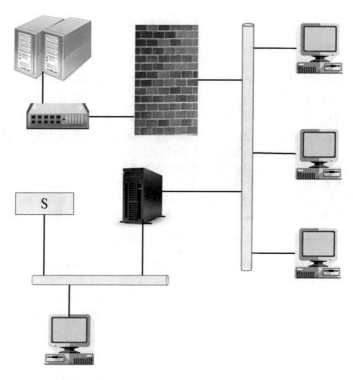

Fig. 5.168 Graphic representation of decomposition transformation on series structure

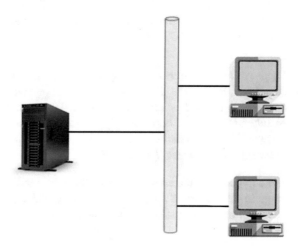

Fig. 5.169 Graphic representation of decomposition transformation on series structure-constituent subsystem 1

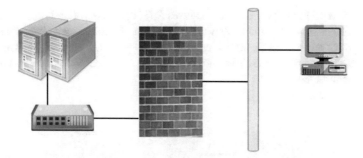

Fig. 5.170 Graphic representation of decomposition transformation on series structure-constituent subsystem 2

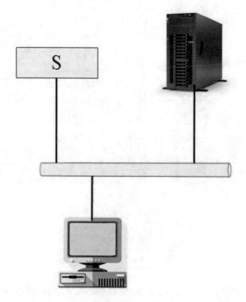

Fig. 5.171 Graphic representation of decomposition transformation on series structure-constituent subsystem 3

Theorem 5.7 *Suppose that the features GY_j of system S have no additivity with respect to the features GY_{ji} $(i=1, 2, \ldots, n)$ of its subsystem S_i-i.e., there exists the following relationship: $S(GY_j)=S(S_1(GY_{j1}\ (a_1, b_1)),\ S_2(GY_{j2}\ (a_2, b_2)),\ \ldots,$ $S_i(GY_{ji}\ (a_i, b_i)),\ \ldots,\ S_n(GY_{jn}\ (a_n, b_n)))$, where (a_i, b_i) stands for the value range of the features of GY_{ji} $(i=1, 2, \ldots, n)$ of its ith subsystem S_i, subsystem S_i and its corresponding features $S_i(GY_{ji}\ (a_i, b_i))$ are obtained once decomposition transformation is conducted on system S, errors with respect to GY_j in subsystems other than subsystem S_i will not be affected as long as the feature of new subsystem S_k is still within the range of $GY_{jk}\ (a_k, b_k))$ if decomposition transformation is conducted on S_k.*

Proof Based on the hypothesis, the features $S_i(GY_{ji}(a_i, b_i))$ (i = 1, 2, ..., n, and i $\neq$ k) in subsystems other than subsystem S_i will not be influenced since changes in the feature of subsystem S_k do not exceed the range of $GY_{jk}(a_k, b_k)$). Therefore, the errors with respect to GY_j in subsystems other than subsystem S_i in system **S** will not be affected.
Proof is finished.

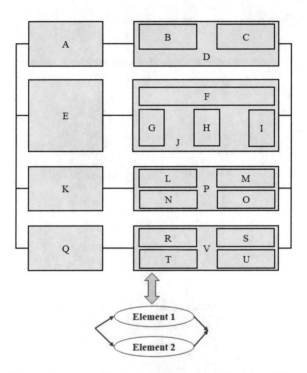

Fig. 5.172 Graphic representation of decomposition transformation on parallel structure

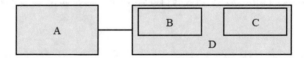

Fig. 5.173 Graphic representation of decomposition transformation on parallel structure-constituent subsystem 1

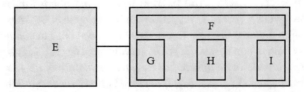

Fig. 5.174 Graphic representation of decomposition transformation on parallel structure-constituent subsystem 2

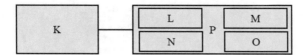

Fig. 5.175 Graphic representation of decomposition transformation on parallel structure-constituent subsystem 3

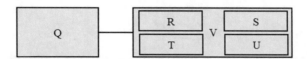

Fig. 5.176 Graphic representation of decomposition transformation on parallel structure-constituent subsystem 4

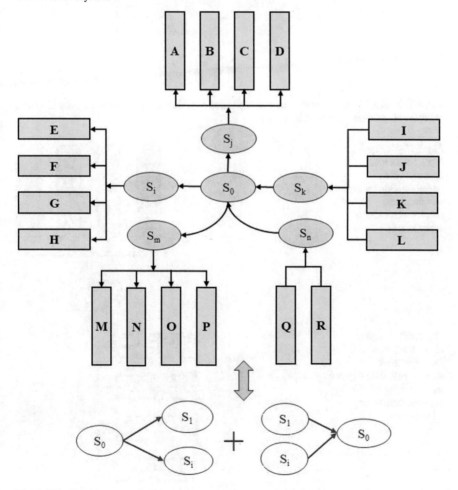

Fig. 5.177 Graphic representation of decomposition transformation on expanding and shrinking structure

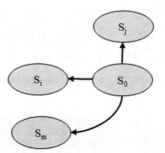

Fig. 5.178 Graphic representation of decomposition transformation on expanding and shrinking structure-constituent subsystem 1

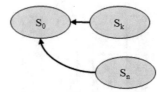

Fig. 5.179 Graphic representation of decomposition transformation on expanding and shrinking structure-constituent subsystem 2

Fig. 5.180 Graphic representation of decomposition transformation on expanding and shrinking structure-constituent subsystem 3

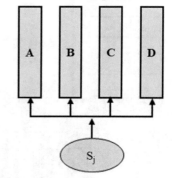

Fig. 5.181 Graphic representation of decomposition transformation on expanding and shrinking structure-constituent subsystem 4

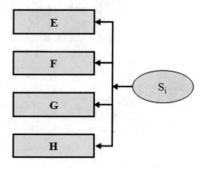

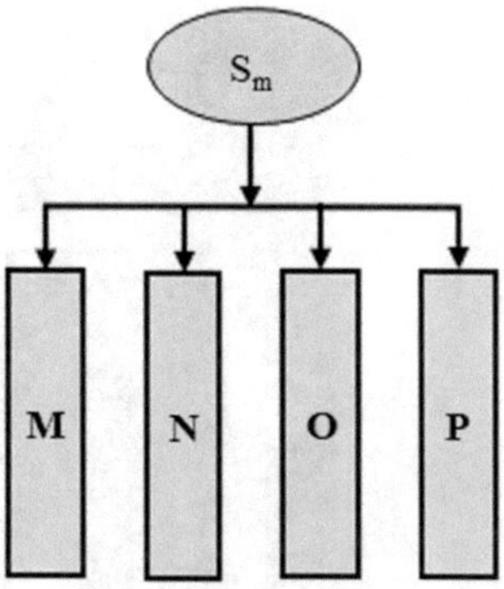

Fig. 5.182 Graphic
representation of
decomposition
transformation on expanding
and shrinking
structure-constituent
subsystem 5

Fig. 5.183 Graphic
representation of
decomposition
transformation on expanding
and shrinking
structure-constituent
subsystem 6

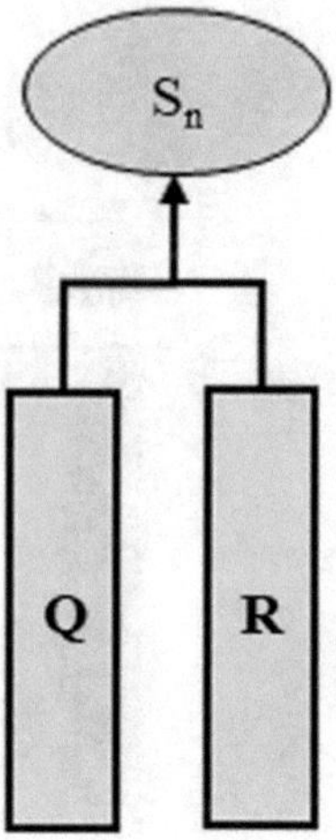

Fig. 5.184 Graphic
representation of
decomposition
transformation on expanding
and shrinking
structure-constituent
subsystem 7

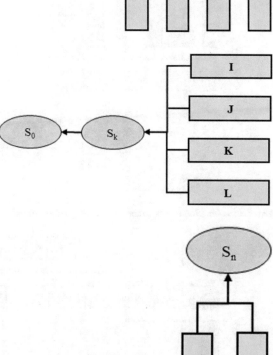

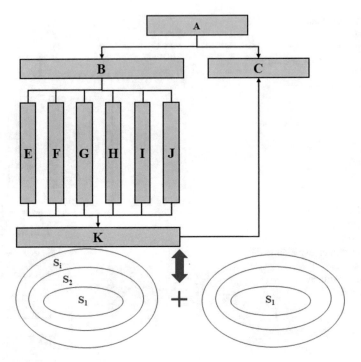

Fig. 5.185 Graphic representation of decomposition transformation on inclusion structure

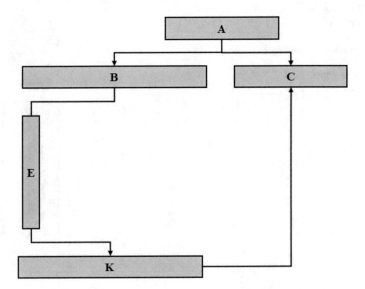

Fig. 5.186 Graphic representation of decomposition transformation on inclusion structure-constituent subsystem 1

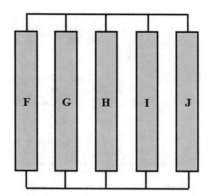

Fig. 5.187 Graphic representation of decomposition transformation on inclusion structure-constituent parallel structure

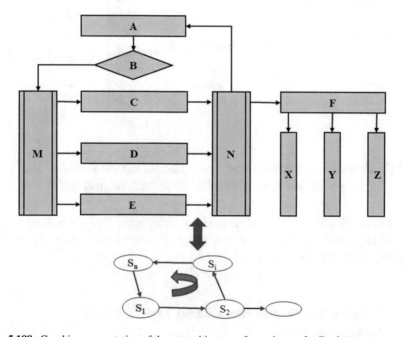

Fig. 5.188 Graphic representation of decomposition transformation on feedback structure

Fig. 5.189 Graphic representation of decomposition transformation on feedback structure-constituent series structure

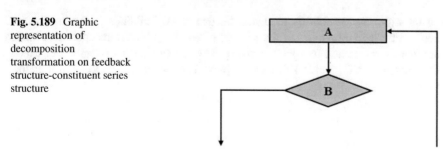

Fig. 5.190 Graphic representation of decomposition transformation on feedback structure-constituent parallel structure

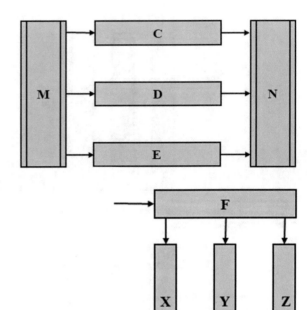

Fig. 5.191 Graphic representation of decomposition transformation on feedback structure-constituent expanding structure

5.4.3   Inverse Decomposition Transformation on Error System Structure-"Combination Transformation"

As "combination transformation" is the inverse decomposition transformation, the combination operation on the error system structure can be achieved through inverse decomposition transformation.

5.5   Destruction Transformation on Error System Structures

Using explosive demolition to dismantle dangerous buildings is the concept of destruction transformation. In the process of repairing broken machine and home appliances, replacing worn-out or broken parts with new spare parts can also be called destruction transformation although we previously called it displacement transformation.

5.5.1   Expression for Logical Proposition of Destruction Transformation on Error System Structure

The expression for the logical proposition of destruction transformation is:

Suppose that $A((U(t), S(t), \vec{p}(t), T(t), L(t)), x(t) = f(u(t), G(t)))$ is an error logical variable defined under judging rule $G(t)$ on universe of discourse $U(t)$, if $T(A((U(t), S(t), \vec{p}(t), T(t), L(t)), x(t) = f(u(t), G(t)))) = A((\emptyset, \emptyset, \emptyset, \emptyset, \emptyset), \emptyset = \emptyset((\emptyset, \emptyset))$, then T is called a destruction transformation connective regarding $A((U(t), S(t), \vec{p}(t), T(t), L(t)), x(t) = f(u(t), G(t)))$ and judging rule $G(t)$ defined on universe of discourse $U(t)$ noted by T_h. If $T(A((U(t), S(t), \vec{p}(t), T(t), L(t)), x(t) = f(u(t), G(t)))) = A((U(t), \emptyset, \vec{p}(t), T(t), L(t)), x(t) = f(u(t), G(t)))$, then T has conducted destruction transformation on subject regarding $A((U(t), S(t), \vec{p}(t), T(t), L(t)), x(t) = f(u(t), G(t)))$ and judging rule $G(t)$ defined on universe of discourse $U(t)$ noted by T_{hsw}. The meaning of subject destruction is: T_{hsw} (subject destruction) $\rightarrow$ subject does not exist $\rightarrow$ no subject is available (there is no need to discuss the subject of interests)(or subject was removed, eradicated, annihilated, fired, sold out, discarded, and moved away, etc.).

5.5.2   Destruction Transformation on Error System Structures

1. Destruction transformation on series structure

    (1) Graphic representation of transformation (referring to Fig. 5.192)
    (2) Representation for logical proposition of destruction transformation
    $$T_h\{A_i((U_i(t), CS(s_1, s_2, \ldots, s_n)_{iu}(t), \vec{p}_{iu}(t), T_{iu}(t), L_{iu}(t)), x_i(t) = f_i(u_i(t), G_{iA}(t))), i = 1, 2, \ldots, n\} = \{B_j((V_j(t), \emptyset, \ldots, s_n)_{jv}t, \vec{p}_{jv}(t), T_{jv}(t), L_{jv}(t)), y_j(t) = g_j(v_j(t), G_{jB}(t))), j = 1, 2, \ldots, m\}.$$
    (3) Symbolic representation for destruction transformation
    $$S = S(s_1, s_2, \ldots, CS(s_1, s_2, \ldots, s_k(t))iu, s_i, s_n) \longrightarrow S(s_1, s_2, \ldots, \emptyset, s_i, s_n).$$
    (4) Representation for the error function of destruction transformation
    $$CS(s_1, s_2, \ldots, CSC(s_1, s_2, \ldots, s_k(t))iu, s_i, s_n) \longrightarrow CS(s_1, s_2, \ldots, \emptyset, s_i, s_n).$$

2. Destruction transformation on parallel structure

    (1) Graphic representation of transformation (referring to Fig. 5.193)
    (2) Representation for logical proposition of destruction transformation
    $$T_h\{A_i((U_i(t), BS(s_1, s_2, \ldots, s_n)_{iu}(t), \vec{p}_{iu}(t), T_{iu}(t), L_{iu}(t)), x_i(t) = f_i(u_i(t), G_{iA}(t))), i = 1, 2, \ldots, n\} = \{B_j((V_j(t), \emptyset, \ldots, s_n)_{jv}t, \vec{p}_{jv}(t), T_{jv}(t), L_{jv}(t)), y_j(t) = g_j(v_j(t), G_{jB}(t))), j = 1, 2, \ldots, m\}.$$

(3) Symbolic representation for destruction transformation

$S = S(s_1, s_2, \ldots, BS(s_1, s_2, \ldots, s_k(t))iu, s_i, s_n) \longrightarrow S(s_1, s_2, \ldots, \emptyset, s_i, s_n).$

(4) Representation for the error function of destruction transformation

$CS(s_1, s_2, \ldots, BSC(s_1, s_2, \ldots, s_k(t))iu, s_i, s_n) \longrightarrow CS(s_1, s_2, \ldots, \emptyset, s_i, s_n).$

3. Destruction transformation on expanding and shrinking structure

(1) Graphic representation of transformation (referring to Fig. 5.194)

(2) Representation for logical proposition of destruction transformation

$T_h\{A_i((U_i(t), KS(s_1, s_2, \ldots, s_n)_{iu}(t), \overrightarrow{p_{iu}}(t), T_{iu}(t), L_{iu}(t)), x_i(t) = f_i(u_i(t),$
$G_{iA}(t))), i = 1, 2, \ldots, n\} = \{B_j((V_j(t), \emptyset, \ldots, s_n)_{jv}t, \overrightarrow{p_{jv}}(t), T_{jv}(t), L_{jv}(t)),$
$y_j(t) = g_j(v_j(t), G_{jB}(t))), j = 1, 2, \ldots, m\}.$

(3) Symbolic representation for destruction transformation

$S = S(s_1, s_2, \ldots, KS(s_1, s_2, \ldots, s_k(t))iu, s_i, s_n) \longrightarrow S(s_1, s_2, \ldots, \emptyset, s_i, s_n).$

(4) Representation for the error function of destruction transformation

$CS(s_1, s_2, \ldots, KSC(s_1, s_2, \ldots, s_k(t))iu, s_i, s_n) \longrightarrow CS(s_1, s_2, \ldots, \emptyset, s_i, s_n).$

4. Destruction transformation on inclusion structure

(1) Graphic representation of transformation (referring to Fig. 5.195)

(2) Representation for logical proposition of destruction transformation

$T_h\{A_i((U_i(t), YS(s_1, s_2, \ldots, s_n)_{iu}(t), \overrightarrow{p_{iu}}(t), T_{iu}(t), L_{iu}(t)), x_i(t) = f_i(u_i(t),$
$G_{iA}(t))), i = 1, 2, \ldots, n\} = \{B_j((V_j(t), \emptyset, \ldots, s_n)_{jv}t, \overrightarrow{p_{jv}}(t), T_{jv}(t), L_{jv}(t)),$
$y_j(t) = g_j(v_j(t), G_{jB}(t))), j = 1, 2, \ldots, m\}.$

(3) Symbolic representation for destruction transformation

$S = S(s_1, s_2, \ldots, YS(s_1, s_2, \ldots, s_k(t))iu, s_i, s_n) \longrightarrow S(s_1, s_2, \ldots, \emptyset, s_i, s_n).$

(4) Representation for the error function of destruction transformation

$CS(s_1, s_2, \ldots, YSC(s_1, s_2, \ldots, s_k(t))iu, s_i, s_n) \longrightarrow CS(s_1, s_2, \ldots, \emptyset, s_i, s_n).$

5. Destruction transformation on feedback structure

(1) Graphic representation of transformation (referring to Fig. 5.196)

(2) Representation for logical proposition of destruction transformation

$T_h\{A_i((U_i(t), FS(s_1, s_2, \ldots, s_n)_{iu}(t), \overrightarrow{p_{iu}}(t), T_{iu}(t), L_{iu}(t)), x_i(t) = f_i(u_i(t),$
$G_{iA}(t))), i = 1, 2, \ldots, n\} = \{B_j((V_j(t), \emptyset, \ldots, s_n)_{jv}t, \overrightarrow{p_{jv}}(t), T_{jv}(t), L_{jv}(t)),$
$y_j(t) = g_j(v_j(t), G_{jB}(t))), j = 1, 2, \ldots, m\}.$

(3) Symbolic representation for destruction transformation

$S = S(s_1, s_2, \ldots, FS(s_1, s_2, \ldots, s_k(t))iu, s_i, s_n) \longrightarrow S(s_1, s_2, \ldots, \emptyset, s_i, s_n).$

(4) Representation for the error function of destruction transformation

$CS(s_1, s_2, \ldots, FSC(s_1, s_2, \ldots, s_k(t))iu, s_i, s_n) \longrightarrow CS(s_1, s_2, \ldots, \emptyset, s_i, s_n).$

Fig. 5.192 Graphic representation of destruction transformation on series structure

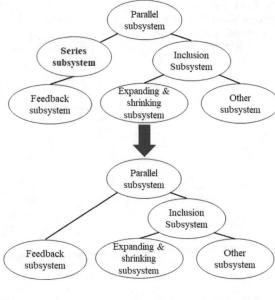

Fig. 5.193 Graphic representation of destruction transformation on parallel structure

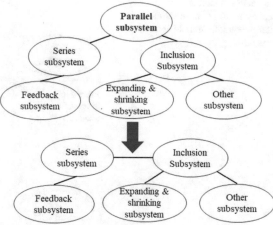

5.5.3   Inverse Destruction Transformation on Error System Structure-"Generation Transformation"

As "generation transformation" is the inverse destruction transformation, the generation operation on the error system structure can be achieved through inverse decomposition transformation. The representation for the error function of destruction transformation $SC(s_1, s_2, \ldots, \emptyset, s_i, s_n) \longrightarrow CS(s_1, s_2, \ldots, S_i(s_1, s_2, \ldots, s_k(t))iu, s_i, s_n)$.

Fig. 5.194 Graphic
representation of destruction
transformation on expanding
and shrinking structure

Fig. 5.195 Graphic
representation of destruction
transformation on inclusion
structure

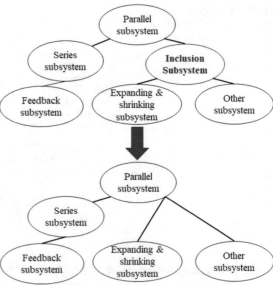

Fig. 5.196 Graphic representation of destruction transformation on feedback structure

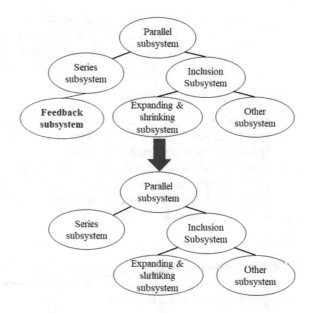

Theorem 5.8 *Suppose that the features GY_j of system S have additivity with respect to the features GY_{ji} ($i=1, 2, \ldots, n$) of its subsystem S_i, and the features GY_{ji} ($i=1, 2, \ldots, n$) of subsystem S_i are independent, errors with respect to GY_j in subsystems other than subsystem S_i will not be affected if destruction transformation is conducted on S_i. However, the features GY_{ji} ($i=1, 2, \ldots, n$) of subsystem S_i can not be achieved since S_i had be been destructed.*

Proof Based on the hypothesis, the errors in other subsystems will not be affected because any change in the features GY_{ji} ($i=1, 2, \ldots, n$) of subsystem S_i will not influence the features of other subsystems in the system **S**. Therefore, errors with respect to GY_j in subsystems other than subsystem S_i will not be changed. Due to the destruction of subsystem S_i, its features GY_{ji} ($i=1, 2, \ldots, n$) can not be achieved. Proof is over.

Theorem 5.9 *Suppose that the features GY_j of system S have no additivity with respect to the features GY_{ji} ($i=1, 2, \ldots, n$) of its subsystem S_i-i.e., there exists the following relationship: $S(GY_{ji})=S(S_1(GY_{j1} (a_1, b_1)), S_2(GY_{j2} (a_2, b_2)), \ldots, S_i(GY_{ji} (a_i, b_i)), \ldots, S_n(GY_{jn} (a_n, b_n)))$, where (a_i, b_i) stands for the value range of the features of GY_{ji} ($i=1, 2, \ldots, n$) of its ith subsystem S_i, then if destruction transformation is conducted on subsystem S_i.*

1. *Two cases are discussed here:*

   (1) The features GY_j of system S can not be realized if S_k is the critical subsystem of S and the S becomes a system with errors.

(2) *The features GY_{ji} $(i=1, 2, \ldots, n)$ of other subsystem $S_i(GY_{ji}(a_i, b_i))$ $(i=1, 2, \ldots, n, i \neq k)$ will not be affected if S_k is not the critical subsystem of S and not all the features GY_j of system S can be realized. Therefore, S is still a system with errors.*

Proof   Based on the hypothesis:

(1) Since subsystem S_k is the critical subsystem of **S**, all the features GY_j of **S** can not be realized if S_k is removed/destructed. According the definition of error system, **S** becomes a system with errors.
(2) Since subsystem S_k is not the critical subsystem of **S**, features GY_{ji} $(i=1, 2, \ldots, n)$ of other subsystem $S_i(GY_{ji}(a_i, b_i))$ $(i=1, 2, \ldots, n, i \neq k)$ will not be influenced. Nevertheless, not all the features GY_j of **S** can be realized. **S** is still a system with errors.

Proof is over.

In the above session, we have discussed the case of removing one subsystem in a system **S**. By using the same norm, one can investigate the case on destruction of multiple subsystems at the same time.

5.5.4   Stable Structures of Error Systems

Here we list some stable structures of error systems.

1. Conditional stability: system reaches stable when certain conditions are met.
2. Unconditional stability: system is stable under any condition.
3. Stable within a range: system is stable within certain range.

The stability of structure is not a basic structure but a certain feature of a system.

Chapter 6
System Acting Forces and Their Applications to Error Systems

Abstract This chapter investigates the relationship between system structure and system functions (features). It also explores the strength, spatial dynamics, direction, and sequence of acting of mutual interacting forces of system structure and how they affect the attainment of system global optimum. For the sake of gaining better understanding on how interactions of system (or subsystem) structures affect system features, we propose the concepts of system structural acting force ("Shi" or 'potential' or 'quan'), acting forces of system elements, chained structural acting force, accumulated acting forces. Thereafter, transformation of system structural acting force, relationship between system structural acting force and system features, and relationship between system structural acting force and optimization of error systems are then examined. Relevant issues, contents, and examples of system acting force are provided accordingly.

6.1 Structural Acting Forces of System

6.1.1 Background

The essence of systems science is its way of holistic thinking. System optimization indicates the process where the whole system reaches global optimum with respect to the objective features contained in the system's intrinsic features. System thinking has been widely used in all walks of life in the society with the purpose of developing high leverage interventions (solutions or policies) to prevent policy resistance(system backfires). System global optimum can not be attained without the synergistic actions of system's subsystems. How can we examine their interactions, in spite of the fact that there exist complicated relationships?

Have done many practical analyses on natural and man-made systems (Sterman, 2000), we found that the strengths of mutual actions of subsystems are influenced by spatial distance, acting direction, and acting sequence. Therefore, in order to explore the relationship between system structure and features (or behaviors), and optimization of systems, we have to investigate the impacts of acting distance, acting

© Springer Nature Switzerland AG 2020

K. Guo and S. Liu, *Error Systems: Concepts, Theory and Applications*,
Studies in Systems, Decision and Control 275,
https://doi.org/10.1007/978-3-030-40760-5_6

direction, and acting sequence of inter-structural forces on the system's features (behaviors).

Before we discuss system acting forces and their applications to error systems, we propose the basic concept, error system structures, and transformations on them. In Chaps. 3, 4, and 5, we have provided a series of concepts: (1) Critical subsystems, major subsystems, and important subsystems; (2) Critical structures, major structures, and important structures; (3) Critical elements, major elements, important elements, and subsystem independence. Next, we discuss the theory and method for conducting system optimization. The steps of system optimization are listed as follows: (1) Determine if system has error; (2) Error should be eliminated if system has error. In some instances, for error occurred in a system, one can start addressing errors from the bottom layer of a system and then move up to higher hierarchies according to acting forces of different subsystem. The method for investigating system error is then confirmed and used to eliminate error in the whole system. (3) For a system without errors (or a system tolerating errors), a programming model is built in which: the expected overall intrinsic features are the objectives; each intrinsic feature contained in objective features is defined as an independent variable; and the system's conditions are constraints. By solving the programming model, the values or the range of values of each intrinsic feature is obtained. (4) Feature-based system optimization is then conducted given the values or value range of each intrinsic feature.

6.1.2   Basics Concepts

By mapping system's elements and interacting direction & sequence to the nodes and directional edges in topology, respectively, a system structural diagram emerges.

Definition 6.1 System structural acting force dictates the strength, direction, and sequence of mutual actions of system's elements, which is a vector used to quantify the effect of one element's action on another one denoted by ω.

In an example of metro system (referring to Fig. 6.1), both the distance between any two stations and metro car's speed within the interval of any stations are structural acting forces of the metro network structure.

Suppose that the object of interests is the transportation network (in Fig. 6.1), it can be represented by the topological portrayal (referring to Fig. 6.2). The purpose of this type of simplification is, without loss of generality, to facilitate data visualization and system analysis.

Definition 6.2 Suppose that system structural diagram is represented by $G = (P, E)$, where $p_0, e_1, p_1, e_2, p_2, e_3, p_3, \ldots, p_{n-2}, e_{n-1}, p_{n-1}, e_n, p_n$ is one structure chain of $G = (P, E)$, w_i ($i = 1, 2, 3, \ldots$, n) stands for the acting forces of corresponding edge with respect to certain variable (or parameter), the "total acting force" of one structural chain is the "sum" of acting forces of all edges of that chain

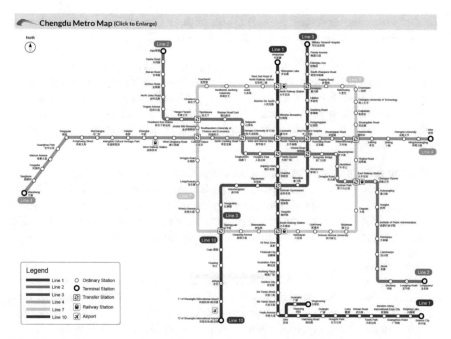

Fig. 6.1 Graphic representation of Chengdu's metro system structure

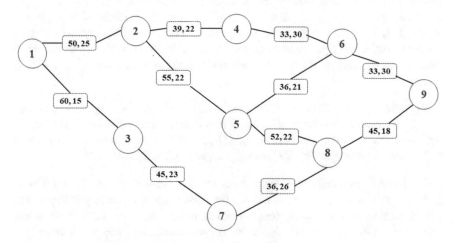

Fig. 6.2 Graphic representation of adding series structure to existing series structure

denoted by $w = \sum w_i$ (if there exists additivity) and other aggregation mechanisms (if there exists no additivity). In $G = (P, E)$, $p = \{p_1 = 1, p_2 = 2, \ldots, p_9 = 9\}$, $E = \{(1, 2), (1, 3), \ldots, (5, 8), (7, 8), (6, 9), (8, 9)\}$; in the rectangle block with two numbers, the first number stands for distance between two adjacent nodes (two adjacent metro stations) denoted by $D = \{d_{ij}, i = 1, 2, 3, \ldots, 9; j = 1, 2, 3, \ldots, 9\}$ and the second number represents the average speed that metro ran between two adjacent nodes denoted by $V = \{v_{ij}, i = 1, 2, 3, \ldots, 9; j = 1, 2, 3, \ldots, 9\}$.

In Fig. 6.2 ($G = (P, E)$), $v_1, e_{12}, v_2, e_{25}, v_5, e_{58}, v_8, \ldots, e_{89}, v_9$ is one structural chain in current metro system, the "sum of acting forces with respect to distance" is $w = \sum d_{ij} = 50 + 55 + 52 + 45 = 202$ and the "sum of acting forces with respect to average velocity" is $w = \sum v_{ij} = 25 + 22 + 22 + 18 = 87$ (The purpose here is just to show the concept since finding the sum of average velocity may not be reasonable in common sense).

Definition 6.3 Suppose that system structural diagram is $G = (P, E)$, $p_0, e_1, p_1, e_2, p_2, e_3, p_3, \ldots, p_{n-2}, e_{n-1}, p_{n-1}, e_n, p_n$ is one structure chain of $G = (P, E)$, w_i ($i = 1, 2, 3, \ldots, n$) stands for the acting forces of corresponding edge with respect to certain variable (or parameter), the minimum acting force with respect to certain variable (or parameter) along certain structural chain is $w_{min} = \min\{w_i, i = 1, 2, \ldots, n\}$

Referring to Fig. 6.2 ($G = (P, E)$), in the structural chain $v_1, e_{12}, v_2, e_{25}, v_5, e_{58}, v_8, \ldots, e_{89}, v_9$, the "minimum of acting forces with respect to distance" is $w = \min\{50, 55, 52, 45\} = 45$ and the "minimum of acting forces with respect to average velocity" in the above structural chain is $w_{min} = \min\{25, 22, 22, 18\} = 18$.

Definition 6.4 Suppose that system structural diagram is $G = (P, E)$, $p_0, e_1, p_1, e_2, p_2, e_3, p_3, \ldots, p_{n-2}, e_{n-1}, p_{n-1}, e_n, p_n$ is one structure chain of $G = (P, E)$, w_i ($i = 1, 2, 3, \ldots, n$) stands for the acting forces of corresponding edge with respect to certain variable (or parameter), the maximum acting force with respect to certain variable (or parameter) along certain structural chain is $w_{max} = \max\{w_i, i = 1, 2, \ldots, n\}$

Referring to Fig. 6.2 ($G = (P, E)$), in the structural chain $v_1, e_{12}, v_2, e_{25}, v_5, e_{58}, v_8, \ldots, e_{89}, v_9$, the "maximum of acting forces with respect to distance" in the above structural chain is $w = \max\{50, 55, 52, 45\}$ 55 and the "maximum of acting forces with respect to average velocity" is $w_{max} = \max\{25, 22, 22, 18\} = 25$.

Definition 6.5 Suppose that system structural diagram is $G = (P, E)$, w_{ij} ($i = 1, 2, 3, \ldots, n; j = 1, 2, \ldots, n$) stands for the acting forces of corresponding edges with respect to certain variable (or parameter), where i represents the ith structural chain and j stands for the jth edge of the ith structural chain, the maximum acting force with respect to certain variable (or parameter) in the whole system structure is $w_{Tmax} = \max\{w_{ij}, i = 1, 2, \ldots, n; j = 1, 2, \ldots, n)\}$.

Referring to Fig. 6.2 ($G = (P, E)$), the "maximum of acting forces with respect to distance" in the whole system structure is $w_{Tmax} = \max\{50, 39, 33, 33, 45, 36, 45, 60, 55, 52, 36\} = 55$; the "maximum of acting forces with respect to average velocity" in the whole system structure is $w_{Tmax} = \max\{25, 22, 30, 30, 18, 26, 23, 15, 22, 22, 21\} = 30$.

Definition 6.6 Suppose that system structural diagram is $G = (P, E)$, w_{ij} ($i = 1$, 2, 3, ..., n; $j = 1$, 2, ..., n) stands for the acting forces of corresponding edges with respect to certain variable (or parameter), where i represents the ith structural chain and j stands for the jth edge of the ith structural chain, the minimum acting force with respect to certain variable (or parameter) in the whole system structure is $w_{Tmin} = \min\{w_{ij}, i = 1, 2, \ldots, n; j = 1, 2, \ldots, n)\}$.

Referring to Fig. 6.2 ($G = (P, E)$), the "maximum of acting forces with respect to distance" in the whole system structure is $w_{Tmin} = \min\{50, 39, 33, 33, 45, 36, 45, 60, 55, 52, 36\} = 33$; the "maximum of acting forces with respect to average velocity" in the whole system structure is $w_{Tmin} = \min\{25, 22, 30, 30, 18, 26, 23, 15, 22, 22, 21\} = 15$.

Definition 6.7 Suppose that system structural diagram is represented by $G = (P, E)$, w_{ij} ($i = 1, 2, 3, \ldots$, n; $j = 1, 2, \ldots$, n) stands for the acting forces of corresponding edges with respect to certain variable (or parameter), where i represents the ith structural chain and j stands for the jth edge of the ith structural chain, the "total acting force" of corresponding edges with respect to certain variable (or parameter) in the whole system is the sum of acting forces of all edges denoted by $w_{Tsum} = \{\sum w_{ij}, i = 1, 2, 3, \ldots, n; j = 1, 2, \ldots, n\}$.

In Fig. 6.2 ($G = (P, E)$), the "total acting force" of corresponding edges with respect to distance" in the metro system is $w_{Tsum} = \sum w_{ij} = 50 + 39 + 33 + 33 + 45 + 36 + 45 + 60 + 55 + 52 + 36 = 484$ and the "total acting force" of corresponding edges with respect to average velocity" in the metro system is $w_{Tsum} = \sum w_{ij} = 25 + 22 + 30 + 30 + 18 + 26 + 23 + 15 + 22 + 22 + 21 = 254$.

6.1.3   Transformations on System Structural Acting Forces

1. Basic concepts of transformations on system structural acting forces

Definition 6.8 Suppose that the corresponding structural acting forces of system structure are represented by w ($i = 1, 2, \ldots$, n), w_i ($i = 1, 2, \ldots$, n) standing for respective acting forces of ith substructure, if $T(w_i) = \lambda\, w_i$, λ is a real number, T has conducted similarity transformation on acting forces w_i denoted by $T_{qx}(w_i)$.

Definition 6.9 Suppose that the corresponding structural acting forces of system structure are represented by w ($i = 1, 2, \ldots$, n), w_i ($i = 1, 2, \ldots$, n) stands for respective acting forces of ith substructure, if $T(w_i) = w_i'$, T has conducted displacement transformation on acting forces w_i denoted by $T_{qz}(w_i)$.

Definition 6.10 Suppose that the corresponding structural acting forces of system structure are represented by w ($i = 1, 2, \ldots$, n), w_i ($i = 1, 2, \ldots$, n) stands for respective acting forces of ith substructure, if $T(w_i) = w_{i1} + w_{i2} + w_{i3} +, \ldots, w_{ik}$, $k \geq 2$, then T has conducted decomposition transformation on acting forces w_i

denoted by $T_{qf}(w_i)$. If $T(w_{i1} + w_{i2} + w_{i3} +, \ldots, w_{ik}) = w_i$, $k \geq 2$, then T has conducted combination transformation on acting forces w_i denoted by $T_{qf}^{-1}(w_i)$ or denoted by $T_{qzu}(w_i)$.

Definition 6.11 Suppose that the corresponding structural acting forces of system structure are represented by w (i $= 1, 2, \ldots$, n), w_i (i $= 1, 2, \ldots$, n) standing for respective acting forces of ith substructure, if $T(w_i) = w_i + a$, a is a value with the same characteristics as w_i, then T has conducted addition transformation on acting forces w_i denoted by $T_{qzj}(w_i)$. If $T(w_i) = w_i - a$, a is a value with the same characteristics as w_i, then T has conducted reduction transformation on acting forces w_i denoted by $T_{qjs}(w_i)$.

Definition 6.12 Suppose that the corresponding structural acting forces of system structure are represented by w (i $= 1, 2, \ldots$, n), w_i (i $= 1, 2, \ldots$, n) stands for respective acting forces of ith substructure, if $T(w_i) = 0$, then T has conducted destruction transformation on acting forces w_i denoted by $T_{qh}(w_i)$. If $T(0) = w_i$, then T has conducted generation transformation on acting forces w_i denoted by $T'_{qh}(w_i)$ or denoted by $T_{qc}(w_i)$.

Definition 6.13 Suppose that the corresponding structural acting forces of system structure are represented by w (i $= 1, 2, \ldots$, n), w_i (i $= 1, 2, \ldots$, n) stands for respective acting forces of ith substructure, if $T(w_i) = w_i$, then T has conducted unit transformation on acting forces w_i denoted by $T_{qd}(w_i)$.

2. Basic transformations on system structural acting forces

   Because it has not been proved that "any system can be decomposed into the five basic structures": (a) series structure, (b) parallel structure, (c) feedback structure, (d) expanding and shrinking structure, (e) inclusion structure. Therefore, we assume there are (a), (b), (c), (d), (e), and other structures. Among which, (a), (b), and (d) can be represented by $m \times n$ format where m stands for the number of parallel series structures from starting point(observing from left to right) and n stands for the number of parallel series structures at the ending point (observing from left to right). Type (a) is $m = n = 1$ (b) is $m = n \geq 2$ and in (d), expanding is the case of $m = 1$ and $n \geq 2$ and shrinking is the case of $m \geq 2$ and $n = 1$. Therefore, the basic types of system are (1) $m \times n$ type, (2) feedback structure, (3) inclusion structure, and (4) other types. Regarding the acting forces of basic structures, five basic transformations can be conducted on them and corresponding inverse transformations and comprehensive transformations can be conducted on them, too.

3. Examples of transformations on system structural acting forces

   In Fig. 6.2, G $=$ (P, E), in the structural chain $v_1, e_{12}, v_2, e_{25}, v_5, e_{58}, v_8, \ldots, e_{89}$, v_9, the "acting forces with respect to distance" is $\{w_i\} = \{50, 55, 52, 45\}$, then $T_{qz}\{w_i\} = \{50, 55, 52, 45\} = \{50, 55, 33, 30\}$ is the displacement transformation conducted on the acting forces with respect to distance in the designated structural chain.

In Fig. 6.2, $G = (P, E)$, in the structural chain $v_1, e_{12}, v_2, e_{25}, v_5, e_{58}, v_8, \ldots, e_{89},$ v_9, the "acting forces with respect to average speed" is $\{w_i\} = \{25, 22, 22, 18\}$, then $T_{qh}\{w_i\} = \{25, 22, 22, 18\} = \{0, 0, 0, 0\}$ is the destruction transformation conducted on the acting forces with respect to average speed in the designated structural chain.

In the aforementioned urban transportation system, the average arrival time along a particular road (structural chain in Fig. 6.2) is $t = \sum \frac{d_i}{v_i}$. Then, $t = \sum \frac{d_i}{v_i}$ $= \frac{50}{25} + \frac{55}{22} + \frac{52}{22} + \frac{45}{18} \approx 9.4$. With the transformation conducted on acting forces with respect to distance, the average arrival time becomes $t = \sum \frac{d_i}{v_i} = \frac{50}{25} + \frac{55}{22} +$ $\frac{33}{22} + \frac{30}{18} \approx 7.7$.

6.1.4 Relevant Research Regarding System Acting Forces

1. Relationship between system structural acting forces and system features;
2. Relationship between system structural acting forces and optimization of system structure;
3. Relationship between system structural acting forces and optimization of system features;
4. Relationship between transformations on system structural acting forces and system features;
5. Relationship between transformations on system structural acting forces and optimization of system;
6. Relationship between five basic transformations on system structural acting forces and system features;
7. Relationship between five basic transformations on system structural acting forces and optimization of system structure;
8. Relationship between five basic transformations on system structural acting forces and optimization of system features;
9. Relationship between five basic transformations on system structural acting forces and optimization of system;
10. Theory of system structural acting forces and its practical applications;
11. Theory, methodology, and applications of system element acting forces;
12. Theory, methodology, and applications of subsystem acting forces.

The optimization of system means the pursuit of global optimum of the system. The global optimization of system denotes the optimization of intrinsic features that are consistent with objective features of the system. The research on optimization of individual feature and global optimization of the system should be addressed separately. For the optimization of individual feature of a system, two cases-i.e., additivity of features and non-additivity of features are discussed in this book. Under the circumstance of additivity of features, relevant laws are investigated and three theorems are put forward accordingly. Regarding the case of non-additivity of features in a system, in the process of system design and management, we should consider the states that the system and its subsystems exhibit when the system attains its its global optimum based on system's structure. Moreover, we should also consider all critical and important subsystems. Last but not the least, in the case that the scenarios for achieving local optimums of different subsystems S_i are different when the system's global optimum is attained, scenario with the minimum total cost should be chosen and implemented.

As for the situation where the features GY_j ($j = 1, 2, \ldots, n$) of system **S** have no additivity with respect to the features GY_{ji} ($i = 1, 2, \ldots, n$) of its subsystem S_i-i.e., it is necessary to know what features GY_{ji} does system **S** require its subsystem S_i to have and the value range of the features GY_{ji} ($i = 1, 2, \ldots, m$) when considering optimizing certain feature GY_i of system **S**. Due to non-additivity on the features GY_{ji} ($i = 1, 2, \ldots, n$) of its subsystem S_i, the relationship between system **S** and its subsystems S_i can be represented by the following equation: $S(GY_j) = S(S_1(GY_{j1} (a_1, b_1)),$ $S_2(GY_{j2} (a_2, b_2)), \ldots, S_i(GY_{ji} (a_i, b_i)), \ldots, S_n(GY_{jn} (a_n, b_n))) = S(S_1(GY_{j1}$ $(a_1(\lambda_1, \lambda_2, \ldots, \lambda_k), b_1(\lambda_1, \lambda_2, \ldots, \lambda_k))), S_2(GY_{j2} (a_2(\lambda_1, \lambda_2, \ldots, \lambda_k), b_2(\lambda_1, \lambda_2, \ldots,$ $\lambda_k))), \ldots, S_i(GY_{ji} (a_i(\lambda_1, \lambda_2, \ldots, \lambda_k), b_i(\lambda_1, \lambda_2, \ldots, \lambda_k))), \ldots, S_n(GY_{jn} (a_n(\lambda_1,$ $\lambda_2, \ldots, \lambda_k), b_n(\lambda_1, \lambda_2, \ldots, \lambda_k))))$, where (a_i, b_i) stands for the value range of the features GY_{ji} ($i = 1, 2, \ldots, m$) of its ith subsystem S_i, $(\lambda_1, \lambda_2, \ldots, \lambda_k)$ is a group of acting foces of system structure. Given the above discussion, a model must be built to incorporate factors of system structure, system acting forces and system features when considering the optimization of system. With our proposed concept of system acting forces, extensive work need to be done to advance relevant in-depth research.

6.2   Acting Forces of System Elements

Based on previous definition on system acting forces, this session discusses the concept of acting forces of system elements. Transformation on acting forces of system elements, relationship between system feature and acting forces of system elements, and relationship between acting forces of system elements and optimization of error system are then discussed. And then, examples are provided to illustrate the above concepts. At last, research issues and contents regarding acting forces of system elements are proposed.

6.2.1   Motivation for Studying Acting Forces of System Elements

There are many different structures in soccer formation tactics such as 4-4-2, 4-3-3, 3-4-3, 3-6-1, 3-4-2-1, 5-4-1, and 4-5-1, etc (Referring to Figs. 6.3, 6.4, 6.5, 6.6, 6.7, 6.8, and 6.9). For instance, in a flat 4-4-2 formation structure (or its variation

Fig. 6.3 Structure of 4-4-2 soccer formation tactics

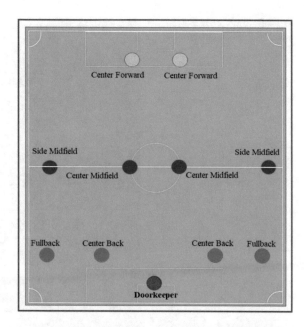

Fig. 6.4 Structure of 4-3-3 soccer formation tactics

Fig. 6.5 Structure of 3-4-3
soccer formation tactics

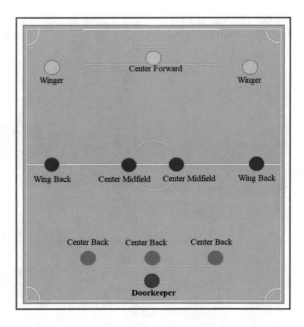

Fig. 6.6 Structure of 3-6-1
soccer formation tactics

Fig. 6.7 Structure of 3-4-2-1 soccer formation tactics

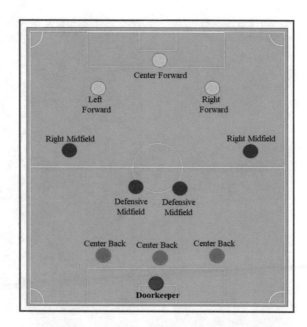

Fig. 6.8 Structure of 5-4-1 soccer formation tactics

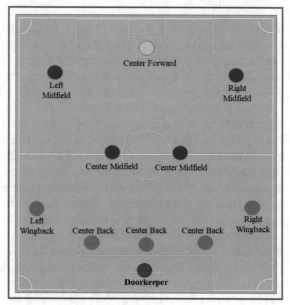

Fig. 6.9 Structure of 4-5-1
soccer formation tactics

Fig. 6.9 Structure of 4-5-1
soccer formation tactics

of diamond 4-4-2), there are four defenders, four midfielders, and two forwards. This is a formation tactics that emphasizes a balance between attacking and defending. For the formation tactics 4-5-1, there are four defenders and five midfielders, and 1 forward, which is a defensive tactics. In the formation tactics 4-3-3, there are four defenders and three midfielders, and three forwards, which is an offensive tactics.

In 4-4-2 formation tactics, the midfielders need to run extensively to provide support to both defense and attack. That is to say that one of the central midfielders has to go up-field to support the attack actions of two forwards and the other central midfielder needs to playing an anchoring role to shield the defense. The left and right midfielders are responsible for moving up along the flanks to the goal line in attacks and protect the full-backs in defense. As far as the team formation system is concerned, what characteristics of team players determine the feature of soccer formation system. Soccer playing techniques are the totalities of all allowed actions and movements used in a game. From the purpose of using them, they are categorized into offensive and defensive techniques. The commonly used techniques include long pass, short pass, scissor pass, volley pass, rolling pass, slide tackle, long drive, grazing shot, off-side trap, diving header, deceptive movement, dribbling, body check, block tackle, overhead kick, chest-high ball, header, and throw-in, etc. As the last defense line, goalkeeper also has many techniques, e.g., finger-tip save and clean catching. In a game, it is very common that one of teams won even both teams adopted the same formation tactics (same structure). This phenomenon indicates that systems' features are different as long as system elements (their valued or acting forces) are different even systems' structures are seemingly the same.

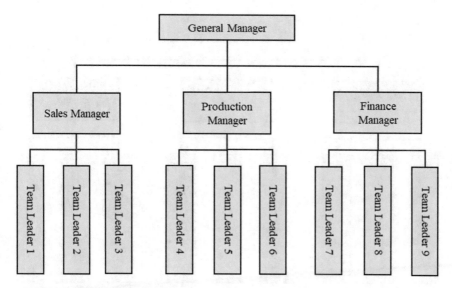

Fig. 6.10 Graphical representation of a typical organization structure

In enterprise management, an organization chart is a simplified management system structure. In a typical organization chart such as Fig. 6.10, it has one general manager and three middle level managers each with four front-line supervisors. Due to the difference in credentials, experiences, charisma, management style and skills, and working attitude of two company's managers and employees, their performance in terms of throughput or profit is different even they are producing the same product using the same process and organization structure.

From the above two examples, we can see how importance system elements play in realizing system's goal (feature) and achieving global optimum. Therefore, for a system of interest, feature-based system optimization is pursued according to values or the range of values of each intrinsic feature under optimal conditions. Having done the optimization for the whole system, system elements are then optimized based on contribution demanded by each optimized feature for the purpose of making the best possible use of things (people or materials).

6.2.2 Concept of Acting Forces of System Elements

Definition 6.14 Acting force of system elements refers to the value w of certain property of system elements; w_{ij} denotes the acting force of jth property of ith system element.

Referring to Fig. 6.11, in the element representation of the management system, the degree to which how well the general manager's style enforces the company's performance-i.e., acting force with respect to management styles is denoted

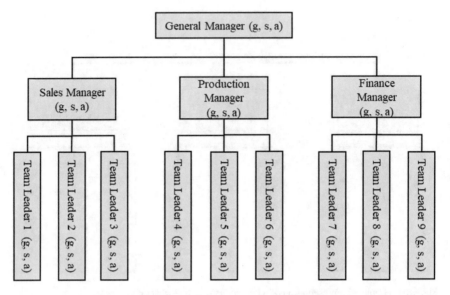

Fig. 6.11 Graphic representation for the elements in a management system

by $g = b$ (b is a numerical value) which is one of properties of elements of the management system. Similarly, for any element in this company, they have corresponding properties such as management style g, management skills (qualifications) $s = \{s_1, s_2, \ldots, s_k\}$, and attitude toward their work, where s_k stands for the kth skill of the selected element (e.g., sales skill, financial management skill, skill in engineering design).

Figure 6.12 illustrates the actual acting forces with respect to the three properties of each element. For convenience, [0, 10] is used to represent corresponding acting force. In practice, it is possible to use [0, 1] or other range expression to represent the acting force. For $s = \{s_1, s_2, \ldots, s_k\}$, the following alternatives, e.g., $s = \min\{s_1, s_2, \ldots, s_k\}$, $s = \max\{s_1, s_2, \ldots, s_k\}$, $s = \sum \frac{s_i}{k}$, or $s = \sum \frac{q_i s_i}{k}$ can be employed according to actual situation.

6.2.3 Transformations on Acting Forces of System Elements

1. Concepts related to transformations on acting forces of system elements

Definition 6.15 Suppose that w_{ij} denotes the acting force of jth property of ith system element, if $T(w_{ij}) = \lambda w_{ij}$, λ is a real number, then T has conducted similarity transformation on w_{ij} denoted by $T_{qx}(w_{ij})$.

In Fig. 6.12, the acting forces with respect to management styles g, management skills $s = \{s_1, s_2, \ldots, s_k\}$, and attitude toward work of sales manager are represented

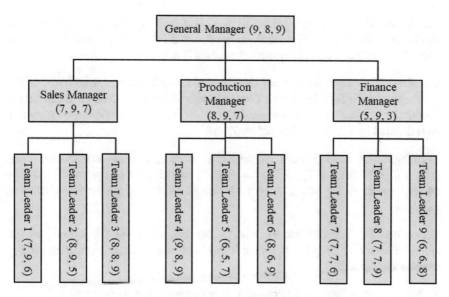

Fig. 6.12 Graphic representation for acting forces of the elements in a management system

by $(g, s, a) = (7, 9, 8)$, if $T_{qx}(7, 9, 8) = (7\lambda, 9\lambda, 8\lambda)$, then T_{qx} has conducted similarity transformation on the relevant acting forces of sales manager.

Definition 6.16 Suppose that w_{ij} denotes the acting force of jth property of ith system element, if $T(w_{ij}) = w'_{ij}$, then T has conducted displacement transformation on w_{ij} denoted by $T_{qz}(w_{ij})$.

Similarly, in Fig. 6.12, the acting forces with respect to management style g, management skills $s = \{s_1, s_2, \ldots, s_k\}$, and attitude toward work of production manager are represented by $(g, s, a) = (8, 9, 7)$, if $T_{qz}(8, 9, 7) = (9, 10, 6)$, then T_{qz} has conducted displacement transformation on the relevant acting forces of production manager.

Definition 6.17 Suppose that w_{ij} denotes the acting force of jth property of ith system element, if $T(w_{ij}) = \{w_{ij1}, w_{ij2}, \ldots, w_{ijk}, k \geq 2\}$, then T has conducted decomposition transformation on w_{ij} denoted by $T_{qf}(w_{ij})$. While $T(\{w_{ij1}, w_{ij2}, \ldots, w_{ijk}, k \geq 2\}) = w_{ij}$, T has done combination transformation on $\{w_{ij1}, w_{ij2}, \ldots, w_{ijk}, k \geq 2\}$ denoted by $T_{qf}^{-1}(\{w_{ij1}, w_{ij2}, \ldots, w_{ijk}, k \geq 2\})$ or denoted by $T_{qzu}(\{w_{ij1}, w_{ij2}, \ldots, w_{ijk}, k \geq 2\})$.

Similarly, in Fig. 6.12, the acting force with respect to management skill is s, $T_{qx}(s) = \{s_1, s_2, \ldots, s_k\}$, then T_{qx} has conducted decomposition transformation on the acting force with respect to management skills of production manager.

Definition 6.18 Suppose that w_{ij} denotes the acting force of jth property of ith system element, if $T(w_{ij}) = w_{ij} + d$, where d is a numerical value, then T has conducted addition transformation on w_{ij} denoted by $T_{qzj}(w_{ij})$. If $T(w_{ij}) = w_{ij}-d$, then T has conducted reduction transformation on w_{ij} denoted by $T_{qjs}(w_{ij})$ (this is actually a special case of addition transformation where d is negative).

Definition 6.19 Suppose that w_{ij} denotes the acting force of jth property of ith system element, if $T(w_{ij}) = 0$, then T has conducted destruction transformation on w_{ij} denoted by $T_{qh}(w_{ij})$. If $T(0) = w_{ij}$, then T has conducted generation transformation on **0** denoted by $T_{qh}^{-1}(0)$ or denoted by $T_{qc}(0)$.

In Fig. 6.12, the acting forces with respect to relevant properties of finance manager is $(g, s, a) = (9, 8, 8)$, if $T_{qh}(9, 8, 8) = (0, 0, 0)$, then T_{qh} has conducted destruction transformation on the acting force with respect to relevant properties of finance manager.

Definition 6.20 Suppose that w_{ij} denotes the acting force of jth property of ith system element, if $T(w_{ij}) = w_{ij}$, then T has conducted unit transformation on w_{ij} denoted by $T_{qd}(w_{ij})$.

2.   Applications of transformations on acting forces of system elements

In Fig. 6.12, among all the acting forces (9, 8, 9) of system element of general manager, acting force with respect to management styles is 9, acting force with respect to management skills is 8, and acting force with respect to attitude toward work is 9. The management skills and attitude toward work are the best of all properties of general manager. Although management skills are not the best and management styles are more important than management skills and attitude toward work. Management skills can be improved given that the general manager has very good attitude toward work. As a result, this system element is qualified for normal operation of this system. By contrast, the acting forces of system element-finance manager are (5, 9, 3). Apparently, he/she is not qualified for the operation of this system. Due to the critical importance of management skills as finance manager, it is not feasible to replace this element in short term. However, with his/her inferior management styles and even worse attitude toward work, this element must be replaced with a more qualified one when chance is ready.

By conducting transformation on system elements and acting force with respect to certain properties of system elements, the system's performance can be apparently improved. Referring to Fig. 6.13, we can see that finance manager, the 2nd production supervisor, and the 1st finance supervisor were replaced with new elements. The attitude toward work of sales manager was apparently improved with appropriate education and training. By conducting expansion transformation on management skills of 3rd production supervisor, 2nd and 3rd production supervisors, their skills were expanded.

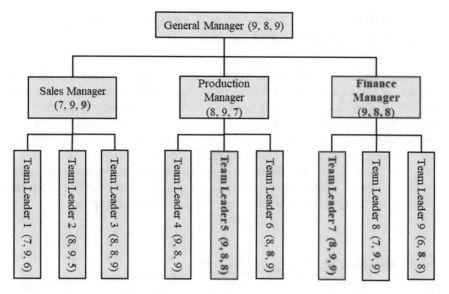

Fig. 6.13 Graphic representation for transformation on acting forces of the elements in a management system

6.2.4 *Relevant Research Regarding Acting Forces of System Elements*

1. Relationship between acting forces of system elements and system features;
2. Relationship between acting forces of system elements and optimization of system structure;
3. Relationship between acting forces of system elements and optimization of system features;
4. Relationship between transformations on acting forces of system elements and system features;
5. Relationship between transformations on acting forces of system elements and optimization of system;
6. Relationship between 5 basic transformations on acting forces of system elements and system features;
7. Relationship between 5 basic transformations on acting forces of system elements and optimization of system structure;
8. Relationship between 5 basic transformations on acting forces of system elements and optimization of system features;
9. Relationship between 5 basic transformations on acting forces of system elements and optimization of system;
10. Theory of acting forces of system elements and its practical applications.

6.3   Relationships Between Acting Forces of System Elements and System Features

In the process of investigating relationships between acting forces of system elements and system structure and system features, there exists certain functional relationship (i.e., $S_G = S(y_{w1}, y_{w2}, \ldots, y_{wn})$ between acting forces of system elements and system features if structure is given. Have done optimization on system structure and normalization on system features and acting forces of system elements, function $S_G = S(y_{w1}, y_{w2}, \ldots, y_{wn})$ has the following properties-i.e., $S_G = S(0, 0, \ldots, 0) = 0$ and $S_G = S(1, 1, \ldots, 1) = 1$. We first propose some basic concepts and laws applied to critical elements, important elements, and acting forces of these elements. Several theorems are then presented regarding acting forces of critical elements.

6.3.1   Motivation on the Investigation

A common commercial airplane is primarily composed of six subsystems i.e., wings, fuselage, tail wings and rudder, landing gears, engines, and control. The major function of the commercial airplane is to safely deliver passengers from departure location to destination, which must be achieved through the synergistic interactions of the six subsystems. During the flying process, airflow on the top of the wing is faster than that of airflow underneath the wing, which generates lower air pressure on the top of the wing and higher air pressure underneath the wing. Therefore, the difference of air pressure produces the lift. Moreover, the spoilers, ailerons, and winglets on the wings offer the function of lift, drag and roll actions. The wings also provide installation locations for turbine engines (2 or 4). The fuselage of an airplane is the backbone to install wing, tail, landing gears, cockpit (control circuit and equipment), and carry passengers and payload as well. The tail wing is used as horizontal stabilizer and to change and control pitch. While the tail rudder is used as vertical stabilizer and to change yaw. Landing gears are used to support airplane on the ground and allow airplane to take off, land, and taxi without damage. Turbine engines generate thrust and also provide power supply for control equipment, air conditioning, and lighting for airplane normal operation. Given the above-mentioned descriptions, the functions of subsystems have essential difference with the system's function of "safely deliver passengers from departure location to destination" and they do not have additivity towards the realization of the whole system's function. The global optimum of this kind of system can not be obtained by simply optimizing its subsystems' features, which needs system **S**'s element to provide certain features or a value range of particular feature. This is to say that each element of system **S** must provide certain acting forces with respect to peculiar properties of each element in order to achieve the global optimum of this system.

Let's recall the example of flashlight system in Chap. 1 (referring to Fig. 1.3). In the flashlight system, light bulb, battery, housing, and switch are all critical elements.

Due to the existence of difference in the acting forces of these elements, the features of system might exhibit difference although these elements are installed in the system structure (same system design).

6.3.2   Relevant Concepts

In the example of Fig. 1.3, the output (Lumen-Lux) of light bulb required by flashlight system is this element's acting force denoted by $w_{lightbulb}$ = [lower limit of Lumen output, upper limit of Lumen output] $\in$ [0, 1], where 0 and 1 are normalized values. While the output (Voltage V) of battery required by flashlight system is this element's acting force denoted by $w_{battery}$ = [lower limit of voltage output, upper limit of voltage output] $\in$ [0, 1], where 0 and 1 are normalized values. By analyzing the relationship between system's features and system structure and system elements, one can obviously know that the features of flashlight can not be realized at all if $w_{lightbulb} = w_{battery} = w_{swithch} = w_{housing} = 0$; the system's features can completely function if $w_{lightbulb} = w_{battery} = w_{swithch} = w_{housing} = w_{reflector} = 1$.

In Fig. 6.11, the degree to which how well the general manager's style enforce the company's performance, i.e., acting force with respect to management styles is denoted by $g = b$ (b is a numerical value) which is one of properties of elements of the management system. Similarly, for any element in this company, they have corresponding properties such as management style g, management skills (qualifications) $s = \{s_1, s_2, \ldots, s_k\}$, and attitude toward their work, where s_k stands for the kth skill of element (e.g., sales skill, financial management skill, skill in engineering design). In our representation, the acting force with respect to certain property of certain element w_{ij} = [lower limit of voltage output, upper limit of voltage output] $\in$ [0, 1], where i stands for the ith element of system **S**, j represents the jth property of the ith element, 0 and 1 are normalized values that stand for the lower and upper limits of the corresponding acting forces. The number in Figs. 6.12 and 6.13 are the original values of their corresponding acting forces (not being normalized). In Figs. 6.14 and 6.15, we use normalized values to represent corresponding acting forces. In Fig. 6.14, for given system structure, all the intrinsic features of the company can not be realized when the acting forces of all system elements are 0. In contrary, for given system structure, all the intrinsic features of the company can be realized when the acting forces of all system elements are 1 (referring to Fig. 6.15).

Theorem 6.1 *In system S, regarding intrinsic feature GY_j of S, removing element y_i or the set of elements $\{y_i, i = 1, 2, \ldots, n\}$ is equivalent to the situation where following situations hold: $w_i = 0$ (the necessary acting force of element for achieving S's intrinsic features) or $\{w_i, i = 1, 2, \ldots, n\} = \{0, 0, \ldots, 0\}$ (necessary acting force set of the element set $\{y_i, i = 1, 2, \ldots, n\}$).*

Proof Proof is omitted here.

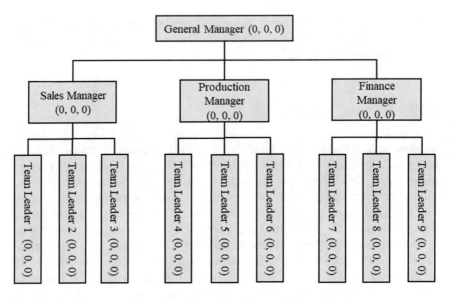

Fig. 6.14 Graphic representation for transformation on acting forces of the elements in a management system-1

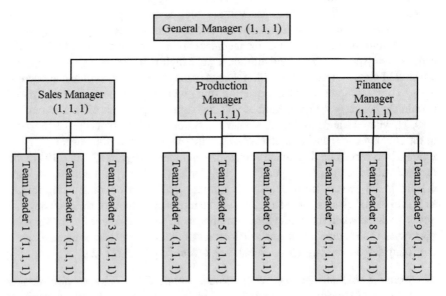

Fig. 6.15 Graphic representation for transformation on acting forces of the elements in a management system-2

6.3.3   Critical and Important Elements in a System

In Chap. 1, we mentioned the definitions of critical and important elements in a system. Suppose that element y_i in a system S is removed, the intrinsic feature GY_j of S can only be partially realized i.e., $[0, a\%]$, where $0 \leq a \leq 100$, system element y_i has the contribution (importance) of $100 - a$ to the intrinsic feature GY_j of S. Especially, when $a = 0$, system element y_i is the critical element to the intrinsic feature GY_j of S. And when $a = 100$, y_i is a surplus(redundant) element to the intrinsic feature GY_j of S. Similarly, suppose that element set $\{y_i, i = 1, 2, \ldots, n\}$ in a system S is removed, the intrinsic feature GY_j of S can only be partially realized i.e., $[0, a\%]$, where $0 \leq a \leq 100$, element set $\{y_i, i = 1, 2, \ldots, n\}$ has the contribution (importance) of $100 - a$ to the intrinsic feature GY_j of S. Especially, when $a = 0$, element set $\{y_i, i = 1, 2, \ldots, n\}$ is the critical element set to the intrinsic feature GY_j of S. And when $a = 100$, $\{y_i, i = 1, 2, \ldots, n\}$ is a surplus (redundant) element set to the intrinsic feature GY_j of S.

Theorem 6.2   *S is a system with critical element $y_i (i = 1, 2, \ldots, m)$, the features GY_j ($j = 1, 2, \ldots, n$) of S by no means can be achieved if error occurring in y_i makes GY_{ji} ($i = 1, 2, \ldots, m; j = 1, 2, \ldots, n$) of element y_i ($i = 1, 2, \ldots, m$) unattainable.*

Proof   Because element y_i is critical to system S and error occurred in y_i ($i = 1, 2, \ldots, m$) makes its feature GY_{ji} ($i = 1, 2, \ldots, m; j = 1, 2, \ldots, n$) unattainable, according to the definition of critical element, the features GY_j ($j = 1, 2, \ldots, n$) of S can by no means be achieved.
Proof is finished!

Theorem 6.3   *S is a system without critical element, the features GY_j ($j = 1, 2, \ldots, n$) of S can be partially achieved if error occurring in any y_i ($i = 1, 2, \ldots, m$) makes GY_{ji} ($i = 1, 2, \ldots, m; j = 1, 2, \ldots, n$) unattainable.*

Proof   According to the definition on critical element, error occurring in any element y_i ($i = 1, 2, \ldots, m$) can not prevent partial realization of S' features although it may affect the feature GY_{ji} ($i = 1, 2, \ldots, m; j = 1, 2, \ldots, n$) of y_i since there is no critical element in system S.
Proof is finished!

Theorem 6.4   *For a system S, there may exist more than one critical element y_i ($i = 1, 2, \ldots, m$).*

Proof   Proof is omitted here!

Theorem 6.5   *S is a system with critical element set $\{y_i\}(i = 1, 2, \ldots, m)$, the features GY_j ($j = 1, 2, \ldots, n$) of S by no means can be achieved if errors occurring in element set $\{y_i\}(i = 1, 2, \ldots, m)$ make GY_{ji} ($i = 1, 2, \ldots, m; j = 1, 2, \ldots, n$) of element set $\{y_i\}(i = 1, 2, \ldots, m)$ unattainable.*

Proof Because element set $\{y_i\}(i = 1, 2, \ldots, m)$ is critical to system **S** and errors occurred in element set $\{y_i\}(i = 1, 2, \ldots, m)$ makes its feature GY_{ji} ($i = 1, 2, \ldots, m$; $j = 1, 2, \ldots, n$) unattainable, according to the definition of critical element, the features GY_j ($j = 1, 2, \ldots, n$) of **S** can by no means be achieved. Proof is finished!

Theorem 6.6 *S is a system without critical element set $\{y_i\}(i = 1, 2, \ldots, m)$, the features GY_j ($j = 1, 2, \ldots, n$) of S can be partially achieved if errors occurring in any element set $\{y_i\}(i = 1, 2, \ldots, m)$ make GY_i ($i = 1, 2, \ldots, m$) unattainable.*

Proof According to the definition on critical element, errors occurring in any element y_i ($i = 1, 2, \ldots, m$) can not prevent partial realization of **S'** features although it may affect the feature GY_i ($i = 1, 2, \ldots, m$) of y_i since there is no critical element set in system **S**. Proof is finished!

6.3.4 Function for Feature Acting Force of System Element

Having done system optimization, by normalizing the element feature required by optimized system **S**, lower and upper limits are obtained. The acting force of that system element is some value falling in the range [lower limit of element feature, upper limit of element feature], i.e., $w_i \in [0, 1]$. In the example of flashlight, the output (Lumen-Lux) of light bulb required by flashlight system is this element's acting force denoted by $w_{lightbulb} = $ [lower limit of Lumen output, upper limit of Lumen output] $\in$ [0, 1], where 0 and 1 are normalized values. While the output (Voltage V) of battery required by flashlight system is this element's acting force denoted by $w_{battery} = $ [lower limit of voltage output, upper limit of voltage output] $\in [0, 1]$, where 0 and 1 are normalized values. The illumination feature of flashlight $GY_{illumination} = 0$ if any equation regarding the following acting forces holds $w_{housing} = 0$ (electrical conductivity of housing), $w_{swich} = 0$ (reliability of switch), $w_{lightbulb} = 0$ (output of light bulb), and $w_{battery} = 0$ (voltage output of battery) if system structure is kept unchanged. It means all of aforementioned elements are critical to the system. The feature acting force of flashlight $w_{GY_{illumination}} = w_{housing} \wedge w_{swich} \wedge w_{lightbulb} \wedge w_{battery}$. The maximum output (Lumen) of flashlight $w_{GY_{illumination}} = w_{lightbulb}$ if the reflector is removed. It means that reflector is an important instead of critical element. Therefore, the feature acting force of flashlight $w_{GY_{illumination}} = w_{housing} \wedge w_{swich} \wedge w_{lightbulb} \wedge w_{battery} \times (1 + w_{reflector})$. It is assumed that $w_{housing} = 0.9$, $w_{swich} = 0.8$, $w_{lightbulb} = 0.7$, $w_{battery} = 0.99$, and $w_{reflector} = 0.6$. Then $w_{GY_{illumination}} = (0.9 \wedge 0.8 \wedge 0.7 \wedge 0.99) \times (1 + 0.6) = 0.7 \times 1.6 = 1.12$. It states that the addition of reflector to flashlight actually increases the Lumen output of flashlight. The feature acting force of flashlight $w_{GY_{illumination}} = 0$ if any acting force among $w_{housing}$, w_{swich}, $w_{lightbulb}$, $w_{battery}$, and $w_{reflector}$ is **0**.

In order to build the function for feature acting force of system element, it is necessary to know the value range of features GY_{ji} ($i = 1, 2, \ldots, m, j = 1, 2,$

..., n; where j represents the jth feature of system S and i stands for ith system element, and GY_{ji} stands for the feature value of element y_i required by the system feature G_i when optimization on S is completed.) required by S after optimization is conducted on system S. The next step is to analyze the functional relationship between system structural acting forces, acting forces element, and feature acting forces. In the above flashlight example, there exists certain functional relationship between system feature and acting force of system element-i.e., system feature $S_G = S(y_{w1}, y_{w2}, \ldots, y_{wn})$. Have done optimization on system structure and normalization on system features and acting forces of system elements, function $S_G = S(y_{w1}, y_{w2}, \ldots, y_{wn})$ has the following properties-i.e., $S_G = S(0, 0, \ldots, 0) = 0$ and $S_G = S(1, 1, \ldots, 1) = 1$.

6.3.5   Conclusion

The system optimization dictates the process that global optimum of system is achieved and the totality of system features is optimized, too. The system optimization also means that all intrinsic features aligned with objective features of the system are optimized. For the purpose of investigating system global optimization, it is essential to examine system structure, system element, optimization on system structure and element. The research on acting forces of system element provides an ideal theoretical framework for studying system global optimization. There exists certain functional relationship between system feature and acting force of system element-i.e., system feature $S_G = S(y_{w1}, y_{w2}, \ldots, y_{wn})$ when system structure is given. Have done optimization on system structure and normalization on system features and acting forces of system elements, function $S_G = S(y_{w1}, y_{w2}, \ldots, y_{wn})$ has the following properties-i.e., $S_G = S(0, 0, \ldots, 0) = 0$ and $S_G = S(1, 1, \ldots, 1) = 1$. The theory of acting forces of system structure and element offers a different perspective for enhancing the research on error systems, which enables researchers and practitioners to study, analyze, prevent, and eliminate errors in complicated socioeconomic systems in a quantitative approach.

Chapter 7
Transformations on Universe of Discourse and Rules in Error Systems

Abstract This chapter chiefly introduces the transformations on universe of discourse and rules for judging error in error systems. The universe of discourse (domain) here states the permitted range of objects of interest. In our research regarding error logic, error set, and error system, there are several types of domains: (1) the universe of discourse for error set $C = \{(u, x)| x = f(G, u), u \in U, f \subseteq U \times \mathbb{R}\}$ is U_C; (2) the range defined by error system $\mathbf{S} = S(U(t), JG(t), Z(t), YS(t), G(t))$ is called system 's domain; (3) in error system $\mathbf{S}$, the domain for structure JG(t) is U_{jg}; (4) in error system $\mathbf{S}$, the domain for subsystems Z(t) is U_z; (5) in error system $\mathbf{S}$, the domain for system elements YS(t) is U_{ys}; and (6) in error system $\mathbf{S}$, the domain for rules for judging error G(t) is U_g. Universe of discourse (domain) plays a critically important role not only in the research of error set and error system but also in the process of error elimination. Many errors can be eliminated by changing the domain to which they pertain. From the definition on transformations, changes on universe of discourse and rules can lead to the changes in conclusion. For instance, regarding the addition of two numbers e.g. $1 + 1 = 1$, under rules and corresponding domain in the field of binary operation $1 + 1 = 1$ is wrong and it is correct under the rules and corresponding domain in the field of boolean algebra. With changed conclusion, it can inversely lead to the changes in rules for judging errors. In this chapter, we first investigate transformations of individual factors since there exists circular relationship between cause and effect. Then we consider the methods and laws for simultaneous transformations of multiple factors.

7.1 Transformations on Domain in Error Systems

7.1.1 Concept of Transformations on Universe of Discourse

Definition 7.1 Suppose that U is a real number set, a and b are lower and upper limits of U, respectively, if $T_{Lx}(U) = K^{+}U = U'$, U and U' have the same lower limits; upper limit of U' is $b' = b + (k-1)(b-a)$, k is constant, then T_{Lx} has conducted positive similarity transformation on universe of discourse U.

© Springer Nature Switzerland AG 2020

K. Guo and S. Liu, *Error Systems: Concepts, Theory and Applications*,
Studies in Systems, Decision and Control 275,
https://doi.org/10.1007/978-3-030-40760-5_7

Definition 7.2 Suppose that U is a real number set, a and b are lower and upper limits of U, respectively, if $T_{Lx}(U) = K^-U = U'$, U and U' have the same upper limits; lower limit of U' is $a' = a - (k-1)(b-a)$, k is constant, then T_{Lx} has conducted negative similarity transformation on universe of discourse U.

Definition 7.3 Suppose that U is a real number set, a and b are lower and upper limits of U, respectively, if $T_{Lx}(U) = KU = U'$, the lower limit of U' is $a' = a-(k-1)(b-a)$, the upper limit of U' is $b' = b + (k-1)(b-a)$, k is constant, then T_{Lx} has conducted similarity transformation on universe of discourse U.

Definition 7.4 Suppose that U is a real number set, if $T_{Lx}(U) = KU = U'$, K is constant, acting force of U is K times of U', then T_{Lx} has conducted similarity transformation on universe of discourse U.

Definition 7.5 Suppose that U is a real number set, if $T_{Lzj}(U) = U \cup U'$, then T_{Lzj} has conducted addition transformation on universe of discourse U.

Definition 7.6 Suppose that U is a real number set, if $T_{Lf}(U) = U_1 \cup U_2 \cup \ldots \cup U_n$, where $U = U_1 \cup U_2 \cup \ldots \cup U_n$, then T_{Lf} has conducted decomposition transformation on universe of discourse U.

Definition 7.7 Suppose that U is a real number set, if $T_{Lz}(U) = U'$, then T_{Lz} has conducted displacement transformation on universe of discourse U.

Definition 7.8 Suppose that U is a real number set, if $T_{Lh}(U) = \Phi$, then T_{Lh} has conducted destruction transformation on universe of discourse U.

Definition 7.9 Suppose that U is a real number set, if $T_{Ly}(U) = U$, then T_{Ly} has conducted unit transformation on universe of discourse U.

7.1.2　Transformations on Universe of Discourse

Suppose that the object system of interest is denoted by $S = S(U(t), JG(t), Z(t), YS(t), G(t))$, G is rules for judging error, transformation operator $T = (T_{ly}, T_{jg}, T_z, T_{ys}, T_g, T_{sj})$, where $T_{ly}, T_{jg}, T_z, T_{ys}, T_g$, and T_{sj} represent corresponding transformations on domain, structure, subsystem, system element, rule, and time in the designed object system, respectively. The position where the transformation connective is placed indicates corresponding transformation. For example, $T = (T_{ly}, T_{jg}, T_z, T_{ys}, T_g, T_{sj})$ is denoted by $T = (T_{Lx}, T_{Lx}, T_{Lx}, T_{Lx}, T_{Lx}, T_{Lx})$ if similarity transformations are conducted on domain, structure, subsystem, system element, rule, and time at the same time.

1. Similarity transformation on universe of discourse
   Let $T = (T_{Lx}, y, y, y, y, y)$, then
   $T(S) = S(T_{Lx}(U(t)), JG(t), Z(t), YS(t), G(t))$
   Let $T = (y, T_{Lx}, y, y, y, y)$, then
   $T(S) = S(U(t), T_{Lx}(JG(t)), Z(t), YS(t), G(t))$

$$\vdots$$
$$\vdots$$

Let T = (y, y, y, y, T_{Lx}, y), then
T(S) = S(U(t), JG(t), Z(t), YS(t), T_{Lx}(G(t)))
Let T = (y, y, y, y, y, T_{Lx}), then
T(S) = S(U(T_{Lx}(t)), JG(T_{Lx}(t)), Z(T_{Lx}(t)), YS(T_{Lx}(t)), G(T_{Lx}(t)))

$$\cdots\cdots\cdots\cdots\cdots\cdots\cdots\cdots\cdots\cdots\cdots\cdots$$

Let T = (T_{Lx}, T_{Lx}, y, y, y, y), then
T(S) = S(T_{Lx}(U(t)), T_{Lx}(JG(t)), Z(t), YS(t), G(t))
Let T = (y, T_{Lx}, T_{Lx}, y, y, y), then
T(S) = S(U(t), T_{Lx}(JG(t)), T_{Lx}(Z(t)), YS(t), G(t))

$$\vdots$$
$$\vdots$$

Let T = (y, y, y, T_{Lx}, T_{Lx}, y), then
T(S) = S(U(t), JG(t), Z(t), T_{Lx}(YS(t)), T_{Lx}(G(t)))
Let T = (y, y, y, y, T_{Lx}, T_{Lx}), then
T(S) = S(U(T_{Lx}(t)), JG(T_{Lx}(t)), Z(T_{Lx}(t)), YS(T_{Lx}(t)), T_{Lx}(G(T_{Lx}(t))))

$$\cdots\cdots\cdots\cdots\cdots\cdots\cdots\cdots\cdots\cdots\cdots\cdots$$

Let T = (T_{Lx}, T_{Lx}, T_{Lx}, y, y, y), then
T(S) = S(T_{Lx}(U(t)), T_{Lx}(JG(t)), T_{Lx}(Z(t)), YS(t), G(t))
Let T = (y, T_{Lx}, T_{Lx}, T_{Lx}, y, y), then
T(S) = S(U(t), T_{Lx}(JG(t)), T_{Lx}(Z(t)), T_{Lx}(YS(t)), G(t))

$$\vdots$$
$$\vdots$$

Let T = (y, y, T_{Lx}, T_{Lx}, T_{Lx}, y), then
T(S) = S(U(t), JG(t), T_{Lx}(Z(t)), T_{Lx}(YS(t)), T_{Lx}(G(t)))
Let T = (y, y, y, T_{Lx}, T_{Lx}, T_{Lx}), then
T(S) = S(U(T_{Lx}(t)), JG(T_{Lx}(t)), Z(T_{Lx}(t)), T_{Lx}(YS(T_{Lx}(t))), T_{Lx}(G(T_{Lx} (t))))

$$\cdots\cdots\cdots\cdots\cdots\cdots\cdots\cdots\cdots\cdots\cdots\cdots$$

Let T = (T_{Lx}, T_{Lx}, T_{Lx}, T_{Lx}, y, y), then
T(S) = S(T_{Lx}(U(t)), T_{Lx}(JG(t)), T_{Lx}(Z(t)), T_{Lx}(YS(t)), G(t))

Let $T = (y, T_{Lx}, T_{Lx}, T_{Lx}, T_{Lx}, y)$, then
$$T(S) = S(U(t), T_{Lx}(JG(t)), T_{Lx}(Z(t)), T_{Lx}(YS(t)), T_{Lx}(G(t)))$$

$$\vdots$$

Let $T = (y, y, T_{Lx}, T_{Lx}, T_{Lx}, T_{Lx})$, then
$$T(S) = S(U(T_{Lx}(t)), JG(T_{Lx}(t)), T_{Lx}(Z(T_{Lx}(t))), T_{Lx}(YS(T_{Lx}(t))), T_{Lx}(G(T_{Lx}(t))))$$

$$\cdots\cdots\cdots\cdots\cdots\cdots\cdots\cdots\cdots\cdots\cdots\cdots$$

Let $T = (T_{Lx}, T_{Lx}, T_{Lx}, T_{Lx}, T_{Lx}, y)$, then
$$T(S) = S(T_{Lx}(U(t)), T_{Lx}(JG(t)), T_{Lx}(Z(t)), T_{Lx}(YS(t)), G(t))$$

$$\vdots$$

Let $T = (y, T_{Lx}, T_{Lx}, T_{Lx}, T_{Lx}, T_{Lx})$, then
$$T(S) = S(U(T_{Lx}(t)), T_{Lx}(JG(T_{Lx}(t))), T_{Lx}(Z(T_{Lx}(t))), T_{Lx}(YS(T_{Lx}(t))), T_{Lx}(G(T_{Lx}(t))))$$

$$\cdots\cdots\cdots\cdots\cdots\cdots\cdots\cdots\cdots\cdots\cdots\cdots$$

Let $T = (T_{Lx}, T_{Lx}, T_{Lx}, T_{Lx}, T_{Lx}, T_{Lx})$, then
$$T(S) = S(T_{Lx}(U(T_{Lx}(t))), T_{Lx}(JG(T_{Lx}(t))), T_{Lx}(Z(T_{Lx}(t))), T_{Lx}(YS(T_{Lx}(t))), T_{Lx}(G(T_{Lx}(t))))$$

2. Addition transformation on universe of discourse

Let $T = (T_{Lzj}, y, y, y, y, y)$, then
$$T(S) = S(T_{Lzj}(U(t)), JG(t), Z(t), YS(t), G(t))$$
Let $T = (y, T_{Lzj}, y, y, y, y)$, then
$$T(S) = S(U(t), T_{Lzj}(JG(t)), Z(t), YS(t), G(t))$$

$$\vdots$$

Let $T = (y, y, y, y, T_{Lzj}, y)$, then
$$T(S) = S(U(t), JG(t), Z(t), YS(t), T_{Lzj}(G(t)))$$
Let $T = (y, y, y, y, y, T_{Lzj})$, then
$$T(S) = S(U(T_{Lzj}(t)), JG(T_{Lzj}(t)), Z(T_{Lzj}(t)), YS(T_{Lzj}(t)), G(T_{Lzj}(t)))$$

$$\cdots\cdots\cdots\cdots\cdots\cdots\cdots\cdots\cdots\cdots\cdots\cdots$$

Let T $= (T_{Lzj}, T_{Lzj}, y, y, y, y)$, then
T(S) $= S(T_{Lzj}(U(t)), T_{Lzj}(JG(t)), Z(t), YS(t), G(t))$
Let T $= (y, T_{Lzj}, T_{Lzj}, y, y, y)$, then
T(S) $= S(U(t), T_{Lzj}(JG(t)), T_{Lzj}(Z(t)), YS(t), G(t))$

$$\vdots$$
$$\vdots$$

Let T $= (y, y, y, T_{Lzj}, T_{Lzj}, y)$, then
T(S) $= S(U(t), JG(t), Z(t), T_{Lzj}(YS(t)), T_{Lzj}(G(t)))$
Let T $= (y, y, y, y, T_{Lzj}, T_{Lzj})$, then
T(S) $= S(U(T_{Lzj}(t)), JG(T_{Lzj}(t)), Z(T_{Lzj}(t)), YS(T_{Lzj}(t)), T_{Lzj}(G(T_{Lzj}\ (t))))$

$$\cdots\cdots\ \cdots\cdots\cdots\cdots\cdots\cdots\cdots\cdots\cdots\cdots\cdots$$

Let T $= (T_{Lzj}, T_{Lzj}, T_{Lzj}, y, y, y)$, then
T(S) $= S(T_{Lzj}(U(t)), T_{Lzj}(JG(t)), T_{Lzj}(Z(t)), YS(t), G(t))$
Let T $= (y, T_{Lzj}, T_{Lzj}, T_{Lzj}, y, y)$, then
T(S) $= S(U(t), T_{Lzj}(JG(t)), T_{Lzj}(Z(t)), T_{Lzj}(YS(t)), G(t))$

$$\vdots$$
$$\vdots$$

Let T $= (y, y, T_{Lzj}, T_{Lzj}, T_{Lzj}, y)$, then
T(S) $= S(U(t), JG(t), T_{Lzj}(Z(t)), T_{Lzj}(YS(t)), T_{Lzj}(G(t)))$
Let T $= (y, y, y, T_{Lzj}, T_{Lzj}, T_{Lzj})$, then
T(S) $= S(U(T_{Lzj}(t)), JG(T_{Lzj}(t)), Z(T_{Lzj}(t)), T_{Lzj}(YS(T_{Lzj}(t))), T_{Lzj}(G\ (T_{Lzj}(t))))$

$$\cdots\cdots\ \cdots\cdots\cdots\cdots\cdots\cdots\cdots\cdots\cdots\cdots\cdots$$

Let T $= (T_{Lzj}, T_{Lzj}, T_{Lzj}, T_{Lzj}, y, y)$, then
T(S) $= S(T_{Lzj}(U(t)), T_{Lzj}(JG(t)), T_{Lzj}(Z(t)), T_{Lzj}(YS(t)), G(t))$
Let T $= (y, T_{Lzj}, T_{Lzj}, T_{Lzj}, T_{Lzj}, y)$, then
T(S) $= S(U(t), T_{Lzj}(JG(t)), T_{Lzj}(Z(t)), T_{Lzj}(YS(t)), T_{Lzj}(G(t)))$

$$\vdots$$
$$\vdots$$

Let T $= (y, y, T_{Lzj}, T_{Lzj}, T_{Lzj}, T_{Lzj})$, then
T(S) $= S(U(T_{Lzj}(t)), JG(T_{Lzj}(t)), T_{Lzj}(Z(T_{Lzj}(t))), T_{Lzj}(YS(T_{Lzj}(t))), T_{Lzj}(G(T_{Lzj}(t))))$

$$\cdots\cdots\ \cdots\cdots\cdots\cdots\cdots\cdots\cdots\cdots\cdots\cdots\cdots$$

Let T = $(T_{Lzj}, T_{Lzj}, T_{Lzj}, T_{Lzj}, T_{Lzj}, y)$, then
T(S) = $S(T_{Lzj}(U(t)), T_{Lzj}(JG(t)), T_{Lzj}(Z(t)), T_{Lzj}(YS(t)), G(t))$

$$\vdots$$
$$\vdots$$

Let T = $(y, T_{Lzj}, T_{Lzj}, T_{Lzj}, T_{Lzj}, T_{Lzj})$, then
T(S) = $S(U(T_{Lzj}(t)), T_{Lzj}(JG(T_{Lzj}(t))), T_{Lzj}(Z(T_{Lzj}(t))), T_{Lzj}(YS(T_{Lzj}(t))), T_{Lzj}(G(T_{Lzj}(t))))$

$\cdots\cdots \cdots\cdots\cdots\cdots\cdots\cdots\cdots\cdots\cdots\cdots\cdots$

Let T = $(T_{Lzj}, T_{Lzj}, T_{Lzj}, T_{Lzj}, T_{Lzj}, T_{Lzj})$, then
T(S) = $S(T_{Lzj}(U(T_{Lzj}(t))), T_{Lzj}(JG(T_{Lzj}(t))), T_{Lzj}(Z(T_{Lzj}(t))), T_{Lzj}(YS(T_{Lzj}(t))),$
$T_{Lzj}(G(T_{Lzj}(t))))$

3. Displacement transformation on universe of discourse

Let T = (T_{Lz}, y, y, y, y, y), then
T(S) = $S(T_{Lz}(U(t)), JG(t), Z(t), YS(t), G(t))$
Let T = (y, T_{Lz}, y, y, y, y), then
T(S) = $S(U(t), T_{Lz}(JG(t)), Z(t), YS(t), G(t))$

$$\vdots$$
$$\vdots$$

Let T = (y, y, y, y, T_{Lz}, y), then
T(S) = $S(U(t), JG(t), Z(t), YS(t), T_{Lz}(G(t)))$
Let T = (y, y, y, y, y, T_{Lz}), then
T(S) = $S(U(T_{Lz}(t)), JG(T_{Lz}(t)), Z(T_{Lz}(t)), YS(T_{Lz}(t)), G(T_{Lz}(t)))$

$\cdots\cdots \cdots\cdots\cdots\cdots\cdots\cdots\cdots\cdots\cdots\cdots\cdots$

Let T = $(T_{Lz}, T_{Lz}, y, y, y, y)$, then
T(S) = $S(T_{Lz}(U(t)), T_{Lz}(JG(t)), Z(t), YS(t), G(t))$

$$\vdots$$
$$\vdots$$

Let T = $(y, y, y, T_{Lz}, T_{Lz}, y)$, then
T(S) = $S(U(t), JG(t), Z(t), T_{Lz}(YS(t)), T_{Lz}(G(t)))$

Let T $= (y, y, y, y, T_{Lz}, T_{Lz})$, then
T(S) $= S(U(T_{Lz}(t)), JG(T_{Lz}(t)), Z(T_{Lz}(t)), YS(T_{Lz}(t)), T_{Lz}(G(T_{Lz}(t))))$

....

Let T $= (T_{Lz}, T_{Lz}, T_{Lz}, y, y, y)$, then
T(S) $= S(T_{Lz}(U(t)), T_{Lz}(JG(t)), T_{Lz}(Z(t)), YS(t), G(t))$
Let T $= (y, T_{Lz}, T_{Lz}, T_{Lz}, y, y)$, then
T(S) $= S(U(t), T_{Lz}(JG(t)), T_{Lz}(Z(t)), T_{Lz}(YS(t)), G(t))$

$$\vdots$$

Let T $= (y, y, T_{Lz}, T_{Lz}, T_{Lz}, y)$, then
T(S) $= S(U(t), JG(t), T_{Lz}(Z(t)), T_{Lz}(YS(t)), T_{Lz}(G(t)))$
Let T $= (y, y, y, T_{Lz}, T_{Lz}, T_{Lz})$, then
T(S) $= S(U(T_{Lz}(t)), JG(T_{Lz}(t)), Z(T_{Lz}(t)), T_{Lz}(YS(T_{Lz}(t))), T_{Lz}(G(T_{Lz}(t))))$

....

Let T $= (T_{Lz}, T_{Lz}, T_{Lz}, T_{Lz}, y, y)$, then
T(S) $= S(T_{Lz}(U(t)), T_{Lz}(JG(t)), T_{Lz}(Z(t)), T_{Lz}(YS(t)), G(t))$
Let T $= (y, T_{Lz}, T_{Lz}, T_{Lz}, T_{Lz}, y)$, then
T(S) $= S(U(t), T_{Lz}(JG(t)), T_{Lz}(Z(t)), T_{Lz}(YS(t)), T_{Lz}(G(t)))$

$$\vdots$$

Let T $= (y, y, T_{Lz}, T_{Lz}, T_{Lz}, T_{Lz})$, then
T(S) $= S(U(T_{Lz}(t)), JG(T_{Lz}(t)), T_{Lz}(Z(T_{Lz}(t))), T_{Lz}(YS(T_{Lz}(t))), T_{Lz}(G(T_{Lz}(t))))$

....

Let T $= (T_{Lz}, T_{Lz}, T_{Lz}, T_{Lz}, T_{Lz}, y)$, then
T(S) $= S(T_{Lz}(U(t)), T_{Lz}(JG(t)), T_{Lz}(Z(t)), T_{Lz}(YS(t)), T_{Lz}G(t)))$

$$\vdots$$

Let T $= (y, T_{Lz}, T_{Lz}, T_{Lz}, T_{Lz}, T_{Lz})$, then
T(S) $= S(U(T_{Lz}(t)), T_{Lz}(JG(T_{Lz}(t))), T_{Lz}(Z(T_{Lz}(t))), T_{Lz}(YS(T_{Lz}(t))), T_{Lz}(G(T_{Lz}(t))))$

....

Let T $= (T_{Lz}, T_{Lz}, T_{Lz}, T_{Lz}, T_{Lz}, T_{Lz})$, then
T(S) $=$ S(T_{Lz}(U(T_{Lz}(t))), T_{Lz}(JG(T_{Lz}(t))), T_{Lz}(Z(T_{Lz}(t))), T_{Lz}(YS(T_{Lz}(t))), T_{Lz}(G(T_{Lz}(t))))

4. Decomposition transformation on universe of discourse

Let T $= (T_{Lf}, y, y, y, y, y)$, then
T(S) = S(T_{Lf}(U(t)), JG(t), Z(t), YS(t), G(t))
Let T $= (y, T_{Lf}, y, y, y, y)$, then
T(S) = S(U(t), T_{Lf}(JG(t)), Z(t), YS(t), G(t))

$$\vdots$$

Let T $= (y, y, y, y, T_{Lf}, y)$, then
T(S) = S(U(t), JG(t), Z(t), YS(t), T_{Lf}(G(t)))
Let T $= (y, y, y, y, y, T_{Lf})$, then
T(S) = S(U(T_{Lf}(t)), JG(T_{Lf}(t)), Z(T_{Lf}(t)), YS(T_{Lf}(t)), G(T_{Lf}(t)))

. . . .

Let T $= (T_{Lf}, T_{Lf}, y, y, y, y)$, then
T(S) = S(T_{Lf}(U(t)), T_{Lf}(JG(t)), Z(t), YS(t), G(t))
Let T $= (y, T_{Lf}, T_{Lf}, y, y, y)$, then
T(S) = S(U(t), T_{Lf}(JG(t)), T_{Lf}(Z(t)), YS(t), G(t))

$$\vdots$$

Let T $= (y, y, y, T_{Lf}, T_{Lf}, y)$, then
T(S) = S(U(t), JG(t), Z(t), T_{Lf}(YS(t)), T_{Lf}(G(t)))
Let T $= (y, y, y, y, T_{Lf}, T_{Lf})$, then
T(S) = S(U(T_{Lf}(t)), JG(T_{Lf}(t)), Z(T_{Lf}(t)), YS(T_{Lf}(t)), T_{Lf}(G(T_{Lf}(t))))

. . . .

Let T $= (T_{Lf}, T_{Lf}, T_{Lf}, y, y, y)$, then
T(S) = S(T_{Lf}(U(t)), T_{Lf}(JG(t)), T_{Lf}(Z(t)), YS(t), G(t))
Let T $= (y, T_{Lf}, T_{Lf}, T_{Lf}, y, y)$, then
T(S) = S(U(t), T_{Lf}(JG(t)), T_{Lf}(Z(t)), T_{Lf}(YS(t)), G(t))

$$\vdots$$

Let T $= (y, y, T_{Lf}, T_{Lf}, T_{Lf}, y)$, then
T(S) $= S(U(t), JG(t), T_{Lf}(Z(t)), T_{Lf}(YS(t)), T_{Lf}(G(t)))$
Let T $= (y, y, y, T_{Lf}, T_{Lf}, T_{Lf})$, then
T(S) $= S(U(T_{Lf}(t)), JG(T_{Lf}(t)), Z(T_{Lf}(t)), T_{Lf}(YS(T_{Lf}(t))), T_{Lf}(G(T_{Lf}(t))))$

$$\cdots\cdots\cdots\cdots\cdots\cdots\cdots\cdots\cdots\cdots\cdots\cdots\cdots\cdots\cdots\cdots$$

Let T $= (T_{Lf}, T_{Lf}, T_{Lf}, T_{Lf}, y, y)$, then
T(S) $= S(T_{Lf}(U(t)), T_{Lf}(JG(t)), T_{Lf}(Z(t)), T_{Lf}(YS(t)), G(t))$
Let T $= (y, T_{Lf}, T_{Lf}, T_{Lf}, T_{Lf}, y)$, then
T(S) $= S(U(t), T_{Lf}(JG(t)), T_{Lf}(Z(t)), T_{Lf}(YS(t)), T_{Lf}(G(t)))$

$$\vdots$$

Let T $= (y, y, T_{Lf}, T_{Lf}, T_{Lf}, T_{Lf})$, then
T(S) $= S(U(T_{Lf}(t)), JG(T_{Lf}(t)), T_{Lf}(Z(T_{Lf}(t))), T_{Lf}(YS(T_{Lf}(t))), T_{Lf}(G(T_{Lf}(t))))$

$$\cdots\cdots\cdots\cdots\cdots\cdots\cdots\cdots\cdots\cdots\cdots\cdots\cdots\cdots\cdots\cdots$$

Let T $= (T_{Lf}, T_{Lf}, T_{Lf}, T_{Lf}, T_{Lf}, y)$, then
T(S) $= S(T_{Lf}(U(t)), T_{Lf}(JG(t)), T_{Lf}(Z(t)), T_{Lf}(YS(t)), T_{Lf}G(t)))$

$$\vdots$$

Let T $= (y, T_{Lf}, T_{Lf}, T_{Lf}, T_{Lf}, T_{Lf})$, then
T(S) $= S(U(T_{Lf}(t)), T_{Lf}(JG(T_{Lf}(t))), T_{Lf}(Z(T_{Lf}(t))), T_{Lf}(YS(T_{Lf}(t))),$
$T_{Lf}(G(T_{Lf}(t))))$

$$\cdots\cdots\cdots\cdots\cdots\cdots\cdots\cdots\cdots\cdots\cdots\cdots\cdots\cdots\cdots\cdots$$

Let T $= (T_{Lf}, T_{Lf}, T_{Lf}, T_{Lf}, T_{Lf}, T_{Lf})$, then
T(S) $= S(T_{Lf}(U(T_{Lf}(t))), T_{Lf}(JG(T_{Lf}(t))), T_{Lf}(Z(T_{Lf}(t))), T_{Lf}(YS(T_{Lf}(t))),$
$T_{Lf}(G(T_{Lf}(t))))$

5. Destruction transformation on universe of discourse

Let T $= (T_{Lh}, y, y, y, y, y)$, then
T(S) $= S(T_{Lh}(U(t)), JG(t), Z(t), YS(t), G(t))$
Let T $= (y, T_{Lh}, y, y, y, y)$, then
T(S) $= S(U(t), T_{Lh}(JG(t)), Z(t), YS(t), G(t))$

$$\vdots$$
$$\vdots$$

Let $T = (y, y, y, y, T_{Lh}, y)$, then
$T(S) = S(U(t), JG(t), Z(t), YS(t), T_{Lh}(G(t)))$
Let $T = (y, y, y, y, y, T_{Lh})$, then
$T(S) = S(U(T_{Lh}(t)), JG(T_{Lh}(t)), Z(T_{Lh}(t)), YS(T_{Lh}(t)), G(T_{Lh}(t)))$

. . . .

Let $T = (T_{Lh}, T_{Lh}, y, y, y, y)$, then
$T(S) = S(T_{Lh}(U(t)), T_{Lh}(JG(t)), Z(t), YS(t), G(t))$
Let $T = (y, T_{Lh}, T_{Lh}, y, y, y)$, then
$T(S) = S(U(t), T_{Lh}(JG(t)), T_{Lh}(Z(t)), YS(t), G(t))$

$$\vdots$$
$$\vdots$$

Let $T = (y, y, y, T_{Lh}, T_{Lh}, y)$, then
$T(S) = S(U(t), JG(t), Z(t), T_{Lh}(YS(t)), T_{Lh}(G(t)))$
Let $T = (y, y, y, y, T_{Lh}, T_{Lh})$, then
$T(S) = S(U(T_{Lh}(t)), JG(T_{Lh}(t)), Z(T_{Lh}(t)), YS(T_{Lh}(t)), T_{Lh}(G(T_{Lh}(t))))$

. . . .

Let $T = (T_{Lh}, T_{Lh}, T_{Lh}, y, y, y)$, then
$T(S) = S(T_{Lh}(U(t)), T_{Lh}(JG(t)), T_{Lh}(Z(t)), YS(t), G(t))$
Let $T = (y, T_{Lh}, T_{Lh}, T_{Lh}, y, y)$, then
$T(S) = S(U(t), T_{Lh}(JG(t)), T_{Lh}(Z(t)), T_{Lh}(YS(t)), G(t))$

$$\vdots$$
$$\vdots$$

Let $T = (y, y, T_{Lh}, T_{Lh}, T_{Lh}, y)$, then
$T(S) = S(U(t), JG(t), T_{Lh}(Z(t)), T_{Lh}(YS(t)), T_{Lh}(G(t)))$
Let $T = (y, y, y, T_{Lh}, T_{Lh}, T_{Lh})$, then
$T(S) = S(U(T_{Lh}(t)), JG(T_{Lh}(t)), Z(T_{Lh}(t)), T_{Lh}(YS(T_{Lh}(t))), T_{Lh}(G(T_{Lh}(t))))$

. . . .

Let $T = (T_{Lh}, T_{Lh}, T_{Lh}, T_{Lh}, y, y)$, then
$T(S) = S(T_{Lh}(U(t)), T_{Lh}(JG(t)), T_{Lh}(Z(t)), T_{Lh}(YS(t)), G(t))$

Let T $= (y, T_{Lh}, T_{Lh}, T_{Lh}, T_{Lh}, y)$, then
T(S) $= S(U(t), T_{Lh}(JG(t)), T_{Lh}(Z(t)), T_{Lh}(YS(t)), T_{Lh}(G(t)))$

$$\vdots$$

Let T $= (y, y, T_{Lh}, T_{Lh}, T_{Lh}, T_{Lh})$, then
T(S) $= S(U(T_{Lx}(t)), JG(T_{Lx}(t)), T_{Lx}(Z(T_{Lx}(t))), T_{Lx}(YS(T_{Lx}(t))), T_{Lx}(G(T_{Lx}(t))))$

$$\cdots\cdots\cdots\cdots\cdots\cdots\cdots\cdots\cdots\cdots\cdots\cdots$$

Let T $= (T_{Lh}, T_{Lh}, T_{Lh}, T_{Lh}, T_{Lh}, y)$, then
T(S) $= S(T_{Lh}(U(t)), T_{Lh}(JG(t)), T_{Lh}(Z(t)), T_{Lh}(YS(t)), T_{Lh}G(t)))$

$$\vdots$$

Let T $= (y, T_{Lh}, T_{Lh}, T_{Lh}, T_{Lh}, T_{Lh})$, then
T(S) $= S(U(T_{Lh}(t)), T_{Lh}(JG(T_{Lh}(t))), T_{Lh}(Z(T_{Lh}(t))), T_{Lh}(YS(T_{Lh}(t))), T_{Lh}(G(T_{Lh}(t))))$

$$\cdots\cdots\cdots\cdots\cdots\cdots\cdots\cdots\cdots\cdots\cdots\cdots$$

Let T $= (T_{Lh}, T_{Lh}, T_{Lh}, T_{Lh}, T_{Lh}, T_{Lh})$, then
T(S) $= S(T_{Lh}(U(T_{Lh}(t))), T_{Lh}(JG(T_{Lh}(t))), T_{Lh}(Z(T_{Lh}(t))), T_{Lh}(YS(T_{Lh}(t))), T_{Lh}(G(T_{Lh}(t))))$

6. Comprehensive transformations on universe of discourse

(1) Comprehensive transformation with two transformation connectives
   Let T $= (T_{Lx}, T_{Lzj}, y, y, y, y)$, then
   T(S) $= S(T_{Lx}(U(t)), T_{Lzj}(JG(t)), Z(t), YS(t), G(t))$
   Let T $= (y, T_{Lx}, T_{Lzj}, y, y, y)$, then
   T(S) $= S(U(t), T_{Lx}(JG(t)), T_{Lzj}(Z(t)), YS(t), G(t))$

$$\vdots$$

   Let T $= (y, y, y, T_{Lzj}, T_{Lx}, y)$, then
   T(S) $= S(U(t), JG(t), Z(t), T_{Lzj}(YS(t)), T_{Lx}(G(t)))$
   Let T $= (y, y, y, y, T_{Lx}, T_{Lzj})$, then
   T(S) $= S(U(T_{Lzj}(t)), JG(T_{Lzj}(t)), Z(T_{Lzj}(t)), YS(T_{Lzj}(t)), T_{Lx}(G(T_{Lzj}(t))))$

$$\vdots$$

Let $T = (T_{Lx}, y, y, T_{Lx}, y, T_{Lzj})$, then

$T(S) = S(T_{Lx}(U(T_{Lzj}(t))), JG(T_{Lzj}(t)), Z(T_{Lzj}(t)), T_{Lx}(YS(T_{Lzj}(t))), G(T_{Lzj}(t)))$

Let $T = (T_{Lzj}, y, T_{Lx}, y, y, T_{Lx})$, then

$T(S) = S(T_{Lzj}(U(T_{Lx}(t))), JG(T_{Lx}(t)), T_{Lx}(Z(T_{Lx}(t))), YS(T_{Lx}(t)), G(T_{Lx}(t)))$

....

Let $T = (T_{Lx}, T_{Lx}, T_{Lzj}, y, y, y)$, then

$T(S) = S(T_{Lx}(U(t)), T_{Lx}(JG(t)), T_{Lzj}(Z(t)), YS(t), G(t))$

Let $T = (y, T_{Lx}, T_{Lx}, T_{Lzj}, y, y)$, then

$T(S) = S(U(t), T_{Lx}(JG(t)), T_{Lx}(Z(t)), T_{Lzj}(YS(t)), G(t))$

$$\vdots$$

Let $T = (y, y, y, T_{Lx}, T_{Lx}, T_{Lzj})$, then

$T(S) = S(U(T_{Lzj}(t)), JG(T_{Lzj}(t)), Z(T_{Lzj}(t)), T_{Lx}(YS(T_{Lzj}(t))), T_{Lx}(G(T_{Lzj}(t))))$

Let $T = (y, y, y, T_{Lzj}, T_{Lx}, T_{Lx})$, then

$T(S) = S(U(T_{Lx}(t)), JG(T_{Lx}(t)), Z(T_{Lx}(t)), T_{Lzj}(YS(T_{Lx}(t))), T_{Lx}(G(T_{Lx}(t))))$

....

Let $T = (T_{Lx}, T_{Lx}, T_{Lx}, T_{Lzj}, y, y)$, then

$T(S) = S(T_{Lx}(U(t)), T_{Lx}(JG(t)), T_{Lx}(Z(t)), T_{Lzj}(YS(t)), G(t))$

Let $T = (y, T_{Lx}, T_{Lx}, T_{Lx}, T_{Lzj}, y)$, then

$T(S) = S(U(t), T_{Lx}(JG(t)), T_{Lx}(Z(t)), T_{Lx}(YS(t)), T_{Lzj}(G(t)))$

$$\vdots$$

Let $T = (y, y, T_{Lx}, T_{Lx}, T_{Lx}, T_{Lzj})$, then

$T(S) = S(U(T_{Lzj}(t)), JG(T_{Lzj}(t)), T_{Lx}(Z(T_{Lzj}(t))), T_{Lx}(YS(T_{Lzj}(t))), T_{Lx}(G(T_{Lzj}(t))))$

Let $T = (y, y, T_{Lx}, T_{Lzj}, T_{Lx}, T_{Lx})$, then

$T(S) = S(U(T_{Lx}(t)), JG(T_{Lx}(t)), T_{Lx}(Z(T_{Lx}(t))), T_{Lzj}(YS(T_{Lx}(t))), T_{Lx}(G(T_{Lx}(t))))$

....

Let T $= (T_{Lx}, T_{Lx}, T_{Lx}, T_{Lzj}, T_{Lzj}, y)$, then
T(S) $=$ S(T_{Lx}(U(t)), T_{Lx}(JG(t)), T_{Lx}(Z(t)), T_{Lzj}(YS(t)), T_{Lzj}(G(t)))
Let T $= (y, T_{Lx}, T_{Lx}, T_{Lx}, T_{Lzj}, T_{Lzj})$, then
T(S) $=$ S(U(T_{Lzj}(t)), T_{Lx}(JG(T_{Lzj}(t))), T_{Lx}(Z(T_{Lzj}(t))), T_{Lx}(YS(T_{Lzj}(t))),
T_{Lzj} (G(T_{Lzj}(t))))

$$\vdots$$

Let T $= (T_{Lzj}, T_{Lzj}, T_{Lx}, T_{Lx}, T_{Lx}, y)$, then
T(S) $=$ S(T_{Lzj}(U(t)), T_{Lzj}(JG(t)), T_{Lx}(Z(t)), T_{Lx}(YS(t)), T_{Lx} (G(t)))
Let T $= (T_{Lzj}, T_{Lzj}, y, T_{Lx}, T_{Lx}, T_{Lx})$, then
T(S) $=$ S(T_{Lzj}(U(T_{Lx}(t))), T_{Lzj}(JG(T_{Lx}(t))), (Z(T_{Lx}(t)), T_{Lx}(YS (T_{Lx}(t))), T_{Lx}
(G (T_{Lx}(t))))

....

Let T $= (T_{Lx}, T_{Lx}, T_{Lx}, T_{Lx}, T_{Lx}, T_{Lzj})$, then
T(S) $=$ S(T_{Lx}(U(T_{Lzj}(t))), T_{Lx}(JG(t)), T_{Lx}(Z(T_{Lzj}(t))), T_{Lx}(YS(T_{Lzj} (t))), T_{Lx}((G(T_{Lzj}
(t))))

$$\vdots$$

Let T $= (T_{Lzj}, T_{Lx}, T_{Lx}, T_{Lx}, T_{Lx}, T_{Lx})$, then
T(S) $=$ S(T_{Lzj}(U(T_{Lx}(t))), T_{Lx}(JG(T_{Lx}(t))), T_{Lx}(Z(T_{Lx}(t))), T_{Lx}(YS(T_{Lx} (t))),
T_{Lx} (G(T_{Lx}(t))))

....

Let T $= (T_{Lx}, T_{Lx}, T_{Lx}, T_{Lzj}, T_{Lzj}, T_{Lzj})$, then
T(S) $=$ S(T_{Lx}(U(T_{Lzj}(t))), T_{Lx}(JG(T_{Lzj}(t))), T_{Lx}(Z(T_{Lzj}(t))), T_{Lzj}
(YS(T_{Lzj}(t))), T_{Lzj} (G(T_{Lzj} (t))))

Let T $= (T_{Lzj}, T_{Lzj}, T_{Lzj}, T_{Lx}, T_{Lx}, T_{Lx})$, then
T(S) $=$ S(T_{Lzi}(U(T_{Lx}(t))), T_{Lzj}(JG(T_{Lx}(t))), T_{Lzj}(Z(T_{Lx}(t))), T_{Lx}
(YS(T_{Lx}(t))), T_{Lx} (G(T_{Lx} (t))))

(2) Comprehensive transformation with three transformation connectives

Let T $= (T_{Lx}, T_{Lzj}, T_{Lz}, y, y, y)$, then
T(S) $=$ S(T_{Lx}(U(t)), T_{Lzj}(JG(t)), T_{Lz}(Z(t)), YS(t), G(t))

Let T = (y, T_{Lx}, T_{Lzj}, T_{Lz}, y, y), then
T(S) = S(U(t), T_{Lx}(JG(t)), T_{Lzj}(Z(t)), T_{Lz}(YS(t)), G(t))

$$\vdots$$

Let T = (y, y, y, T_{Lzj}, T_{Lx}, T_{Lz}), then
T(S) = S(U(T_{Lz}(t)), JG(T_{Lz}(t)), Z(T_{Lz}(t)), T_{Lzj}(YS(T_{Lz}(t))), T_{Lx}(G(T_{Lz}(t))))
Let T = (y, y, y, T_{Lz}, T_{Lx}, T_{Lzj}), then
T(S) = S(U(T_{Lzj}(t)), JG(T_{Lzj}(t)), Z(T_{Lzj}(t)), T_{Lz}(YS(T_{Lzj}(t))), T_{Lx}(G(T_{Lzj}(t))))

$$\vdots$$

Let T = (T_{Lz}, y, y, T_{Lx}, y, T_{Lzj}), then
T(S) = S(T_{Lz}(U(T_{Lzj}(t))), JG(T_{Lzj}(t)), Z(T_{Lzj}(t)), T_{Lx}(YS(T_{Lzj}(t))), G(T_{Lzj}(t)))
Let T = (T_{Lzj}, y, T_{Lz}, y, y, T_{Lx}), then
T(S) = S(T_{Lzj}(U(T_{Lx}(t))), JG(T_{Lx}(t)), T_{Lz}(Z(T_{Lx}(t))), YS(T_{Lx}(t)), G(T_{Lx}(t)))

······ ·································

Let T = (T_{Lx}, T_{Lx}, T_{Lzj}, T_{Lz}, y, y), then
T(S) = S(T_{Lx}(U(t)), T_{Lx}(JG(t)), T_{Lzj}(Z(t)), T_{Lz}(YS(t)), G(t))
Let T = (y, T_{Lx}, T_{Lx}, T_{Lzj}, T_{Lz}, y), then
T(S) = S(U(t), T_{Lx}(JG(t)), T_{Lx}(Z(t)), T_{Lzj}(YS(t)), T_{Lz}(G(t)))

$$\vdots$$

Let T = (y, y, T_{Lz}, T_{Lx}, T_{Lx}, T_{Lzj}), then
T(S) = S(U(T_{Lzj}(t)), JG(T_{Lzj}(t)), T_{Lz}(Z(T_{Lzj}(t))), T_{Lx}(YS(T_{Lzj}(t))), T_{Lx}(G(T_{Lzj}(t))))
Let T = (y, y, T_{Lz}, T_{Lzj}, T_{Lx}, T_{Lx}), then
T(S) = S(U(T_{Lx}(t)), JG(T_{Lx}(t)), T_{Lz}(Z(T_{Lx}(t))), T_{Lzj}(YS(T_{Lx}(t))), T_{Lx}(G(T_{Lx}(t))))

······ ·································

Let T = (T_{Lx}, T_{Lx}, T_{Lx}, T_{Lzj}, T_{Lz}, y), then
T(S) = S(T_{Lx}(U(t)), T_{Lx}(JG(t)), T_{Lx}(Z(t)), T_{Lzj}(YS(t)), T_{Lz}(G(t)))

Let $T = (T_{Lz}, T_{Lx}, T_{Lx}, T_{Lx}, T_{Lzj}, y)$, then
$T(S) = S(T_{Lz}(U(t)), T_{Lx}(JG(t)), T_{Lx}(Z(t)), T_{Lx}(YS(t)), T_{Lzj}(G(t)))$

$$\vdots$$

Let $T = (y, T_{Lz}, T_{Lx}, T_{Lx}, T_{Lx}, T_{Lzj})$, then
$T(S) = S(U(T_{Lzj}(t)), T_{Lz}(JG(T_{Lzj}(t))), T_{Lx}(Z(T_{Lzj}(t))), T_{Lx}(YS(T_{Lzj}\ (t))),$
$T_{Lzj}(G(T_{Lzj}(t))))$
Let $T = (y, T_{Lz}, T_{Lx}, T_{Lzj}, T_{Lx}, T_{Lx})$, then
$T(S) = S(U(T_{Lx}(t)), T_{Lz}(JG(T_{Lx}(t))), T_{Lx}(Z(T_{Lx}(t))), T_{Lzj}\ (YS(T_{Lx}\ (t))),$
$T_{Lx}(G\ (T_{Lx}\ (t))))$

$$\cdots\cdots\cdots\cdots\cdots\cdots\cdots\cdots\cdots\cdots\cdots\cdots\cdots\cdots$$

Let $T = (T_{Lx}, T_{Lz}, T_{Lz}, T_{Lzj}, T_{Lzj}, y)$, then
$T(S) = S(T_{Lx}(U(t)), T_{Lz}(JG(t)), T_{Lz}(Z(t)), T_{Lzj}(YS(t)), T_{Lzj}(G(t)))$
Let $T = (y, T_{Lx}, T_{Lz}, T_{Lz}, T_{Lzj}, T_{Lzj})$, then
$T(S) = S(U(T_{Lzj}(t)), T_{Lx}(JG(T_{Lzj}(t))), T_{Lz}(Z(T_{Lzj}(t))), T_{Lz}(YS(T_{Lzj}\ (t))),$
$T_{Lzj}\ (G(T_{Lzj}(t))))$

$$\vdots$$

Let $T = (T_{Lzj}, T_{Lz}, T_{Lz}, T_{Lz}, T_{Lx}, y)$, then
$T(S) = S(T_{Lzj}(U(t)), T_{Lz}(JG(t)), T_{Lz}(Z(t)), T_{Lz}(YS(t)), T_{Lx}\ (G(t)))$
Let $T = (T_{Lzj}, T_{Lz}, y, T_{Lz}, T_{Lz}, T_{Lx})$, then
$T(S) = S(T_{Lzj}(U(T_{Lx}(t))), T_{Lz}(JG(T_{Lx}(t))), Z(T_{Lx}(t)), T_{Lz}(YS\ (T_{Lx}(t))),$
$T_{Lz}(G\ (T_{Lx}(t))))$

$$\cdots\cdots\cdots\cdots\cdots\cdots\cdots\cdots\cdots\cdots\cdots\cdots\cdots\cdots$$

Let $T = (T_{Lx}, T_{Lz}, T_{Lz}, T_{Lz}, T_{Lz}, T_{Lzj})$, then
$T(S) = S(T_{Lx}(U(T_{Lzj}(t))), T_{Lz}(JG(t)), T_{Lz}(Z(T_{Lzj}(t))), T_{Lz}(YS(T_{Lzj}(t))),$
$T_{Lz}((G(T_{Lzj}(t))))$

$$\vdots$$

Let $T = (T_{Lzj}, T_{Lz}, T_{Lz}, T_{Lz}, T_{Lz}, T_{Lx})$, then
$T(S) = S(T_{Lzj}(U(T_{Lx}(t))), T_{Lz}(JG(T_{Lx}(t))), T_{Lz}(Z(T_{Lx}(t))), T_{Lz}(YS(T_{Lx}\ (t))),$
$T_{Lz}\ (G(T_{Lx}(t))))$

······ ··

Let $T = (T_{Lx}, T_{Lx}, T_{Lx}, T_{Lz}, T_{Lzj}, T_{Lzj})$, then
$T(S) = S(T_{Lx}(U(T_{Lzj}(t))), T_{Lx}(JG(T_{Lzj}(t))), T_{Lx}(Z(T_{Lzj}(t))), T_{Lz}(YS(T_{Lzj}(t))),$
$T_{Lzj}(G(T_{Lzj}(t))))$

Let $T = (T_{Lzj}, T_{Lz}, T_{Lz}, T_{Lx}, T_{Lx}, T_{Lx})$, then
$T(S) = S(T_{Lzi}(U(T_{Lx}(t))), T_{Lz}(JG(T_{Lx}(t))), T_{Lz}(Z(T_{Lx}(t))), T_{Lx}(YS(T_{Lx}(t))),$
$T_{Lx}(G(T_{Lx}(t))))$

(3) Comprehensive transformation with four transformation connectives

Let $T = (T_{Lx}, T_{Lzj}, T_{Lz}, T_{Lf}, y, y)$, then
$T(S) = S(T_{Lx}(U(t)), T_{Lzj}(JG(t)), T_{Lz}(Z(t)), T_{Lf}(YS(t)), G(t))$
Let $T = (y, T_{Lx}, T_{Lzj}, T_{Lz}, T_{Lf}, y)$, then
$T(S) = S(U(t), T_{Lx}(JG(t)), T_{Lzj}(Z(t)), T_{Lz}(YS(t)), T_{Lf}(G(t)))$

$$\vdots$$

Let $T = (y, y, T_{Lf}, T_{Lzj}, T_{Lx}, T_{Lz})$, then
$T(S) = S(U(T_{Lz}(t)), JG(T_{Lz}(t)), T_{Lf}(Z(T_{Lz}(t))), T_{Lzj}(YS(T_{Lz}(t))), T_{Lx}(G(T_{Lz}(t))))$
Let $T = (y, y, T_{Lf}, T_{Lz}, T_{Lx}, T_{Lzj})$, then
$T(S) = S(U(T_{Lzj}(t)), JG(T_{Lzj}(t)), T_{Lf}(Z(T_{Lzj}(t))), T_{Lz}(YS(T_{Lzj}(t))), T_{Lx}(G(T_{Lzj}(t))))$

$$\vdots$$

Let $T = (T_{Lz}, y, T_{Lf}, T_{Lx}, y, T_{Lzj})$, then
$T(S) = S(T_{Lz}(U(T_{Lzj}(t))), JG(T_{Lzj}(t)), T_{Lf}(Z(T_{Lzj}(t))), T_{Lx}(YS(T_{Lzj}(t))),$
$G(T_{Lzj}(t)))$
Let $T = (T_{Lzj}, y, T_{Lz}, y, T_{Lf}, T_{Lx})$, then
$T(S) = S(T_{Lzj}(U(T_{Lx}(t))), JG(T_{Lx}(t)), T_{Lz}(Z(T_{Lx}(t))), YS(T_{Lx}(t)), T_{Lf}(G(T_{Lx}(t))))$

······ ··

Let $T = (T_{Lx}, T_{Lx}, T_{Lzj}, T_{Lz}, T_{Lf}, y)$, then
$T(S) = S(T_{Lx}(U(t)), T_{Lx}(JG(t)), T_{Lzj}(Z(t)), T_{Lz}(YS(t)), T_{Lf}(G(t)))$
Let $T = (T_{Lf}, T_{Lx}, T_{Lx}, T_{Lzj}, T_{Lz}, y)$, then
$T(S) = S(T_{Lf}(U(t)), T_{Lx}(JG(t)), T_{Lx}(Z(t)), T_{Lzj}(YS(t)), T_{Lz}(G(t)))$

$$\vdots$$

$$\vdots$$

Let T = $(y, T_{Lf}, T_{Lz}, T_{Lx}, T_{Lx}, T_{Lzj})$, then

T(S) = $S(U(T_{Lzj}(t)), T_{Lf}(JG(T_{Lzj}(t))), T_{Lz}(Z(T_{Lzj}(t))), T_{Lx}(YS(T_{Lzj}(t))), T_{Lx}(G(T_{Lzj}(t))))$

Let T = $(y, T_{Lf}, T_{Lz}, T_{Lzj}, T_{Lx}, T_{Lx})$, then

T(S) = $S(U(T_{Lx}(t)), T_{Lf}(JG(T_{Lx}(t))), T_{Lz}(Z(T_{Lx}(t))), T_{Lzj}(YS(T_{Lx}(t))), T_{Lx}(G(T_{Lx}(t))))$

· ·

Let T = $(T_{Lx}, T_{Lf}, T_{Lf}, T_{Lzj}, T_{Lz}, y)$, then

T(S) = $S(T_{Lx}(U(t)), T_{Lf}(JG(t)), T_{Lf}(Z(t)), T_{Lzj}(YS(t)), T_{Lz}(G(t)))$

Let T = $(T_{Lz}, T_{Lx}, T_{Lf}, T_{Lf}, T_{Lzj}, y)$, then

T(S) = $S(T_{Lz}(U(t)), T_{Lx}(JG(t)), T_{Lf}(Z(t)), T_{Lf}(YS(t)), T_{Lzj}(G(t)))$

$$\vdots$$
$$\vdots$$

Let T = $(y, T_{Lz}, T_{Lx}, T_{Lf}, T_{Lf}, T_{Lzj})$, then

T(S) = $S(U(T_{Lzj}(t)), T_{Lz}(JG(T_{Lzj}(t))), T_{Lx}(Z(T_{Lzj}(t))), T_{Lf}(YS(T_{Lzj}(t))),$
$T_{Lf}(G(T_{Lzj}(t))))$

Let T = $(y, T_{Lz}, T_{Lx}, T_{Lzj}, T_{Lf}, T_{Lf})$, then

T(S) = $S(U(T_{Lf}(t)), T_{Lz}(JG(T_{Lf}(t))), T_{Lx}(Z(T_{Lf}(t))), T_{Lzj}(YS(T_{Lf}(t))),$
$T_{Lf}(G(T_{Lf}(t))))$

· ·

Let T = $(T_{Lx}, T_{Lz}, T_{Lz}, T_{Lf}, T_{Lzj}, y)$, then

T(S) = $S(T_{Lx}(U(t)), T_{Lz}(JG(t)), T_{Lz}(Z(t)), T_{Lf}(YS(t)), T_{Lzj}(G(t)))$

Let T = $(y, T_{Lx}, T_{Lz}, T_{Lz}, T_{Lf}, T_{Lzj})$, then

T(S) = $S(U(T_{Lzj}(t)), T_{Lx}(JG(T_{Lzj}(t))), T_{Lz}(Z(T_{Lzj}(t))), T_{Lz}(YS(T_{Lzj}(t))),$
$T_{Lf}(G(T_{Lzj}(t))))$

$$\vdots$$
$$\vdots$$

Let T = $(T_{Lzj}, T_{Lz}, T_{Lz}, T_{Lz}, T_{Lx}, T_{Lf})$, then

T(S) = $S(T_{Lzj}(U(T_{Lf}(t))), T_{Lz}(JG(T_{Lf}(t))), T_{Lz}(Z(T_{Lf}(t))), T_{Lz}(YS(T_{Lf}(t))),$
$T_{Lx}(G(T_{Lf}(t))))$

Let T = $(T_{Lzj}, T_{Lz}, T_{Lf}(t), T_{Lz}, T_{Lz}, T_{Lx})$, then

$T(S) = S(T_{Lzj}(U(T_{Lx}(t))), T_{Lz}(JG(T_{Lx}(t))), T_{Lf}(Z(T_{Lx}(t))), T_{Lz}(YS(T_{Lx}(t))), T_{Lz}(G(T_{Lx}(t))))$

$\cdots\cdots\ \cdots\cdots\cdots\cdots\cdots\cdots\cdots\cdots\cdots\cdots\cdots\cdots$

Let $T = (T_{Lx}, T_{Lf}, T_{Lf}, T_{Lf}, T_{Lz}, T_{Lzj})$, then

$T(S) = S(T_{Lx}(U(T_{Lzj}(t))), T_{Lf}(JG(t)), T_{Lf}(Z(T_{Lzj}(t))), T_{Lf}(YS(T_{Lzj}(t))), T_{Lz}((G(T_{Lzj}(t))))$

$\vdots$
$\vdots$

Let $T = (T_{Lzj}, T_{Lf}, T_{Lz}, T_{Lf}, T_{Lz}, T_{Lf})$, then

$T(S) = S(T_{Lzj}(U(T_{Lf}(t))), T_{Lf}(JG(T_{Lf}(t))), T_{Lz}(Z(T_{Lf}(t))), T_{Lf}(YS(T_{Lf}(t))), T_{Lz}(G(T_{Lf}(t))))$

(4) Comprehensive transformation with five transformation connectives

Let $T = (T_{Lx}, T_{Lzj}, T_{Lz}, T_{Lf}, T_{Lh}, y)$, then
$T(S) = S(T_{Lx}(U(t)), T_{Lzj}(JG(t)), T_{Lz}(Z(t)), T_{Lf}(YS(t)), T_{Lh}(G(t)))$
Let $T = (y, T_{Lx}, T_{Lzj}, T_{Lz}, T_{Lf}, T_{Lh})$, then
$T(S) = S(U(T_{Lh}(t)), T_{Lx}(JG(T_{Lh}(t))), T_{Lzj}(Z(T_{Lh}(t))), T_{Lz}(YS(T_{Lh}(t))), T_{Lf}(G(T_{Lh}(t))))$

$\vdots$
$\vdots$

Let $T = (y, T_{Lh}, T_{Lf}, T_{Lzj}, T_{Lx}, T_{Lz})$, then
$T(S) = S(U(T_{Lz}(t)), T_{Lh}(JG(T_{Lz}(t))), T_{Lf}(Z(T_{Lz}(t))), T_{Lzj}(YS(T_{Lz}(t))), T_{Lx}(G(T_{Lz}(t))))$
Let $T = (y, T_{Lh}, T_{Lf}, T_{Lz}, T_{Lx}, T_{Lzj})$, then
$T(S) = S(U(T_{Lzj}(t)), T_{Lh}(JG(T_{Lzj}(t))), T_{Lf}(Z(T_{Lzj}(t))), T_{Lz}(YS(T_{Lzj}(t))), T_{Lx}(G(T_{Lzj}(t))))$

$\vdots$
$\vdots$

Let $T = (T_{Lz}, T_{Lh}, T_{Lf}, T_{Lx}, y, T_{Lzj})$, then
$T(S) = S(T_{Lz}(U(T_{Lzj}(t))), T_{Lh}(JG(T_{Lzj}(t))), T_{Lf}(Z(T_{Lzj}(t))), T_{Lx}(YS(T_{Lzj}(t))), G(T_{Lzj}(t)))$
Let $T = (T_{Lzj}, y, T_{Lz}, T_{Lh}, T_{Lf}, T_{Lx})$, then
$T(S) = S(T_{Lzj}(U(T_{Lx}(t))), JG(T_{Lx}(t)), T_{Lz}(Z(T_{Lx}(t))), T_{Lh}(YS(T_{Lx}(t))), T_{Lf}(G(T_{Lx}(t))))$

7.1.3 Theorem of Transformation on Universe of Discourse

Theorem 7.1 *Suppose U is universe of discourse, G is a group of rules defined on U, if $U_1 \subseteq U$ and $U_2 \subseteq U$, $U_1 \subseteq U_2$, $C_1 = \{(u, x) \mid u \in U_1, x = f(u, G), x \in A\}$, $C_2 = \{(u, y) \mid u \in U_2, y = g(u, G), y \in B\}$, $a = \inf(x)$ (limit inferior), $b = \sup(x)$ (limit superior), $c = \inf(y)$ (limit inferior), $d = \inf(y)$ (limit superior), then $a \leq c$, $b \leq d$, $A = \{x \mid (u, x) \in C_1\}$, $B = \{y \mid (u, y) \in C_2\}$.*

Proof Proof is omitted here.

7.2 Transformations on Rules of Error Systems

7.2.1 Necessity in Studying Rules for Judging Errors

During the course of grading homework, teachers always put some symbols or written comments on the homework sheet, which were used to judge if some questions or the steps of certain question were correct or not. How are those symbols or comments obtained? Of course, they are a group of rules obtained from axioms, theorems, and laws. Judge, in court, needs to decide if the defendant is guilty and what kind of penalty needs to be put on a criminal. In an organization, the under-performing employees are published based on the company's regulations or rules. Parents disciple children according to the social norms or values, law, regulations, and disciplines in that family. In summary, rules are needed to determine whether the political system, policies, decisions, theoretic systems in a country or organization are correct or appropriate or not. Therefore, a group of rules need to be defined before one makes judgment.

7.2.2 Properties of Rules for Judging Errors

1. The necessity and inevitability of changes in rules for judging errors
   In order to meet the very basic physiological needs, human beings must have some actions and activities. Regarding the results of certain activity, how should we know if it is wrong or right? This is easier said than done. The following items must be clarified before answering the question:

   (1) where did this activity happen?
   (2) when did this activity happen?
   (3) which field does this activity belong to?
   (4) what was the purpose of addressing this activity?
   (5) what kinds of knowledge, skills and techniques were involved to start this activity?

Item (5) is included in items (1) and (3).

Example 1 It is impossible to tell if it is right or wrong if a driver is driving his car on the right side of the road before the actual location is confirmed. In China, according to the traffic rules and regulations, his action has no problem. However, he is wrong by breaking the rules if he is driving in the UK because all vehicles must keep on the left side of the road.

Example 2 Someone did a calculation and got the following result $1 + 1 = 1$. Is this correct? According to laws in binary computing (Table 7.1):

The result is incorrect if this person is performing binary computation. On the contrary, the computing result is correct if this person is doing Boolean calculation. Because the Boolean calculation is as (Table 7.2):

Example 3 Tying knots and scratching marks on stone are advanced tally mark and numerical system. Is this proposition correct? In 5000 years ago, as tally mark, tying knots, and scratching marks on stones were advanced and the above proposition was correct in that sense. However, many advanced numerical systems have been developed to handle this nowadays. Comparatively, the primitive techniques are outdated and the proposition is false in the contemporary world.

By analyzing the above three examples, example **1** tells us that the location change caused the changes in rules for judging errors. And example **2** exhibits that different rules in different fields are needed to evaluate error in corresponding areas or fields. While example **3** shows that rules are contingent on the historic period and technology development. Moreover, rules for judging errors also change with various purposes of using them.

Table 7.1 Computation of binary numbers

Binary number 1	Binary number 2	Sum
0	0	0
0	1	1
1	1	10

Table 7.2 The computation in Boolean algebra

Boolean number 1	Boolean number 2	Sum
0	0	1
0	1	0
1	1	1

Example 4 In evaluating the functionality of a bicycle, residents in urban areas require a light and aesthetic style while farmers in village use it as transport tool demanding heavy-duty and all-terrain-use style.

2. Definitions of rule function

Definition 7.10 Suppose that $S = [S_0, S_1]$, $K = \{(X, Y, Z) \mid a \le x \le b, c \le y \le d, e \le z \le f\}$, $Z = \{z_1, z_2, \ldots \ldots, z_n)$, $M = \{(m_1, m_2, \ldots \ldots, m_i)\}$, where S is the temporal set; K is the spatial set; Z is the set composed of different fields; M is the purpose set. Suppose that G is a group of rules, $D = S \times Z \times M \times K$, if $f : D \to G$, then f is called a function for rules of judging errors simplified as rule function, noted by $G = f(D)$, $G = f(s, k, z, m)$, or $G = G(s, k, z, m)$.

Special forms of rule function:

(a) When space, field, and purpose are given, the rule is the function with respect to time, i.e., if $k = k_0$, $z = z_0$, and $m = m_0$, $G = f(s, k_0, z_0, m_0) = f(s)$ holds.
(b) When time, field, and purpose are given, the rule is the function with respect to space, i.e., if $s = s_0$, $z = z_0$, and $m = m_0$, $G = f(s_0, k, z_0, m_0) = f(k)$ holds. Similarly, the following relationships hold.
(c) If $s = s_0$, $k = k_0$, and $m = m_0$, $G = f(s_0, k_0, z, m_0) = f(z)$ holds.
(d) If $s = s_0$, $k = k_0$, and $z = z_0$, $G = f(s_0, k_0, z_0, m) = f(m)$ holds.
(e) If $s = s_0$ and $k = k_0$, $G = f(s_0, k_0, z, m) = f(z, m)$ holds.
(f) If $k = k_0$ and $z = z_0$, $G = f(s, k_0, z_0, m) = f(s, m)$ holds.
(g) If $m = m_0$ and $z = z_0$, $G = f(s, k, z_0, m_0) = f(s, k)$ holds.
(h) If $s = s_0$ and $m = m_0$, $G = f(s_0, k, z, m_0) = f(k, z)$ holds.
(i) If $s = s_0$ and $z = z_0$, $G = f(s_0, k, z_0, m) = f(k, m)$ holds.
(j) If $k = k_0$ and $m = m_0$, $G = f(s, k_0, z, m_0) = f(s, z)$ holds.
(k) If $s = s_0$, $G = f(s_0, k, z, m) = f(k, z, m)$ holds.
(l) If $k = k_0$, $G = f(s, k_0, z, m) = f(s, z, m)$ holds.
(m) If $z = z_0$, $G = f(s, k, z_0, m) = f(s, k, m)$ holds.
(n) If $m = m_0$, $G = f(s, k, z, m_0) = f(s, k, z)$ holds.
(o) If $s = s_0$, $k = k_0$, $z = z_0$, and $m = m_0$, $G = f(s_0, k_0, z_0, m_0)$ holds. It is called constant rule function noted by $G = f_0$ or $G = G_0$ or G_0.

3. Definitions of transformations on rules

Definition 7.11 Suppose that G_1 and G_2 are two group of rules, if $g_1 \ne \varnothing$, $g_2 \ne \varnothing$, $g_1 \Longleftrightarrow g_2$, $g_1 \subseteq G_1$, $g_2 \subseteq G_2$, then $G_1 \sim G_2$ (G_1 is similar to G_2). If $g_1 \ne G_1$ and $g_2 \ne G_2$, the G_1 is partially similar to G_2. If $g_1 \ne G_1$ and $g_2 \ne G_2$, then $G_1 \Longleftrightarrow G_2$ (G_1 is equal to G_2). If $g_1 \Longleftrightarrow g_2$ does not hold as $g_1 \ne \varnothing$, $g_2 \ne \varnothing$, $g_1 \subseteq G_1$, and $g_2 \subseteq G_2$, then G_1 and G_2 are mutually independent of each other.

Definition 7.12 Suppose that $T_{gzj}(G) = G \cup G_1$, then T_{gzj} has done addition transformation on rule G.

Definition 7.13 Suppose that $T_{gf}(G) = G_1 \cup G_2 \cup \ldots \cup G_n$, $G = G_1 \cup G_2 \cup \ldots \cup G_n$, then T_{gf} has done decomposition transformation on rule G.

Definition 7.14 Suppose that $T_{gz}(G) = G'$, then T_{gz} has done displacement transformation on rule G.

Definition 7.15 Suppose that $T_{gh}(G) = \varnothing$, then T_{gh} has done destruction transformation on rule G.

Definition 7.16 Suppose that $T_{gd}(G) = G$, then T_{gd} has done unit transformation on rule G.

7.2.3 Transformations on Rules for Judging Errors in Error Systems

Transformation system is composed of: similarity transformation $T_x \subseteq \{T_{xly}, T_{xjg}, T_{xz}, T_{xys}, T_{xg}, T_{xsj}\}$ (T_x has done similarity transformation on domain, structure, subsystem, system element, rule, and time; same pattern follows for other transformations), addition transformation $T_{zj} \subseteq \{T_{zjly}, T_{zjjg}, T_{zjz}, T_{zjys}, T_{zjg}, T_{zjsj}\}$, displacement transformation $T_z \subseteq \{T_{zly}, T_{zjg}, T_{zz}, T_{zys}, T_{zg}, T_{zsj}\}$, decomposition transformation $T_f \subseteq \{T_{fly}, T_{fjg}, T_{fz}, T_{fys}, T_{fg}, T_{fsj}\}$, destruction transformation $T_h \subseteq \{T_{hly}, T_{hjg}, T_{hz}, T_{hys}, T_{hg}, T_{hsj}\}$, and unit transformation $T_d \subseteq \{T_{dly}, T_{djg}, T_{dz}, T_{dys}, T_{dg}, T_{dsj}\}$; and the conjunction, disjunction, and inverse operations on transformations.

Suppose that the rule system of interest is denoted by $G = \{G_{sj}, G_{kj}, G_z, G_{md}\}$, G is rules for judging error, transformation operator $T = (T_{sj}, T_{lj}, T_z, T_{md})$, where T_{sj}, T_{lj}, T_z, and T_{md} represent corresponding transformation on time, space, field applied, and purpose for applying rules in the designed object system, respectively. The position where the transformation operator is placed indicates corresponding transformation. For example, $T = (T_{sj}, T_{lj}, T_z, T_{md})$ is denoted by $T = (T_x, T_x, T_x, T_x)$ if similarity is conducted.

1. Similarity transformation on rules
   Let $T = (T_x, y, y, y)$, then
   $T(G) = (T_x(S), KJ, ZY, MD)$
   Let $T = (y, T_x, y, y)$, then
   $T(G) = (S, T_x(KJ), ZY, MD)$

$$\vdots$$

   Let $T = (y, y, y, T_x)$, then
   $T(G) = (S, KJ, ZY, T_x(MD))$

   $\cdots\cdots\cdots\cdots\cdots\cdots\cdots\cdots\cdots\cdots\cdots\cdots\cdots\cdots\cdots\cdots\cdots\cdots$

   Let $T = (T_x, T_x, y, y)$, then
   $T(G) = (T_x(S), T_x(KJ), ZY, MD)$ Let $T = (y, T_x, T_x, y)$, then

$T(G) = (S, T_x(KJ), T_x(ZY), MD)$

$$\vdots$$

Let $T = (y, y, T_x, T_x)$, then
$T(G) = (S, KJ, T_x(ZY), T_x(MD))$

. . . .

Let $T = (T_x, T_x, T_x, y)$, then
$T(G) = (T_x(S), T_x(KJ), T_x(ZY), MD)$
Let $T = (y, T_x, T_x, T_x)$, then
$T(G) = (S, T_x(KJ), T_x(ZY), T_x(MD))$

$$\vdots$$

Let $T = (T_x, y, T_x, T_x)$, then
$T(G) = (T_x(S), KJ, T_x(ZY), T_x(MD))$

. . . .

Let $T = (T_x, T_x, T_x, T_x)$, then
$T(G) = (T_x(S), T_x(KJ), T_x(ZY), T_x(MD))$

2. Addition transformation on rules
   Let $T = (T_{zj}, y, y, y)$, then
   $T(G) = (T_{zj}(S), KJ, ZY, MD)$
   Let $T = (y, T_{zj}, y, y)$, then
   $T(G) = (S, T_{zj}(KJ), ZY, MD)$

$$\vdots$$

Let $T = (y, y, y, T_{zj})$, then
$T(G) = (S, KJ, ZY, T_{zj}(MD))$

. . . .

Let $T = (T_{zj}, T_{zj}, y, y)$, then
$T(G) = (T_{zj}(S), T_{zj}(KJ), ZY, MD)$

Let $T = (y, T_{zj}, T_{zj}, y)$, then
$T(G) = (S, T_{zj}(KJ), T_{zj}(ZY), MD)$

$$\vdots$$
$$\vdots$$

Let $T = (y, y, T_{zj}, T_{zj})$, then
$T(G) = (S, KJ, T_{zj}(ZY), T_{zj}(MD))$

....

Let $T = (T_{zj}, T_{zj}, T_{zj}, y)$, then
$T(G) = (T_{zj}(S), T_{zj}(KJ), T_{zj}(ZY), MD)$
Let $T = (y, T_{zj}, T_{zj}, T_{zj})$, then
$T(G) = (S, T_{zj}(KJ), T_{zj}(ZY), T_{zj}(MD))$

$$\vdots$$
$$\vdots$$

Let $T = (T_{zj}, y, T_{zj}, T_{zj})$, then
$T(G) = (T_{zj}(S), KJ, T_{zj}(ZY), T_{zj}(MD))$

....

Let $T = (T_{zj}, T_{zj}, T_{zj}, T_{zj})$, then
$T(G) = (T_{zj}(S), T_{zj}(KJ), T_{zj}(ZY), T_{zj}(MD))$

3. Displacement transformation on rules
   Let $T = (T_z, y, y, y)$, then
   $T(G) = (T_z(S), KJ, ZY, MD)$
   Let $T = (y, T_z, y, y)$, then
   $T(G) = (S, T_z(KJ), ZY, MD)$

$$\vdots$$
$$\vdots$$

Let $T = (y, y, y, T_z)$, then
$T(G) = (S, KJ, ZY, T_z(MD))$

....

Let T = (T_z, T_z, y, y), then
T(G) = $(T_z(S), T_z(KJ), ZY, MD)$
Let T = (y, T_z, T_z, y), then
T(G) = $(S, T_z(KJ), T_z(ZY), MD)$

$$\vdots$$

Let T = (y, y, T_z, T_z), then
T(G) = $(S, KJ, T_z(ZY), T_z(MD))$

$\cdots\cdots \cdots\cdots\cdots\cdots\cdots\cdots\cdots\cdots\cdots\cdots$

Let T = (T_z, T_z, T_z, y), then
T(G) = $(T_z(S), T_z(KJ), T_z(ZY), MD)$
Let T = (y, T_z, T_z, T_z), then
T(G) = $(S, T_z(KJ), T_z(ZY), T_z(MD))$

$$\vdots$$

Let T = (T_z, y, T_z, T_z), then
T(G) = $(T_z(S), KJ, T_z(ZY), T_z(MD))$

$\cdots\cdots \cdots\cdots\cdots\cdots\cdots\cdots\cdots\cdots\cdots\cdots$

Let T = (T_z, T_z, T_z, T_z), then
T(G) = $(T_z(S), T_z(KJ), T_z(ZY), T_z(MD))$

4. Decomposition transformation on rules
   Let T = (T_f, y, y, y), then
   T(G) = $(T_f(S), KJ, ZY, MD)$
   Let T = (y, T_f, y, y), then
   T(G) = $(S, T_f(KJ), ZY, MD)$

$$\vdots$$

Let T = (y, y, y, T_f), then
T(G) = $(S, KJ, ZY, T_f(MD))$

$\cdots\cdots \cdots\cdots\cdots\cdots\cdots\cdots\cdots\cdots\cdots\cdots$

Let T $= (T_f, T_f, \text{y}, \text{y})$, then
T(G) $= (T_f(\text{S}), T_f(\text{KJ}), \text{ZY}, \text{MD})$
Let T $= (\text{y}, T_f, T_f, \text{y})$, then
T(G) $= (\text{S}, T_f(\text{KJ}), T_f(\text{ZY}), \text{MD})$

$$\vdots$$

Let T $= (\text{y}, \text{y}, T_f, T_f)$, then
T(G) $= (\text{S}, \text{KJ}, T_f(\text{ZY}), T_f(\text{MD}))$

....

Let T $= (T_f, T_f, T_f, \text{y})$, then
T(G) $= (T_f(\text{S}), T_f(\text{KJ}), T_f(\text{ZY}), \text{MD})$
Let T $= (\text{y}, T_f, T_f, T_f)$, then
T(G) $= (\text{S}, T_f(\text{KJ}), T_f(\text{ZY}), T_f(\text{MD}))$

$$\vdots$$

Let T $= (T_f, \text{y}, T_f, T_f)$, then
T(G) $= (T_f(\text{S}), \text{KJ}, T_f(\text{ZY}), T_f(\text{MD}))$

....

Let T $= (T_f, T_f, T_f, T_f)$, then
T(G) $= (T_f(\text{S}), T_f(\text{KJ}), T_f(\text{ZY}), T_f(\text{MD}))$

5. Destruction transformation on rules
   Let T $= (T_h, \text{y}, \text{y}, \text{y})$, then
   T(G) $= (T_h(\text{S}), \text{KJ}, \text{ZY}, \text{MD})$
   Let T $= (\text{y}, T_h, \text{y}, \text{y})$, then
   T(G) $= (\text{S}, T_h(\text{KJ}), \text{ZY}, \text{MD})$

$$\vdots$$

Let T $= (\text{y}, \text{y}, \text{y}, T_h)$, then
T(G) $= (\text{S}, \text{KJ}, \text{ZY}, T_h(\text{MD}))$

....

Let T $= (T_h, T_h, y, y)$, then
T(G) $= (T_h(S), T_h(KJ), ZY, MD)$
Let T $= (y, T_h, T_h, y)$, then
T(G) $= (S, T_h(KJ), T_h(ZY), MD)$

$$\vdots$$

Let T $= (y, y, T_h, T_h)$, then
T(G) $= (S, KJ, T_h(ZY), T_h(MD))$

$\cdots\cdots \cdots\cdots\cdots\cdots\cdots\cdots\cdots\cdots\cdots\cdots\cdots$

Let T $= (T_h, T_h, T_h, y)$, then
T(G) $= (T_h(S), T_h(KJ), T_h(ZY), MD)$
Let T $= (y, T_h, T_h, T_h)$, then
T(G) $= (S, T_h(KJ), T_h(ZY), T_h(MD))$

$$\vdots$$

Let T $= (T_h, y, T_h, T_h)$, then
T(G) $= (T_h(S), KJ, T_h(ZY), T_h(MD))$

$\cdots\cdots \cdots\cdots\cdots\cdots\cdots\cdots\cdots\cdots\cdots\cdots\cdots$

Let T $= (T_h, T_h, T_h, T_h)$, then
T(G) $= (T_h(S), T_h(KJ), T_h(ZY), T_h(MD))$

6. Comprehensive transformation on rules

   (1) Comprehensive transformation with two transformation connectives
       Let T $= (T_x, T_{zj}, y, y)$, then
       T(G) $= (T_x(S), T_{zj}(KJ), ZY, MD)$
       Let T $= (y, T_h, T_{zj}, y)$, then
       T(G) $= (S, T_h(KJ), T_{zj}(ZY), MD)$

$$\vdots$$

       Let T $= (y, T_x, y, T_{zj})$, then
       T(G) $= (S, T_x(KJ), ZY, T_{zj}(MD))$

$\cdots\cdots \cdots\cdots\cdots\cdots\cdots\cdots\cdots\cdots\cdots\cdots\cdots\cdots\cdots$

Let $T = (T_x, T_{zj}, T_x, y)$, then
$T(G) = (T_x(S), T_{zj}(KJ), T_x(ZY), MD)$
Let $T = (T_x, T_{zj}, y, T_x)$, then
$T(G) = (T_x(S), T_{zj}(KJ), ZY, T_x(MD))$

$\vdots$
$\vdots$

Let $T = (T_{zj}, y, T_x, T_x)$, then
$T(G) = (T_{zj}(S), KJ, T_x(ZY), T_x(MD))$

$\cdots\cdots \cdots\cdots\cdots\cdots\cdots\cdots\cdots\cdots\cdots\cdots\cdots\cdots$

Let $T = (T_x, T_{zj}, T_x, T_x)$, then
$T(G) = (T_x(S), T_{zj}(KJ), T_x(ZY), T_x(MD))$
Let $T = (T_x, T_x, T_{zj}, T_x)$, then
$T(G) = (T_x(S), T_x(KJ), T_{zj}(ZY), T_x(MD))$

$\vdots$
$\vdots$

Let $T = (T_x, T_x, T_x, T_{zj})$, then
$T(G) = (T_x(S), T_x(KJ), T_x(ZY), T_{zj}(MD))$

$\cdots\cdots \cdots\cdots\cdots\cdots\cdots\cdots\cdots\cdots\cdots\cdots\cdots\cdots$

Let $T = (T_x, T_x, T_{zj}, T_{zj})$, then
$T(G) = (T_x(S), T_x(KJ), T_{zj}(ZY), T_{zj}(MD))$

(2) Comprehensive transformation with three transformation connectives
Let $T = (T_x, T_{zj}, T_f, y)$, then
$T(G) = (T_x(S), T_{zj}(KJ), T_f ZY), MD)$
Let $T = (y, T_h, T_{zj}, T_f)$, then
$T(G) = (S, T_h(KJ), T_{zj}(ZY), T_f MD))$

$\vdots$
$\vdots$

Let $T = (T_f, T_x, y, T_{zj})$, then
$T(G) = (T_f(S), T_x(KJ), ZY, T_{zj}(MD))$

$\cdots\cdots\ \cdots\cdots\cdots\cdots\cdots\cdots\cdots\cdots\cdots\cdots$

Let $T = (T_x, T_{zj}, T_x, T_f)$, then
$T(G) = (T_x(S), T_{zj}(KJ), T_x(ZY), T_f(MD))$
Let $T = (T_x, T_{zj}, T_f, T_x)$, then
$T(G) = (T_x(S), T_{zj}(KJ), T_f(ZY), T_x(MD))$

$\vdots$

$\vdots$

Let $T = (T_{zj}, T_f, T_x, T_x)$, then
$T(G) = (T_{zj}(S), T_f(KJ), T_x(ZY), T_x(MD))$

$\cdots\cdots\ \cdots\cdots\cdots\cdots\cdots\cdots\cdots\cdots\cdots\cdots$

Let $T = (T_x, T_{zj}, T_f, T_f)$, then
$T(G) = (T_x(S), T_{zj}(KJ), T_f(ZY), T_f(MD))$
Let $T = (T_f, T_f, T_{zj}, T_x)$, then
$T(G) = (T_f(S), T_f(KJ), T_{zj}(ZY), T_x(MD))$

$\vdots$

$\vdots$

Let $T = (T_f, T_x, T_{zj}, T_{zj})$, then
$T(G) = (T_f(S), T_x(KJ), T_{zj}(ZY), T_{zj}(MD))$

$\cdots\cdots\ \cdots\cdots\cdots\cdots\cdots\cdots\cdots\cdots\cdots\cdots$

Let $T = (T_{zj}, T_{zj}, T_f, T_x)$, then
$T(G) = (T_{zj}(S), T_{zj}(KJ), T_f(ZY), T_x(MD))$

(3) Comprehensive transformation with four transformation connectives
Let $T = (T_x, T_{zj}, T_f, T_h)$, then
$T(G) = (T_x(S), T_{zj}(KJ), T_f(ZY), T_h(MD))$
Let $T = (T_h, T_{zj}, T_f, T_x)$, then
$T(G) = (T_h(S), T_{zj}(KJ), T_f(ZY), T_x(MD))$

$\vdots$

$\vdots$

Let $T = (T_f, T_x, T_z, T_{zj})$, then
$T(G) = (T_f(S), T_x(KJ), T_z(ZY), T_{zj}(MD))$

$$\vdots$$
$$\vdots$$

Let $T = (T_h, T_z, T_f, T_x)$, then
$T(G) = (T_h(S), T_z(KJ), T_f(ZY), T_x(MD))$

Proposition 7.1 *Suppose that G is a group of qualified rules for judging errors defined on U in error system, if u_1, $u_2 \in U$, the error value of u_1 under rule G is $\boldsymbol{a}$ and the error value of u_2 under rule G is $\boldsymbol{b}$, if $u_1 = u_2$, then $\boldsymbol{a} = \boldsymbol{b}$*

Proof If $\mathbf{a} \neq \mathbf{b}$, there are two error values for same object under the same rule G defined on universe of discourse U, which is contradicted to the assumption that G is a group of qualified rules defined on U.
Proof is completed here!

Proposition 7.2 *Suppose that G_1 and G_2 are two group of qualified rules for judging errors defined on U in error system, if $u \in U$ is non-erroneous under rule G_1 and is erroneous under rule G_2, then either G_1 or G_2 is erroneous, otherwise, G_1 and G_2 have different scope of applications.*

Proof If G_1 and G_2 are two group of qualified rules for judging errors defined on U in error system, if the error values of $u \in U$ under rules G_1 and G_2 are $\mathbf{a}$ and $\mathbf{b}$, respectively, if $\mathbf{a} \neq \mathbf{b}$, G_1 and G_2 are not the same rules according to Proposition 7.1. This proposition is correct.
Proof is completed here!

Chapter 8
Error Function

Abstract This chapter introduces the concept of error function in order to quantitatively investigate errors. Error function is categorized according to the differences in the range of error values and the object of interests. Scalar and vector functions are then examined. The process of building error function is discussed based on different rules in judging errors. Last but not the least, the relationships between values of error functions under different rules of judging errors are examined. Three propositions are proposed accordingly. This chapter mainly seeks to quantify the errors in error system models, which makes it possible for the error system models to be widely applied in practice.

8.1 Concept of Error Function

Function is a mapping from one set to the other, which depicts the relationship between two sets. In order to capture error in a quantitative manner, it is necessary to figure out the relationships between each element in the object systems and certain error values. For achieving this, one needs to identify the relationship between object system and domain of real numbers $\Re$. As the object system can be represented by an object set, certain deterministic relationships between object set and real set $\Re$ is the "error function" that we are meaning to address in this chapter.

8.1.1 Definition for Error Function

Definition 8.1 Suppose that U is an object set, G is a set of rules for judging error, $V = \{ (u, G) \mid u \in U \}$, $f : V \to \Re$, then f is called an error function under the rule of judging errors G defined on U denoted by $x = f(G, u)$ i.e., $f(u)$, $\Re$ represents the universe of discourse of real numbers, x is the error value of object u defined under judging rules of G.

© Springer Nature Switzerland AG 2020

251

K. Guo and S. Liu, *Error Systems: Concepts, Theory and Applications*,
Studies in Systems, Decision and Control 275,
https://doi.org/10.1007/978-3-030-40760-5_8

From the definition, f is the mapping from $V = \{ (u, G) \mid u \in U \}$ to domain of real numbers $\mathfrak{R}$. There are two independent variables in function f: (1) element in object set U; (2) judging rules for error G, which together correspond to a real number-i.e., error value of object $u \in U$ under the rules of G. There are two types of dynamics where u changes within U and G changes with respect to different factors on U.

8.1.2   Categorization of Error Functions

Given the differences in specific requirements, error functions can be categorized based on their characteristics. In this session, error functions are categorized based on value range of error function and partial domain.

1. To categorize error function according to the value difference $ran(f)$ of $f : V \to \mathfrak{R}$

   (1) If $ran(f) = \{0, 1\}$, then f is called a classic error function defined on U, apparently it is not continuous;
   (2) If $ran(f) = [0, 1]$, then f is called a fuzzy error function defined on U;
   (3) If $ran(f) = (-\infty, +\infty)$, then f is called an error function with critical points defined on U;
   (4) If $ran(f) = [0, +\infty)$, then f is called a non-negative error function defined on U.

2. To categorize error function according to the domain difference $dom(f)$ of $f : V \to \mathfrak{R}$

   (1) If $dom(f) = \{(u, G_m) \mid u \in U\}$, where U is domain for the objective features of system **S**, G_m are rules for judging if there exists error in objective feature of **S**, then f is called an error function for objective features of **S**;
   (2) If $dom(f) = \{(u, G_{gm}) \mid u \in U\}$, where U is domain for the objective features of system **S**, G_{gm} are rules for judging judging the attainability of objective features of **S**, then f is called an error function for judging level of realization on objective features of **S**;
   (3) If $dom(f) = \{(u, G_h) \mid u \in U\}$, where U is the domain for intrinsic features of system **S**, G_m are rules for judging the attainability of intrinsic features of **S**, then f is called an error function judging level of realization on intrinsic features of **S**;
   (4) If $dom(f) = \{(u, G_p) \mid u \in P(U)\}$, where $P(U)$ is the power set of system domain U, G_p are rules for judging error of object domain U in system **S**, then f is called an association error function defined on U.

3. Based on the above two categorizing mechanisms, 12 error functions are obtained as below:

   (1) Classic error function for objective features in system **S**;
   (2) Fuzzy error function for objective features in system **S**;
   (3) Error function with critical points for objective features in system **S**;
   (4) Non-negative error function for objective features in system **S**;
   (5) Classic error function for realization of objective features in system **S**;
   (6) Fuzzy error function for realization of objective features in system **S**;
   (7) Error function with critical points for realization of objective features in system **S**;
   (8) Non-negative error function for realization of objective features in system **S**;
   (9) Classic error function for system **S** behavior;
   (10) Fuzzy error function for system **S** behavior;
   (11) Error function with critical points for system **S** behavior;
   (12) Non-negative error function for system **S** behavior.

8.2 Forms of Error Functions

In mathematics, different types of functions have distinct forms and even the same type of function has various forms due to pre-given conditions in reality. Error function, with no exception, has different forms in different situations. Two special error functions are discussed in this session.

8.2.1 Definitions of Error Function

Definition 8.2 Suppose that $f_1(G_1, u)$ and $f_2(G_2, u)$ are defined on domain U, if $G_1 \nRightarrow u$ and $G_2 \nRightarrow u$, $\forall u \in U$ such that $f_1(G_1, u) = f_2(G_2, u)$ holds, then $f_1(G_1, u)$ is said to be equal to $f_2(G_2, u)$ denoted by $f_1(G_1, u) = f_2(G_2, u)$.

8.2.2 Scalar Error Functions

Definition 8.3 Suppose that U is the domain of an error function, $V = \{ v \mid v = f(G, u), u \in U \}$, $f(G, X) = f(X) = \sum_{u \in X} v_u$, where X is a subset of U, and when $v = f(G, u) < 0$, let $v = 0$, then $f(X)$ is called the scalar error function denoted by $f_h(X)$.

1. Relationships between scalar error functions

(1) Equivalence of scalar error functions

Definition 8.4 Suppose that $f_{h1}(X)$ and $f_{h2}(X)$ are defined on domain U, $\forall X \subseteq U$ such that $f_{h1}(X)=f_{h2}(X)$, $f_{h1}(X)$ is said to be equal to $f_{h2}(X)$; when $G_1 = G_2$ such that $f_{h1}(X) \equiv f_{h2}(X)$, $f_{h1}(X)$ is said to be identically equal to $f_{h2}(X)$.

(2) Magnitude of scalar error functions

Definition 8.5 Suppose that $f_{h1}(X)$ and $f_{h2}(X)$ are defined on domain U, $\forall X \subseteq U$ such that $f_{h1}(X) \leq f_{h2}(X)$, $f_{h1}(X)$ is said to be less than and equal to $f_{h2}(X)$; particularly, $\forall X \subseteq U$ such that $f_{h1}(X) < f_{h2}(X)$, $f_{h1}(X)$ is said to be less than $f_{h2}(X)$.

2. Properties of scalar error function

Proposition 8.1 *Suppose that $f_h(X)$ is defined on domain U and $X \subseteq U$, function $f_h(X)$ has the property of non-negativity on domain U.*

Proof Since $f_h(X) = \sum_{u \in X} v_u$, from the definition for $f_h(X)$, $\forall u \in X$ such that $v_u \geq 0$, hence $\sum_{u \in X} v_u \geq 0$, $\therefore f_h(X) \geq 0$.
Proof is completed!

Proposition 8.2 *Suppose that $f_h(X)$ is defined on domain U, $X_1 \subseteq U$, $X_2 \subseteq U$, $X_1 \cap X_2 = \Phi$, $X_1 \cup X_2 = X$, and $f_h(X)$ is not association error function, then $f_h(X) = f_h(X_1) + f_h(X_2)$.*

Proof Based on given conditions, $f_h(X) = \sum_{u \in X} v_u = \sum_{u \in X_1} v_u + \sum_{u \in X_2} v_u = f_h(X_1) + f_h(X_2)$.
Proof is completed!

Proposition 8.3 *Suppose that $f_h(X)$ is defined on domain U, $X_1 \subseteq U$, $X_2 \subseteq U$, $X_1 \subseteq X_2$, then $f_h(X_1) \leq f_h(X_2)$.*

Proof Since $X_1 \subseteq X_2$, $f_h(X_2) = f_h(X_1) + f_h(X_2 - X_1) + f_h(X_1, X_2 - X_1)$. From proposition 8.2, it is known that $f_h(X_2 - X_1) \geq 0$, $f_h(X_1, X_2 - X_1) \geq 0$; thus $f_h(X_1) \leq f_h(X_2)$.
Proof is completed!

3. Operations of scalar error functions

(1) Union operation on scalar error functions

Definition 8.6 Suppose that $V = \{ v \mid v = f(G, u), u \in U \}$ is defined on domain U, X is a subset of U, $f_{h1}(X) = f_{h1}(G_1, X) = \sum_{u \in X} v_{1u}$, $f_{h2}(X) = f_{h2}(G_2, X) = \sum_{u \in X} v_{2u}$. If $\forall X \subseteq U$ such that $f_{h3}(X) = f_{h3}(G_3, X) = \sum_{u \in X} max(v_{1u}, v_{2u})$, $f_{h3}(X)$ is the union operation on $f_{h1}(X)$ and $f_{h2}(X)$ denoted by $f_{h3}(X) = f_{h1}(X) \oplus f_{h2}(X)$.

Proposition 8.4 *Suppose that $f_h(X_1)$ and $f_h(X_2)$ are scalar error functions defined on domain U, then $f_{h1}(X) \oplus f_{h2}(X) = f_{h2}(X) \oplus f_{h1}(X)$.*

Proof Proof is omitted here.

(2) Intersection operation on scalar error functions

Definition 8.7 Suppose that $V = \{ v \mid v = f(G, u), u \in U \}$ is defined on domain U, X is a subset of U, $f_{h1}(X) = f_{h1}(G_1, X) = \sum_{u \in X} v_{1u}$, $f_{h2}(X) = f_{h2}(G_2, X) = \sum_{u \in X} v_{2u}$. If $\forall X \subseteq U$ such that $f_{h3}(X) = f_{h3}(G_3, X) = \sum_{u \in X} min(v_{1u}, v_{2u})$, $f_{h3}(X)$ is the intersection operation on $f_{h1}(X)$ and $f_{h2}(X)$ denoted by $f_{h3}(X) = f_{h1}(X) \otimes f_{h2}(X)$.

Proposition 8.5 *Suppose that $f_h(X_1)$ and $f_h(X_2)$ are scalar error functions defined on domain U, then $f_{h1}(X) \otimes f_{h2}(X) = f_{h2}(X) \otimes f_{h1}(X)$.*

Proof Proof is omitted here.

Proposition 8.6 *Suppose that $f_h(X_1)$ and $f_h(X_2)$ are scalar error functions defined on domain U, if $\forall X \subseteq U$ and $f_{h1}(X) \leq f_{h2}(X)$, there exists:*

(a) $f_{h1}(X) \oplus f_{h2}(X) = f_{h2}(X)$;
(b) $f_{h1}(X) \otimes f_{h2}(X) = f_{h1}(X)$.

Proof $\because \forall X \subseteq U, f_h(X_1) \leq f_h(X_2)$
$\therefore f_{h1}(X) \oplus f_{h2}(X) = \sum_{u \in X} max(v_{1u}, v_{2u}) = \sum_{u \in X} v_{2u} = f_{h2}(X)$.
Proof is omitted here!

Similarly, (b) can be proved.

Proposition 8.7 *Suppose that $f_h(X_1)$, $f_h(X_2)$, and $f_h(X_3)$ are three scalar error functions defined on domain U, there exists:*

(a) $f_{h1}(X) \oplus [f_{h2}(X) \otimes f_{h3}(X)] = [f_{h1}(X) \oplus f_{h2}(X)] \otimes [f_{h1}(X) \oplus f_{h3}(X)]$;
(b) $f_{h1}(X) \otimes [f_{h2}(X) \oplus f_{h3}(X)] = [f_{h1}(X) \otimes f_{h2}(X)] \oplus [f_{h1}(X) \otimes f_{h3}(X)]$.

Proof Proof is omitted here!

(3) Extension transformation operation on scalar error functions

Definition 8.8 Suppose that $f_h(X)$ is a scalar error function defined on domain U, if $T[f_h(X)] = g_h(Y)$, then T is said to have conducted extension transformation on $f_h(X)$, T is called extension transformation operator.

4. Weighted scalar error functions

Definition 8.9 Suppose that $V = \{v \mid v = f(G, u), u \in U\}$ is defined on domain U, $X = \{u_1, u_2, \ldots, u_n\}$ is a subset of U, there exists a coefficient set $C \subseteq \{(c_1, c_2,$

$$\ldots, c_n) \mid c_1 + c_2 + \ldots + c_n = 1, c_i \geq 0\}; \text{ if } f(G, X) = f(X) = \sum_{i=1}^{n} c_i v_i (G, u_i),$$

and when $v_i(G, u_i) = f(G, u_i) < 0$, let $v_i = 0$, then $f(X)$ is called weighted scalar error function noted by $f_{qh}(X)$.

Weighted scalar error function exercises high effectiveness in fitting multiple error values. In order to achieve that, several concerns need to be given consideration: (1) the so-called "weight" represents relative significance of certain information used for different errors, identification of this type of information must reflect decision-makers' weight on the importance of each error because there exist differences in feedback from different errors and reliability of different error values. Weight can be obtained through Delphi approach, analytic hierarchy process (AHP), comparative laws, entropy model, and programming techniques. (2) when employing weighted scalar error function, one needs to make sure that different errors are independent and error values should possess comparability. Error values should be normalized if they are not comparable in their existing states.

8.2.3   Vector Type Error Function

Definition 8.10 Suppose that $V = V_1 \times V_2 \times \ldots \times V_n$ is defined on domain $U = U_1 \times U_2 \times \ldots \times U_n$, where $V_i = (G_i, U_i)$, $f_i(G_i \nRightarrow \mu_i)$ $(i = 1, 2, \ldots, n)$; $f(u) = \{v_1 = f_1(G_1 \nRightarrow u_1), v_2 = f_2(G_2 \nRightarrow u_2), \ldots, v_n = f_n(G_n \nRightarrow u_n)\}$, $\vec{x} = (v_1, v_2, \ldots, v_n)$, $u = (u_1, u_2, \ldots, u_n)$, $u_i \in U_i$, $i = 1, 2, \ldots, n$; then $f(u)$ is called a vector type error function.

1. Relationships and operations of vector type error functions;
   In math, two vectors are equal if both direction and magnitude are equal. Suppose that $\vec{x_1} = (x_{11}, x_{12}, \ldots, x_{1n})$, $\vec{x_2} = (x_{21}, x_{22}, \ldots, x_{2n})$, if $x_{1i} \leq x_{2i}$, $i = 1, 2, \ldots, n$, thus vector $\vec{x_1}$ is less than and equal to vector $\vec{x_2}$ denoted by $\vec{x_1} \leq \vec{x_2}$. With no exception, vector type error function adheres to theories related to conventional vector theory.

Definition 8.11 Suppose that vector type error functions $f_1(\vec{x})$ and $f_2(\vec{x})$ are defined on domain $U = U_1 \times U_2 \times \ldots \times U_n$, if $\forall \vec{x} \in U$ such that $f_1(\vec{x}) = f_1(G_1, \vec{x}) = f_2(G_2, \vec{x}) = f_2(\vec{x})$, then vector type error functions $f_1(\vec{x})$ and $f_2(\vec{x})$ are said to be equal. Vector type error function $f_1(\vec{x})$ is said to be identically equal to $f_2(\vec{x})$ when $G_1 = G_2$ denoted by $f_1(\vec{x}) \equiv f_2(\vec{x})$

Definition 8.12 Suppose that vector type error functions $f_1(\vec{x})$ and $f_2(\vec{x})$ are defined on domain $U = U_1 \times U_2 \times \ldots \times U_n$, if $\forall \vec{x} \in U$ such that $f_1(\vec{x}) = f_1(G_1, \vec{x}) \leq f_2(G_2, \vec{x}) = f_2(\vec{x})$, then vector type error function $f_1(\vec{x})$ is less than and equal to $f_2(\vec{x})$ denoted by $f_1(\vec{x}) \leq f_2(\vec{x})$. Particularly, if $\forall \vec{x} \in U$ such that $f_1(\vec{x}) = f_1(G_1, \vec{x}) < f_2(G_2, \vec{x}) = f_2(\vec{x})$, then vector type error function $f_1(\vec{x})$ is less than $f_2(\vec{x})$ denoted by $f_1(\vec{x}) < f_2(\vec{x})$.

Proposition 8.8 *Suppose that $f(G, \vec{x_1})$ and $f(G, \vec{x_2})$ are defined on domain U, where $\vec{x_1}$ and $\vec{x_2}$ have no correlation, $\vec{x} = \vec{x_1} + \vec{x_2}$ such that $f(\vec{x}) = f(\vec{x_1}) + f(\vec{x_2})$.*

Proof Proof is omitted here!

Definition 8.13 Suppose that vector type error functions $f_1(\vec{x})$ and $f_2(\vec{x})$ are defined on domain $U = U_1 \times U_2 \times \ldots \times U_n$, $f_1(\vec{x}) = f_1(G_1, \vec{x}) = (v_{11}, v_{12}, \ldots, v_{1n})$, $f_2(\vec{x}) = f_2(G_2, \vec{x}) = (v_{21}, v_{22}, \ldots, v_{2n})$; if $\forall \vec{x} \in U$ such that $f(\vec{x}) = f(G, \vec{x}) = (max\ (v_{11}, v_{21}), max\ (v_{12}, v_{22}), \ldots, max\ (v_{1n}, v_{2n}))$, thus $f(\vec{x})$ is called the union operation on $f_1(\vec{x})$ and $f_2(\vec{x})$ denoted by $f(\vec{x}) = f_1(\vec{x}) \oplus f_2(\vec{x})$.

Proposition 8.9 *Suppose that vector type error functions $f_1(\vec{x})$ and $f_2(\vec{x})$ are defined on domain $U = U_1 \times U_2 \times \ldots \times U_n$, there exists $f_1(\vec{x}) \oplus f_2(\vec{x}) = f_2(\vec{x}) \oplus f_1(\vec{x})$.*

Proof Proof is omitted here!

Definition 8.14 Suppose that vector type error functions $f_1(\vec{x})$ and $f_2(\vec{x})$ are defined on domain $U = U_1 \times U_2 \times \ldots \times U_n$, $f_1(\vec{x}) = f_1(G_1, \vec{x}) = (v_{11}, v_{12}, \ldots, v_{1n})$, $f_2(\vec{x}) = f_2(G_2, \vec{x}) = (v_{21}, v_{22}, \ldots, v_{2n})$; if $\forall \vec{x} \in U$ such that $f(\vec{x}) = f(G, \vec{x}) = (min\ (v_{11}, v_{21}), min\ (v_{12}, v_{22}), \ldots, min\ (v_{1n}, v_{2n}))$, thus $f(\vec{x})$ is called the intersection operation on $f_1(\vec{x})$ and $f_2(\vec{x})$ denoted by $f(\vec{x}) = f_1(\vec{x}) \otimes f_2(\vec{x})$.

Proposition 8.10 *Suppose that vector type error functions $f_1(\vec{x})$ and $f_2(\vec{x})$ are defined on domain $U = U_1 \times U_2 \times \ldots \times U_n$, there exists $f_1(\vec{x}) \otimes f_2(\vec{x}) = f_2(\vec{x}) \otimes f_1(\vec{x})$.*

Proof Proof is omitted here!

Proposition 8.11 *Suppose that vector type error functions $f_1(\vec{x})$ and $f_2(\vec{x})$ are defined on domain $U = U_1 \times U_2 \times \ldots \times U_n$, if $\forall \vec{x} \in U$ such that $f_1(\vec{x}) \leq f_2(\vec{x})$, then*

(a) $f_1(\vec{x}) \oplus f_2(\vec{x}) = f_2(\vec{x})$;

(b) $f_1(\vec{x}) \otimes f_2(\vec{x}) = f_1(\vec{x})$.

Proof Proof is omitted here!

Proposition 8.12 *Suppose that vector type error functions $f_1(\vec{x})$, $f_2(\vec{x})$, and $f_3(\vec{x})$ are defined on domain $U = U_1 \times U_2 \times \ldots \times U_n$, then*

(a) $f_1(\vec{x}) \oplus [f_2(\vec{x}) \otimes f_3(\vec{x})] = [f_1(\vec{x}) \oplus f_2(\vec{x})] \otimes [f_1(\vec{x}) \oplus f_3(\vec{x})]$;

(b) $f_1(\vec{x}) \otimes [f_2(\vec{x}) \oplus f_3(\vec{x})] = [f_1(\vec{x}) \otimes f_2(\vec{x})] \oplus [f_1(\vec{x}) \otimes f_3(\vec{x})]$.

Proof Proof is omitted here!

Definition 8.15 Suppose that vector type error function $f_1(\vec{x})$ is defined on domain U, there exists $T\,[f_1(\vec{x})] = g_2(\vec{y})$, then T is said to have conducted transformation on $f_1(\vec{x})$ within domain U, T is called an operator.

Definition 8.16 Suppose that vector type error function $f(\vec{x})$ is defined in domain U, there exists $T[\,f(\vec{x})] = f(\vec{y})$, then T is said to have conducted transformation on independent variable of $f(\vec{x})$ within domain U, T is called a transformation operator for independent variable.

2. Scalar vector type error function.

Definition 8.17 Suppose that vector type error function $f(\vec{X})$ is defined on domain $U = U_1 \times U_2 \times \ldots \times U_n$, where $V = V_1 \times V_2 \times \ldots \times V_n$, $V = \{v_{iu} \mid v_{iu} = f_j(G_j, u_j), u_j \in U_j\}, (i = 1, 2, \ldots, n)$; $f(\vec{x}) = f(G, \vec{X}) = \{\, f_h(\vec{X}_1), f_h(\vec{X}_2), \ldots, f_h(\vec{X}_n)\,\} = \{\, \sum_{u_1 \in X_1} v_{1u_1}, \sum_{u_2 \in X_2} v_{2u_2}, \ldots, \sum_{u_n \in X_n} v_{nu_n}\,\}$, $\vec{X} = \{\, \vec{X}_1, \vec{X}_2, \ldots, \vec{X}_n\,\}$, X_i is a subset of U_i, $i = 1, 2, \ldots, n$, i.e., $\vec{X} \subseteq U$; let $v_{ij} = 0$, then $f(\vec{X})$ is called scalar vector type error function.

Definition 8.18 Suppose that scalar vector type error functions $f_1(\vec{X})$ and $f_2(\vec{X})$ are defined on domain $U = U_1 \times U_2 \times \ldots \times U_n$, if $\forall\, \vec{X} \subseteq U$, $f_1(\vec{X}) = f_1(G_1, \vec{X}) = f_2(G_2, \vec{X}) = f_2(\vec{X})$, then $f_1(\vec{X})$ is said to be equal to $f_2(\vec{X})$; especially, when $G_1 = G_2$, $f_1(\vec{X})$ is said to be identically equal to $f_2(\vec{X})$ denoted by $f_1(\vec{X}) \equiv f_2(\vec{X})$

Definition 8.19 Suppose that scalar vector type error functions $f_1(\vec{X})$ and $f_2(\vec{X})$ are defined on domain $U = U_1 \times U_2 \times \ldots \times U_n$, if $\forall\, \vec{X} \subseteq U$, $f_1(\vec{X}) = f_1(G_1, \vec{X}) \leq f_2(G_2, \vec{X}) = f_2(\vec{X})$, then $f_1(\vec{X})$ is said to be less than and equal to $f_2(\vec{X})$ $f_1(\vec{X}) \leq f_2(\vec{X})$. Particularly, $\forall\, \vec{X} \subset U$, values of function meet $f_1(G_1, \vec{X}) < f_2(G_2, \vec{X})$.

Proposition 8.13 *Scalar vector type error function* $f(\vec{X})$ *has the property of non-negativity on domain U.*

Proof Proof is omitted here!

Proposition 8.14 *Suppose that scalar vector type error functions* $f(\vec{X}_1)$ *and* $f(\vec{X}_2)$ *are defined on domain* $U = U_1 \times U_2 \times \ldots \times U_n$, $\vec{X}_1 = \{\ \vec{X}_{11}, \vec{X}_{12}, \ldots, \vec{X}_{1n}\ \}$, $\vec{X}_2 = \{\ \vec{X}_{21}, \vec{X}_{22}, \ldots, \vec{X}_{2n}\ \}$; *if* $X_{1i} \cap X_{2i} = \Phi$, $X_{1i} \cup X_{2i} = X_i$, $i = 1, 2, \ldots, n$, *and* $\vec{X}_1$ *and* $\vec{X}_2$ *are not correlated, thus* $f(\vec{X}) = f(\vec{X}_1) + f(\vec{X}_2)$.

Proof Proof is omitted here!

Proposition 8.15 *Suppose that scalar vector type error functions* $f(\vec{X}_1)$ *and* $f(\vec{X}_2)$ *are defined on domain* $U = U_1 \times U_2 \times \ldots \times U_n$, $\vec{X}_1 = \{\ \vec{X}_{11}, \vec{X}_{12}, \ldots, \vec{X}_{1n}\ \}$, $\vec{X}_2 = \{\ \vec{X}_{21}, \vec{X}_{22}, \ldots, \vec{X}_{2n}\ \}$; *if* $X_{1i} \subseteq X_{2i}$, $i = 1\ 2,, \ldots, n$, *thus* $f(\vec{X}_1) \leq f(\vec{X}_2)$.

Proof Proof is omitted here!

Scalar vector type error function also has union, intersection and extension operations, which also have similar characteristics as vector type error function.

8.3 Relationships Between Error Function and Judging Rules for Errors

Based on definition for error function f, it has two independent variables-i.e., element u in object set U and judging rule G for errors. Therefore, error function f is determined by judging rules G if u is kept constant. The values of error function are changing with respect to G, which might exhibit certain regularity.

8.3.1 Effects of Judging Rules on Error Function f

Judging rules, in reality, are requirements and conditions for evaluating and judging the object of interests. Generally, there are two types of indicators-i.e., qualitative and quantitative ones. For instance, when procuring fight jet, maximum velocity, flight radius, maximum take-off weight, and price are quantitative indicators. While flight agility, maneuverability, and maintainability, etc. are qualitative indicators.

Quantitative indicators can be divided into multiple types including but are not limited to efficiency, cost, fixed value, and range, etc. Efficiency indicators refer to the case that smaller value means larger error; cost indicators refer to the case that higher cost means larger error; fixed value indicates the phenomenon that a particular value will be located (or optimal) to meet the requirements, i.e., the farther an actual value is off the fixed point, the larger the error value is; range indicators refer to the case that the chosen values must fall in certain predefined range. Except for the above four indicators, two more indicators are defined as blow.

Definition 8.20 The indicator that the maximum error value is attained when it approaches to a fixed value, which is called fixed value approximation indicator.

Definition 8.21 The indicator that the maximum error value can be found when it approaches to certain range, which is called range approximation indicator.

For example, in the process of extracting salt from seawater, two methods can be used: evaporation and crystallization. In the extracting process, it is better to make the operating environment deviate from the temperature that increases the solubility of salt. Soluble level is the deviation indicator because the further the seawater pool's temperature is away from the optimal temperature for achieving the maximum soluble level the easier for it to extract crystal salt. Evaporation approach is adopted when the temperature is higher than the optimal temperature and crystallization is conducted when temperature is lower than the optimal temperature.

As for qualitative indicators, quantification approaches (such as Bipolar measure) can be employed to make some of them quantifiable. Others that are not apt for quantification can still be used in evaluating the error functions by using natural language processing techniques(NLP), modeling techniques of linguistic variables, and textual script analysis.

Because judging rules for errors might include above-mentioned indicators or their combination thereof, error functions can take different forms based on the chosen indicators in the judging rules for errors when element u in object set U is given.

(1) Judging rules including efficiency indicator

It is assumed that judging rules of object system has provided efficiency indicator for certain state s_i, where $s_i \geq a$ and error value gets larger as s_i is much smaller than a. Under this circumstance, error function can take the forms of fuzzy error function, non-negative error function, and error function with critical points. For $s_i \in (-\infty, +\infty)$, $\forall s_i \geq a \geq 0$, if error values are all equal to 0, then error function can take the forms of fuzzy error function and non-negative error function e.g.,

$$f(G, u) = \begin{cases} e^{(1/(s_i - a))} & s_i < a \\ 0 & s_i \geq a \geq 0 \end{cases} \quad or \qquad (8.1)$$

$$f(G, \mu) = \begin{cases} ln(a - s_i + 1) & s_i < a \\ 0 & s_i \geq a \geq 0 \end{cases} \qquad (8.2)$$

And for $s_i \in (-\infty, +\infty)$, $\forall s_i \geq a \geq 0$, error function with critical points takes the lead if error value is less than 0 and gets smaller as s_i increases. For example: $f(G, u) = \alpha (a - s_i)^{\frac{1}{\beta}}$, α is positive constant and β is an odd number that is larger than 1.

(2) Judging rules including cost indicator

It is assumed that judging rules of object system offer cost indicator for certain state s_i, where $s_i \leq a$ and error value gets larger as s_i is much larger than a. Under this circumstance, error function can take the forms of fuzzy error function, non-negative error function, and error function with critical points.

(3) Judging rules including fixed value indicator

It is assumed that judging rules of object system present fixed value indicator for certain state s_i, where $s_i = a$ and error value gets larger as s_i deviates much farther from a. Under this circumstance, error function can take the forms of fuzzy error function and non-negative error function. While for $s_i \in (-\infty, +\infty)$, fuzzy error function can be chosen:

$$f(G, u) = \begin{cases} e^{(1/(a-s_i))} & s_i > a \\ 0 & s_i = a \ or \\ e^{(1/(s_i-a))} & s_i < a \end{cases} \tag{8.3}$$

$$f(G, u) = \begin{cases} ln(s_i - a + 1) & s_i > a \\ 0 & s_i = a \\ ln(a - s_i + 1) & s_i < a \end{cases} \tag{8.4}$$

(4) Judging rules including range indicator

Suppose that judging rules of object system provide range indicator for certain state s_i, where $s_i \in [a, b]$, and error value gets larger as s_i deviates much farther from range $[a, b]$. Under this circumstance, error function can take the forms of fuzzy error function and non-negative error function. While for $s_i \in (-\infty, +\infty)$, following functions can be used:

$$f(G, u) = \begin{cases} e^{(1/(b-s_i))} & s_i > b \\ 0 & a \leq s_i \leq b \ or \\ e^{(1/(s_i-a))} & s_i < a \end{cases} \tag{8.5}$$

$$f(G, u) = \begin{cases} ln(s_i - b + 1) & s_i > b \\ 0 & a \leq s_i \leq b \\ ln(a - s_i + 1) & s_i < a \end{cases} \tag{8.6}$$

(5) Judging rules including fixed value approximation indicator

Suppose that judging rules of object system provide fixed value approximation indicator for certain state s_i, where $s_i \neq a$, and error value gets larger as s_i

approaches to a. Under this circumstance, error function can take the forms of fuzzy error function and non-negative error function. While for $s_i \in (-\infty, +\infty)$, following functions can be used:

$$f(G, u) = \begin{cases} e^{(a-s_i)} & s_i > a \\ 1 & s_i = a \text{ or} \\ e^{(s_i-a)} & s_i < a \end{cases} \tag{8.7}$$

$$f(G, u) = \begin{cases} \alpha(s_i - a)^{-\frac{1}{\beta}} & s_i \geq a \\ \alpha(a - s_i)^{-\frac{1}{\beta}} & s_i < a \end{cases} \tag{8.8}$$

α is positive constant and $\beta \geq 1$.

(6) Judging rules including range approximation indicator

Suppose that judging rules of object system present range approximation indicator for certain state s_i, where $s_i \notin [a, b]$, and error value gets larger as s_i approaches to $[a, b]$. Under this circumstance, error function can take the forms of fuzzy error function and non-negative error function. While for $s_i \in (-\infty, +\infty)$, following functions can be used:

$$f(G, u) = \begin{cases} e^{(b-s_i)} & s_i > b \\ 1 & a \leq s_i \leq b \\ e^{(s_i-a)} & s_i < a \end{cases} \tag{8.9}$$

(7) Judging rules including qualitative indicator

Suppose that judging rules of object system include qualitative indicator, error function can take the forms of classic error function:

$$f(G, u) = \begin{cases} 1 & G \nRightarrow a \\ 0 & G \Rightarrow a \end{cases} \tag{8.10}$$

The above-mentioned examples intend to address how different indicators might exert impacts on error functions. In practice, error functions could have millions of forms. Therefore, the construction and form adoption of error function depend on specific issues and situation. We consider not only the judging rules but also actual needs. The major principle for depicting error function is to objectively evaluate to what extent the object of interests violates the judging rules.

8.3.2 Relationships Between Different Error Values Under Different Judging Rules

Proposition 8.16 *Suppose that G_1 and G_2 are two groups of judging rules and $G_1 \sim G_2$ such that $f(G_1, u) = f(G_2, u)$ holds for any object $u \in U$.*

Proof Proof is omitted here!

Proposition 8.17 *Suppose that G are judging rules defined on domain U and $U_1 \subset U_2 \subset U$ such that $max(f(G \nRightarrow u_1)) \leq max(f(G \nRightarrow u_2))$ holds if $f(G \nRightarrow u) \geq 0$ for any object $u \in U$ and judging rules G.*

Proof $\because U_1 \subset U_2$
$\therefore f(G \nRightarrow U_2) = f(G \nRightarrow U_1) + f(G \nRightarrow U_2 - U_1)$
As $f(G \nRightarrow u) \geq 0$ for any object $u \in U$ and judging rules G.
$\therefore max(f(G \nRightarrow U_1)) \leq max(f(G \nRightarrow U_2))$.

Proposition 8.18 *Suppose that G are judging rules defined on domain U and $U_1 \subset U_2 \subset U$ such that $min(f(G \nRightarrow u_1)) \leq min(f(G \nRightarrow u_2))$ holds if $f(G \nRightarrow u) \geq 0$ for any object u and judging rules G.*

Proof $\because U_1 \subset U_2$
$\therefore f(G \nRightarrow U_2) = f(G \nRightarrow U_1) + f(G \nRightarrow U_2 - U_1)$
$\therefore min(f(G \nRightarrow u_1)) \leq min(f(G \nRightarrow u_2))$.
Proof is completed!

Chapter 9
Applications of Error Systems Theory

Abstract The primary objective of our research is to prevent, identify, and eliminate errors in complicated socioeconomic systems. This chapter introduces the applications of error systems theory in real examples and project. A case of waste water treatment system is presented in this section to illustrate how the steps of identifying and eliminating errors in a system are exercised in reality.

9.1 Preparation for Eliminating Errors

9.1.1 Identifying Error by Analyzing Root Causes

There might exist various causes for errors and the structures of errors also have diverse properties. According to error systems theory, for the object/issue of interest, the first step is to abstract it to be an object system and then rules for judging error are established. Thereafter, the causes for producing errors are investigated from a theoretical angle. Besides, those proposed theories should be validated using evidence and data from the field.

9.1.2 Principles for Eliminating Errors

1. Well-targeted planning and implementing
   In the process of preventing, identifying, reducing, and eliminating errors, a precisely targeted plan must be well designed. Different methods are adopted for dealing with errors generated in distinct space, time, and fields. Root causes need to be scrutinized and figured out in order to reduce, prevent, and eliminate errors, which resembles the philosophy of diagnosing diseases and exercising a permanent cure in Chinese medicine.

© Springer Nature Switzerland AG 2020
K. Guo and S. Liu, *Error Systems: Concepts, Theory and Applications*,
Studies in Systems, Decision and Control 275,
https://doi.org/10.1007/978-3-030-40760-5_9

2. Hierarchical structure

   As system's structure has different hierarchies, so does the rules for judging errors. The process of eliminating error also demands hierarchical actions at different layers.

3. Iterative processes

   The only thing in the world does not change is change. The undertaking of eliminating errors is a continual recurring process in which the rules for judging errors are dynamically varying with the change of elements and structure of the system associated with the identified errors. Generally, there is no one-for-ever panacea in reality.

4. Multistage operations

   During the course of human history, human beings came to understand and unfolded the world gradually. With the help of ever-emerging technologies, human beings are seemingly approaching the coding secret of the world in which we are living than any time in history. Nevertheless, people are confronting more challenges than ever since the universe is unfolding enormous secrets to us. People have different objectives and criteria in understanding the universe as they are knowing about more about the universe and the insufficiency of their knowledge, which renders the error to possess different characteristics during different periods. Thus, the process of eliminating errors demands multistage actions, where different phases could span a long period of time.

5. Minimum costs

   In general, there exist multiple alternatives to prevent or eliminate error. When evaluating different alternatives, one major factor needs to be considered is the cost effectiveness.

9.1.3   Transformations and Their Impacts

In an object system, any change in one element or subsystem will definitely affect the behaviors of other elements or subsystems since everything is connected to everything else and the elements or subsystems are tightly coupled. For example, some conclusions of an object system can be affected when some conditions in the object system change. The change in the element of object system also affects the interacting elements, which extends the impacts to other elements as well. The change in the intrinsic features of object system will generate different conclusions. This section provides diagram for illustrating the transition of the impacts when elements of the object system are changing (Fig. 9.1).

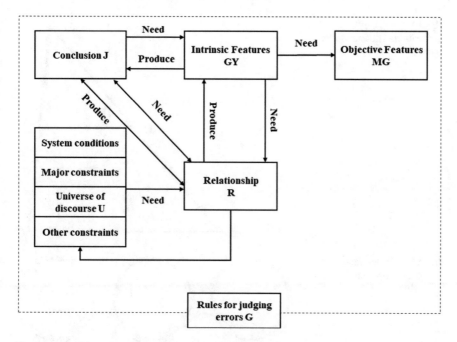

Fig. 9.1 Transition of the impact as elements changing

9.2 Eliminating Errors in Systems Through Similarity Transformation

Generally, eliminating errors with common features or similarities is exercised by using the laws and methods of similarity transformation. This session offers some examples on how the laws and mechanisms in changes of system error when system structure changes are used to eliminated errors with similarities in practice.

9.2.1 Introduction

The morph that scale of a structure is augmented or lessened (miniature or giant) while shape is kept unchanged is called similarity transformation on system structure (referring to Figs. 9.2, 9.3, 9.4, 9.5, and 9.6).

Suppose that W and V are two subjects, W and V are similar as long as they have more than one equivalent parameters.

Fig. 9.2 Simple example of similarity transformation on figure-1

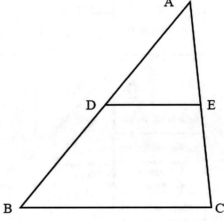

Fig. 9.3 Simple example of similarity transformation on figure-2

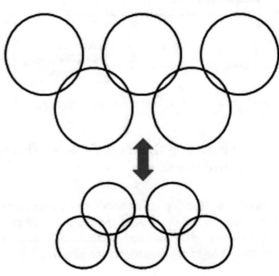

Fig. 9.4 Simple example of similarity transformation on figure-3

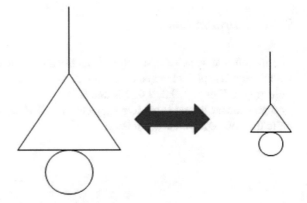

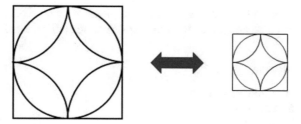

Fig. 9.5 Simple example of similarity transformation on figure-4

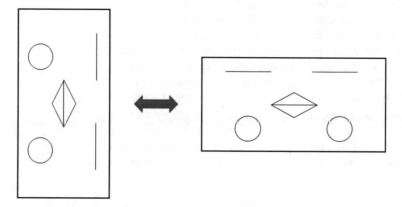

Fig. 9.6 Simple example of similarity transformation on figure-5

9.2.2 Principles in Eliminating Errors Through Transformation

Three principles must be held when considering the transformation in eliminating errors.

(1) Objective needs
(2) Determined by actual conditions
(3) Minimum costs.

9.2.3 Eliminating Error Through Similarity Transformation

Guo and Liu [91] proposed "15-6-3" method for eliminating error. Regarding four basic factors-i.e., universe of discourse, rules for judging errors, time, and system (structure, element, and subsystem or parameters of object such as subject, space, properties, quantifier, error value, and error function), 15 paths are provided as follows. Regarding transformation definitions, please refer to Chap. 5 for details. The 15 paths can also be applied to other transformations discussed in the ensuing sessions.

1. Fifteen for eliminating errors

    (a) Transformation on domain;
    (b) Transformation on system;
    (c) Temporal transformation;
    (d) Transformation on rules;
    (e) Simultaneous transformation on domain and system;
    (f) Simultaneous transformation on domain and time;
    (g) Simultaneous transformation on domain and rules;
    (h) Simultaneous transformation on system and time;
    (i) Simultaneous transformation on system and rules;
    (j) Simultaneous transformation on time and rules;
    (k) Simultaneous transformation on system, domain, and time;
    (l) Simultaneous transformation on system, domain, and rules;
    (m) Simultaneous transformation on domain, time, and rules;
    (n) Simultaneous transformation on system, time, and rules;
    (o) Simultaneous transformation on system, domain, time, and rules.

9.2.4 Examples Illustrating Error Elimination Through Similarity Transformation

1. Case background and error type

(1) *Case origin and short description*:

"*Not all insurgencies have been protracted affairs. It has become a matter of conventional wisdom that insurgencies last an average of 10 years and that the insurgents win about 40% of the time. These statistics have appeared in USA Today, PBS, Pentagon media briefings and on National Public Radio. The insight these numbers are meant to convey is that counterinsurgencies are inherently long and difficult struggles against wily and resilient foes, so it is unrealistic to expect rapid, quantifiable progress in the near term. Fortunately, these statistics are misleading and the associated analysis is wrong.*

    The source of this mistaken conventional wisdom is the prestigious Dupuy Institute, which has been providing rigorous quantitative analysis to the military for more than 40 years. In May 2007, Dupuy researchers published the preliminary results of a study in which they examined 63 modern insurgencies for a variety of factors, including the longevity and the success rate of the conflicts. Given their analytical talent and track record of precision, their statistical computations are undoubtedly accurate. The problem, however, isn't with their math; it's with the initial selection of cases."[28]

(2) **Conclusion**: According to the research from Dupuy Institute, insurgences are protracted affairs with an average of 10 years' duration and only about 60% of them were put down.

(3) **Error type**: error in conclusion

2. Technical analysis on the error

(1) *Concepts*

Based on the definition from the Counterinsurgency Doctrine Manual(FM 3–24), an insurgency is "an organized movement aimed at the overthrow of a constituted government through the use of subversion and armed conflict." In this case, we use this definition as rules for judging if there is insurgence.

(2) *Error analysis*

"*In the past 100 years, there have been considerably more than 63 movements that would fit that definition, so to create a manageable data set, only the most violent and intense conflicts are likely to be included. However, because casualty counts are often a function of time, this method naturally trends toward long wars and excludes cases in which government forces crushed the nascent insurgency before it even got off the ground.*" [28]. The high failure rate of insurgences is obvious since it is very risky of using limited organized guerrilla forces or super power-supported armed proxies to counter against modern constituted government. Therefore, the conclusion of Dupuy Institute-i.e., "insurgences are protracted affairs with an average of 10 years' duration and only about 60% of them were put down" is erroneous (wrong). The theoretical reason for this error here resides in the inconsistency between the universe of discourse for choosing insurgence cases and the scope for insurgence defined in its definition. And the practical reason for this error is that samples collected are not representative for the obtained unbiased conclusion.

3. Approach to eliminate error

In this error, there exists inconsistency between the universe of discourse for choosing insurgence cases and the scope for insurgence defined in its definition. Specifically, the scope for definition of insurgence is larger than the universe of discourse used for selecting samples of insurgences. Suppose that the universe of discourse used for selecting samples of insurgences is U_1 and the scope for definition of insurgence is U, similarity transformation is conducted on U_1, i.e., $T_x(U_1) = kU_1 = U$.

Note: similarity transformation is just one of approaches for eliminating this type of error.

9.3   Eliminating Errors in Systems Through Displacement Transformation

9.3.1   Starting Story

There was a very interesting ancient story named "Measuring elephant's weight by Cao Chong" recorded in "*Records of the Three Kingdoms · Book of Wei · Biographies of Cao Chong*". Sun Quan, the King of Kingdom Wu, gave Cao Cao, the King of Kingdom Wei, an elephant as a gift. Cao Cao and his staffs never saw this kind of giant animal before and he wanted to know the weight of this hunk. However, no one knew how to measure this thing since Kingdom Wei did not have very large weight scale then. Some staff proposed that a very large weight scale must be built in order to measure the weight of this elephant. Another staff opposed the proposal by mentioning that it was not possible to measure a live animal since it was very hard to make it stand still on the built huge weight scale. Therefore, he then proposed that the elephant should be cut into pieces in order to know its weight. All other people including Cao Cao laughed at him since his idea was so stupid. Cao Chong who was Cao Cao's youngest son, about 5 years old, stood out in the crowd and proposed a very smart suggestion. He idea was: (1) prepare a large boat in the adjacent river; (2) allure (using some food such as banana) the elephant to board the large boat; (3) mark the waterline on the boat; (4) drive elephant out of the boat; (5) order solders to move stones to the boat until the waterline was reached; (6) measure the weight of each stone and calculate the total weight of all stones. The weight of all stones used is equal to the weight of the elephant. Of course, one can also use people or sand as items for the replacement in this example. This is a typical example that used displacement transformation to resolve actual problem.

9.3.2   Definition on Displacement Transformation

Definition 9.1   Suppose that $u(t) \in U$ is the object needing error elimination defined under judging rule G on universe of discourse U, if $T(u(t)) = v(t)$, then T has conducted displacement transformation on u(t) under $G(t)$ on U denoted by T_z.

Definition 9.2   Suppose that $(u(t) \in U, (u(t), x(t)) \in C$, C is the error set defined under judging rule G on universe of discourse U, if $T(u(t), x(t)) = f(u(t), \vec{p}, G(t))) = (v(t), y(t) = f(v(t), \vec{p}, G(t)))'$, then T has done displacement transformation on u(t) under $G(t)$ on U denoted by T_z.

Definition 9.3   Suppose that $\{A_i((U_i, S_{iu}(t), \vec{p}_{iu}, T_{iu}(t), L_{iu}(t)), x_i(t) = f_i(u_i(t), G_{iA}(t))), i = 1, 2, \ldots, m\}$ is the object needing error elimination defined under judging rule $G(t)$ on universe of discourse $U(t)$, if $T_z\{A_i((U_i, S_{iu}(t), \vec{p}_{iu}, T_{iu}(t), L_{iu}(t)), x_i(t) = f_i(u_i(t), G_{iA}(t))), i = 1, 2, \ldots, m\} = \{B_j((U_j, S_{jv}(t), \vec{p}_{jv},$

$T_{jv}(t), L_{jv}(t)), y_j(t) = g_j (v_j(t), G_{jB}(t))), j = 1, 2, \ldots, n\}$, then T_z has conducted displacement transformation on $\{A_i((U_i, S_{iu}(t), \overrightarrow{p_{iu}}, T_{iu}(t), L_{iu}(t)), x_i(t) = f_i (u_i(t), G_{iA}(t))), i = 1, 2, \ldots, m\}$ under $G(t)$ on U. Here, $\{B_j((U_j, S_{jv}(t), \overrightarrow{p_{jv}}, T_{jv}(t), L_{jv}(t)), y_j(t) = g_j (v_j(t), G_{jB}(t))), j = 1, 2, \ldots, n\}$ is used to replace $\{A_i((U_i, S_{iu}(t), \overrightarrow{p_{iu}}, T_{iu}(t), L_{iu}(t)), x_i(t) = f_i (u_i(t), G_{iA}(t))), i = 1, 2, \ldots, m\}$. In the transformation, we hope to replace the error value of $\{x_i(t), i = 1, 2, \ldots, m\}$ with the desired error value of $\{y_j(t), j = 1, 2, \ldots, n\}$.

There are more displacement transformations listed as below.

(1) Suppose that $a = max\{x_i(t), i = 1, 2, \ldots, m\}$; $b = min\{x_i(t), i = 1, 2, \ldots, m\}$; $c = max\{y_j(t), j = 1, 2, \ldots, n\}$; $d = min\{y_i(t), j = 1, 2, \ldots, n\}$. If $c < a$ and $d < b$, then T_{zcz} has conducted soothing displacement transformation on error value denoted by T_{zycz};

(2) Suppose that $a = max\{x_i(t), i = 1, 2, \ldots, m\}$; $b = min\{x_i(t), i = 1, 2, \ldots, m\}$; $c = max\{y_j(t), j = 1, 2, \ldots, n\}$; $d = min\{y_i(t), j = 1, 2, \ldots, n\}$. If $c > a$ and $d > b$, then T_{zcz} has conducted worsening displacement transformation on error value denoted by T_{zhcz};

(3) In the displacement transformation $\{x_i(t), i = 1, 2, \ldots, m\} \to \{y_j(t), j = 1, 2, \ldots, n\}$, if $y_i(t) = kx_i(t)$ $(i = 1, 2, \ldots, m, j = 1, 2, \ldots, n)$, then T_{zcz} has conducted error-value amplification displacement transformation denoted by T_{zkcz}.

(a) If $k \geq 1$, then T_{zcz} has conducted positive amplification displacement transformation on error value noted by T_{zkzcz}

(b) If $k \leq -1$, then T_{zcz} has conducted negative amplification displacement transformation on error value noted by T_{zkfcz};

(c) If $0 < k < 1$, then T_{zcz} has conducted positive decreasing displacement transformation on error value denoted by T_{zkzscz};

(d) If $-1 < k < 0$, then T_{zcz} has conducted negative decreasing displacement transformation on error value denoted by T_{zkfscz};

(e) If $k = 0$, then T_{zcz} has conducted error-elimination displacement transformation denoted by T_{zkhlcz}.

9.3.2.1 Types of Displacement Transformation in Error Elimination

In general, the object of interest $u(t)$ not only has vertical and horizontal structure but also contains many hierarchies and relationships. Therefore, it is necessary to conduct displacement transformation at different hierarchies. Suppose that in $\{A_i((U_i, S_{iu}(t), \overrightarrow{p_{iu}}, T_{iu}(t), L_{iu}(t)), x_i(t) = f_i (u_i(t), G_{iA}(t))), i = 1, 2, \ldots, m\}$ and $\{B_j((U_j, S_{jv}(t), \overrightarrow{p_{jv}}, T_{jv}(t), L_{jv}(t)), y_j(t) = g_j (v_j(t), G_{jB}(t))), j = 1, 2, \ldots, n\}$:

(1) $\{u_i(t), i = 1, 2, \ldots, m\} \to \{v_j(t), j = 1, 2, \ldots, n\}$, T_z has conducted domain displacement transformation with respect to $G(t)$ and $\{A_i((V_i, S_{iu}(t), \overrightarrow{p_{iu}}, T_{iu}(t), L_{iu}(t)), x_i(t) = f_i (u_i(t), G_{iA}(t))), i = 1, 2, \ldots, m\}$ denoted by T_{zly}. In this case

$U(t) \cap V(t) \neq \Phi$ or $U(t) \cap V(t) = \Phi$. For example, when discussing issues related to human resource in China, suppose that domain Guangdong province $U(t)$ and domain China $V(t)$ are two domains, $V(t)$ can be used to replace $U(t)$.

(2) $\{S_{iu}(t), i = 1, 2, \ldots, m\} \rightarrow \{S_{jv}(t), j = 1, 2, \ldots, n\}$, T_z has conducted subject displacement transformation with respect to $G(t)$ and $\{A_i((V_i, S_{iu}(t), \overrightarrow{p_{iu}}, T_{iu}(t), L_{iu}(t)), x_i(t) = f_i(u_i(t), G_{iA}(t))), i = 1, 2, \ldots, m\}$ denoted by T_{zsw}. In this case, the displacement transformation is conducted on subject in $(U_i, S_{iu}(t), \overrightarrow{p_{iu}}, T_{iu}(t), L_{iu}(t))$ to achieve the expected goal.

(3) $\{\overrightarrow{p_{iu}}, i = 1, 2, \ldots, m\} \rightarrow \{\overrightarrow{p_{jv}}, j = 1, 2, \ldots, n\}$, T_z has conducted spatial displacement transformation with respect to $G(t)$ and $\{A_i((V_i, S_{iu}(t), \overrightarrow{p_{iu}}, T_{iu}(t), L_{iu}(t)), x_i(t) = f_i(u_i(t), G_{iA}(t))), i = 1, 2, \ldots, m\}$ denoted by T_{zkj}. In this case, the displacement transformation is carried out on the location of subject in $(U_i, S_{iu}(t), \overrightarrow{p_{iu}}, T_{iu}(t), L_{iu}(t))$ to achieve the expected goal.

(4) $\{T_{iu}, i = 1, 2, \ldots, m\} \rightarrow \{T_{jv}, j = 1, 2, \ldots, n\}$, T_z has conducted property displacement transformation with respect to $G(t)$ and $\{A_i((V_i, S_{iu}(t), \overrightarrow{p_{iu}}, T_{iu}(t), L_{iu}(t)), x_i(t) = f_i(u_i(t), G_{iA}(t))), i = 1, 2, \ldots, m\}$ denoted by T_{ztz}. In this case, the displacement transformation is conducted on the property of subject in $(U_i, S_{iu}(t), \overrightarrow{p_{iu}}, T_{iu}(t), L_{iu}(t))$ to achieve the expected goal.

(5) $\{L_{iu}, i = 1, 2, \ldots, m\} \rightarrow \{L_{jv}, j = 1, 2, \ldots, n\}$, T_z has conducted displacement transformation on the quantifier with respect to $G(t)$ and $\{A_i((V_i, S_{iu}(t), \overrightarrow{p_{iu}}, T_{iu}(t), L_{iu}(t)), x_i(t) = f_i(u_i(t), G_{iA}(t))), i = 1, 2, \ldots, m\}$ denoted by T_{zlz}. In this case, the displacement transformation is carried out on the quantifier of property $T(t)$ in $(U_i, S_{iu}(t), \overrightarrow{p_{iu}}, T_{iu}(t), L_{iu}(t))$ to achieve the expected goal. For example, the quantifier of length $T_u(t)$ and the quantifier of width $T_v(t)$ can be exchanged.

(6) $\{x_i(t), i = 1, 2, \ldots, m\} \rightarrow \{y_j(t), j = 1, 2, \ldots, n\}$, T_z has conducted displacement transformation on error value with respect to $G(t)$ and $\{A_i((V_i, S_{iu}(t), \overrightarrow{p_{iu}}, T_{iu}(t), L_{iu}(t)), x_i(t) = f_i(u_i(t), G_{iA}(t))), i = 1, 2, \ldots, m\}$ denoted by T_{zcz}. In this case, the displacement transformation is carried out on the error value in $(U_i, S_{iu}(t), \overrightarrow{p_{iu}}, T_{iu}(t), L_{iu}(t))$ to achieve the expected goal. For example, we hope to use the anticipated error value $\{y_j(t), j = 1, 2, \ldots, n\}$ to replace the undesirable error value $\{x_i(t), i = 1, 2, \ldots, m\}$.

(7) $\{G_{iA}(t), i = 1, 2, \ldots, m\} \rightarrow \{G_{jB}(t), j = 1, 2, \ldots, n\}$, T_z has conducted displacement transformation on rule with respect to $G(t)$ and $\{A_i((V_i, S_{iu}(t), \overrightarrow{p_{iu}}, T_{iu}(t), L_{iu}(t)), x_i(t) = f_i(u_i(t), G_{iA}(t))), i = 1, 2, \ldots, m\}$ denoted by T_{zgz}. In this case, the displacement transformation is carried out on the rule to achieve the expected goal.

(8) $\{f_i, i = 1, 2, \ldots, m\} \rightarrow \{g_j, j = 1, 2, \ldots, n\}$, T_z has conducted displacement transformation on error function with respect to $G(t)$ and $\{A_i((V_i, S_{iu}(t), \overrightarrow{p_{iu}}, T_{iu}(t), L_{iu}(t)), x_i(t) = f_i(u_i(t), G_{iA}(t))), i = 1, 2, \ldots, m\}$ denoted by T_{zhs}. In this case, the displacement transformation is implemented on error function to achieve the expected goal.

(9) $t_A \rightarrow t_B$ has conducted temporal displacement transformation with respect to $G(t)$ and $\{A_i((V_i, S_{iu}(t), \vec{p}_{iu}, T_{iu}(t), L_{iu}(t)), x_i(t) = f_i (u_i(t), G_{iA}(t))), i = 1, 2, \ldots, m\}$ denoted by T_{zsj}. In this case, the displacement transformation is conducted on time to achieve the expected goal.

If displacement transformation(T_z) has been conducted on domain $\{u_i(t), i = 1, 2, \ldots, m\} \rightarrow \{v_j(t), j = 1, 2, \ldots, n\}$, subject $\{S_{iu}(t), i = 1, 2, \ldots, m\} \rightarrow \{S_{jv}(t), j = 1, 2, \ldots, n\}$, space $\{\vec{p}_{iu}, i = 1, 2, \ldots, m\} \rightarrow \{\vec{p}_{jv}, j = 1, 2, \ldots, n\}$, property $\{T_{iu}, i = 1, 2, \ldots, m\} \rightarrow \{T_{jv}, j = 1, 2, \ldots, n\}$, quantifier $\{L_{iu}, i = 1, 2, \ldots, m\} \rightarrow \{L_{jv}, j = 1, 2, \ldots, n\}$, error function $\{f_i, i = 1, 2, \ldots, m\} \rightarrow \{g_j, j = 1, 2, \ldots, n\}$, error value $\{x_i(t), i = 1, 2, \ldots, m\} \rightarrow \{y_j(t), j = 1, 2, \ldots, n\}$, rule $\{G_{iA}(t), i = 1, 2, \ldots, m\} \rightarrow \{G_{jB}(t), j = 1, 2, \ldots, n\}$, and time $t_A \rightarrow t_B$ at the same time, then T_z has conducted comprehensive displacement transformation on the object of interests with respect to $G(t)$ and $\{A_i((V_i, S_{iu}(t), \vec{p}_{iu}, T_{iu}(t), L_{iu}(t)), x_i(t) = f_i (u_i(t), G_{iA}(t))), i = 1, 2, \ldots, m\}$ denoted by T_{zq}. In this situation, the displacement transformation is conducted on domain, subject, space, property, quantifier, error function, error value, time, and rules for judging errors to achieve the expected goal. Here, the displacement transformation connectives are $T_z \subseteq \{T_{zly}, T_{zsw}, T_{zkj}, T_{ztz}, T_{zlz}, T_{zcz}, T_{zgz}, T_{zhs}, T_{zsj}, T_{zq}\}$ (displacement transformation) and T_z^{-1} (inverse displacement transformation connectives).

9.3.3   Examples Illustrating Error Elimination Through Displacement Transformation

Example 1: Error in locating the operation site for a restaurant

1. Case background and error type

(Note: this is an old case occurred in 2005–2006. Therefore, information such as price only reflects that period's actual situation.) The site discussed in this case is to the north of Beijing UME International Cineplex across the street from Beijing Shuang'an Mall, which had an area of $200 \, m^2$ (Referring to Fig. 9.7). In 2005, its lessee was a Japanese style noodle restaurant named Yuanshengyuan. Before the opening of this noodle restaurant, 3 other restaurants were shut down on this site. Due to large passenger volume, this location was surrounded by many chained fast food providers such as McDonald, KFC, Yoshinoya, Yongho King, and A & J Restaurant. Before 2002, the lessee of this site was Mianaimian, which was taken over by another Japanese style restaurant after 2002. At the beginning of 2004, this property was leased to Este Pizza. Within less one year, it was transferred to Yuanshengyuan. It was witnessed that Yuanshengyuan's staffs used loudspeaker to solicit customers even during peak dining time. It obviously meant that the restaurant was then not attractive. It had no doubt that it would be passed to another lessee soon. The question is what happened to this site.

Fig. 9.7 Map showing the site location

This site is connected to Shuang'an mall across street through pedestrian overpass. The east side of Shuang'an has several office buildings namely, digital mansion and Zhongdianxinxi building, etc. The Supermarket Sends is located to the southwest side of this site. Large resident communities named Shuangyushu and Zhichunli are located to the north and east sides of this site, respectively. Dangdai Mall and Renmin University lie to the northwest side of this site. Therefore, shoppers, community residents, work commuters, and university students form the passenger flew around this area. With detailed investigation, it was found that majority of the passengers passing by this site were community residents. Moreover, some work commuters and shopper also came to this site looking for food during dining peak time. This was the pattern of passenger flow around this site.

2. Technical analysis on the error

(1) *Error in selecting objects(potential customers)*
A common feature among former unsuccessful restaurants was that their price was unwisely set higher than the expected costs for community residents (majority of the potential customers). The average consumption per meal was 20RMB, which was then a cost appropriate for white collar commuters and passengers shopping at luxury mall. For nearby residents, they would rather cook at home x than eat at restaurant with this expense for noodle or Donburi(rice topped with dishes). University students preferred to each KFC or McDonald with this price. Due to limited coverage distance of local brand restaurants, potential customers were within a radius of

500 m. Moreover, it was inconvenient for commuters working in the office buildings across street and shoppers lingering around in Shuang'an mall to cross the overpass to eat here. Therefore, the dilemma was that real customers were insufficient in the restaurant settled here although the passenger volume passing by this site was large.

(2) *Error in selecting spatial location*
The second type of error is that they tempted to run a middle (or high) end restaurant in a site with high-density low-end potential customers. The selection of this site made it inconvenient for serving effective customers. That could have been a different story if those restaurants were located at some sites adjacent to Shuang'an mall at the south side of the 3rd ring road.

3. Approach to eliminate error

(1) *Scenario 1*
   In the 1st type of error, displacement transformation can be conducted on the object-i.e., re-positioning target customers if the spatial location can not be changed. Current restaurant should review and revise its pricing tactics to cater to the needs of students and community residents.
(2) *Scenario 2*
   In the 2nd type of error, spatial displacement transformation can be conducted to relocate current restaurant to a site that is convenient to commuters working in high-end office buildings and shoppers if they still wanted to use current pricing tactics.

Example 2: Common-sense error in daily life

In order to perfectly peel the boiled egg's shell off, people often put boiled eggs in cold tap water (impotable water), which shrinks the membrane between egg-white and the egg shell allowing people peel it off in one perfect piece. However, this process created opportunity for bacteria or virus in tap water to enter the gap between membrane and egg shell, which possibly entered consumer's body. It had food hygiene issues especially for baby.

1. Rules for judging errors

   (a) *"Food should be nontoxic and harmless, conform to proper nutritive require-ments and have appropriate sensory properties such as colour, fragrance, and taste."* The Food Hygiene Law of the People's Republic of China § 6 (1995).
   (b) *"Principal and supplemental foods intended specially for infants and preschool children shall conform to the nutritive and hygienic standards promulgated by the administrative department of public health under State Council.":* The Food Hygiene Law of the People's Republic of China § 7 (1995)

2. Error analysis
   Putting boiled egg into cold tap water will possibly bring bacteria or virus into the peeled egg, which violates the requirements in Article 6 of Food Hygiene Law quoted above.
3. Approach to eliminate error
   Displacement transformation can be conducted on the object by replacing cold tap water with cold boiled water.

9.4   Eliminating Errors in Systems Through Decomposition Transformation

9.4.1   Starting Story

Project structure must be carefully analyzed before developing project schedule. Project structural decomposition is done through tool such as Work Breakdown Structure (WBS), which can systematically unfold the project structure to be the smallest, independent, mono-content, and meaningful elements easier to be applied to cost accounting. The specific implementation processes and details are also provided with the deployment of decomposition process. The decomposition process should be able to guarantee that clear connecting logic and precedence among obtained constituent elements are clarified, confirmed, and verified. Each element should be assigned to particular stakeholder and commanding, integrating, and coordinating mechanisms must be established. The bases for developing project schedule are: (1) scope of the project's objectives; (2) project duration (deadline); (3) properties of project; (4) internal and external conditions; (5) properties of decomposed project elements; (6) available resources. In order to guarantee the accomplishment of project objectives, the development of project schedule must take into consideration of cost, quality, safety, objective conditions, and risk as well as their relationships.

The poor performance management of an enterprise can be partially attributed to poorly defined strategies and/or the way that senior management team translated them into plans in lower level hierarchies which was a kind of decomposition transformation on the strategy element of the whole enterprise system. Without well-developed strategies and plans in different departments or units, it is not possible to have effective and efficient operation for each department since they do not know how to align employees' individual goals with the company's objectives. It is not uncommon that a whole department is performing poorly even though its employees are well qualified because the company lacks an effective layer-by-layer breakdown procedures regarding performance indicators.

In order to make their games more attractive to players, many game platforms provides a magic box (inventory) where players can use the function of "disenchanting" (i.e. breakdown or decomposition) to covert armor or weapons into more powerful, magical, and practical gears such as dusts, essences, shards which can be employed enchanting recipes. Let's imagine that soldiers in real battle field can use this kind of concepts(disenchanting and enchanting, decompose and compose) to decompose and recompose their backpack's gears. This is going to significantly reduce the weight they carry and lessen the reliance on resupply convoys, and consequently increase the survival rate in complicated situations.

From the above simple examples, we know that decomposition transformation has practical applications in eliminating errors or dealing with challenging issues.

9.4.2 Definition on Decomposition Transformation

Definition 9.4 Suppose that $u(t) \in U$ is the object needing error elimination defined under judging rule G on universe of discourse U, if $T(u(t)) = \{u_1(t), u_2(t), \dots, u_n(t)\}$, then T has conducted decomposition transformation on u(t) under $G(t)$ on U noted by T_f.

Definition 9.5 Suppose that $(u(t) \in U, (u(t), x(t)) \in C, C$ is the error set defined under judging rule G on universe of discourse U, if $T(u(t), x(t)) = \{(u_1(t), x_1(t) = f_1((u_1(t), \vec{p_1}), G(t))), (u_2(t), x_2(t) = f_2((u_2(t), \vec{p_2}), G_2(t))), \dots \dots, (u_n(t), x_n(t) = f_n((u_n(t), \vec{p})n), G_n(t)))\}$, then T has done decomposition transformation on $(u(t), x(t))$ under $G(t)$ on U denoted by T_f.

Definition 9.6 Suppose that $A((U, S(t), \vec{p}, T(t), L(t)), x(t) = f((u(t), \vec{p}), G(t)))$ is the object needing error elimination defined under judging rule $G(t)$ on universe of discourse $U(t)$, if $T(A((U, S(t), \vec{p}, T(t), L(t)), x(t) = f((u(t), \vec{p}), G(t)))) = \{A_1((U_1, S_1(t), \vec{p_1}, T_1(t), L_1(t)), x_1(t) = f_1((u_1(t), \vec{p_1}), G_1(t))), A_2((U_2, S_2(t), \vec{p_2}, T_2(t), L_2(t)), x_2(t) = f_2((u_2(t), \vec{p_2}), G_2(t))), \dots \dots, A_n((U_n, S_n(t), \vec{p_n}, T_n(t), L_n(t)), x_n(t) = f_n((u_n(t), \vec{p_n}), G_n(t)))\}$, here $u_1(t) \quad h\ u_2(t) \quad \dots \quad u_n(t) = u(t)$, then T has conducted decomposition transformation on $A((U, S(t), \vec{p}, T(t), L(t)), x(t) = f((u(t), \vec{p}), G(t)))$ under judging rule $G(t)$ on universe of discourse $U(t)$ denoted by T_f.

In this case:

(1) In $\{x_i(t), i = 1, 2, \dots, m\}$, if $x_i(t) \geq x(t), i = 1, 2, \dots, n$, then T_f has conducted worsening decomposition transformation on error denoted by T_{fz};
(2) In $\{x_i(t), i = 1, 2, \dots, m\}$, if $x_i(t) \leq 0, i = 1, 2, \dots, n$, then T_f has conducted error-elimination decomposition transformation denoted by T_{fx};
(3) In $\{x_i(t), i = 1, 2, \dots, m\}$, if $x_i(t) = kx(t), i = 1, 2, \dots, n$, then T_f has conducted amplification decomposition transformation on error value denoted by T_{fk}.

  (a) If $k \geq 1$, then T_{fk} has conducted positive amplification decomposition transformation on error value denoted by T_{fzk};

(b) If $k \leq -1$, then T_{fk} has conducted negative amplification decomposition transformation on error value denoted by T_{ffk};

(c) If $0 < k < 1$, then T_{fk} has conducted positive decreasing decomposition transformation on error value denoted by T_{fzs};

(d) If $-1 < k < 0$, then T_{fk} has conducted negative decreasing decomposition transformation on error value denoted by T_{ffs};

(e) If $k = 0$, then T_{fk} has conducted error-elimination decomposition transformation denoted by T_{fhl}.

9.4.3 Types of Decomposition Transformations in Error Elimination

Types of decomposition transformation:

(1) Physical decomposition
(2) Mathematical decomposition
(3) Decomposing according to objective needs and requirements
(4) Comprehensive decomposition: based on the definition for T_f and the element of $A((U, S(t), \vec{p}, T(t), L(t)), x(t) = f((u(t), \vec{p}), G(t)))$, decomposition transformation T_f can be conducted on universe of discourse U, object $u(t)$, error value $x(t)$, error function f, time t, and rule for judging errors of $A((U, S(t), \vec{p}, T(t), L(t)), x(t) = f((u(t), \vec{p}), G(t)))$, therefore, $T_f \subseteq \{T_{fly}, T_{fsw}, T_{fkj}, T_{ftz}, T_{flz}, T_{fcz}, T_{fgz}, T_{fhs}, T_{fsj}, T_{fq}\}$. The type of error logical variable of $B((U, S(t), \vec{p}, T(t), L(t)), x(t) = f((u(t), \vec{p}), G(t)))$ does not change if decomposition transformation T_f is not carried out on error function f.

9.4.4 Examples Illustrating Error Elimination Through Decomposition Transformation

Example 1 Case background and error type

Carrot, a root vegetable, is a good source of vitamin K1, potassium, fiber, antioxidants, and beta carotene [203]. Due to carrot's crunchy, tasty, and nutritious properties, many people liked to eat it. Some other people also ate it when they were drinking liquor.

1. Rules for judging errors

    (a) *"Food should be nontoxic and harmless, conform to proper nutritive requirements and have appropriate sensory properties such as colour, fragrance, and taste."* The Food Hygiene Law of the People's Republic of China § 6 (1995).

(b) Research finding: "Alcohol, vitamin A, and beta-carotene: adverse interactions, including hepatotoxicity and carcinogenicity" [140]

2. Error analysis
   According to the above-listed two rules, eating carrot while drinking liquor (or drinking liquor product with carrot) is erroneous. The beta carotene contained in carrot when consumed together with alcohol will produce toxins in the liver, and thereby damage the function of the liver.
3. Approach to eliminate error
   We can borrow the concept of decomposition transformation by taking them at different times, which eliminates the error in the above case.

9.5 Eliminating Errors in Systems Through Addition Transformation

9.5.1 Starting Story

In SAP software, there are several ways to find enhancements. One of them is described as follows: "(1) to find MODX _ FUNCTION _ ACTIVE _ CHECK using system functions; (2) to add a breakpoint at the end of this FUNCTION code; (3) execute the TCODE that needs to be enhanced. If there is any enhancement, it will automatically jump into the DEBUG interface. In the DEBUG interface, look at the f_tab field. The Smod shown here is a list of all the enhancements for this TCODE. These enhancements are all in the form of EXIT_XXXXXX_XXX"(Anonymous, http://programmersought.com/article/35291363781/;jsessionid=5AE70A7479F3CD9386899B693D36E228). In this case, process of adding breakpoint is the concept of addition transformation.

In another example, with the increased concentration of glass fiber, reinforced Polypropylene (RPP) demonstrates obvious improvement in tensile strength, flexural modulus, and composite impact strength. At the same time, charpy impact, tensile impact, and high speed impact properties apparently increase with fiber length up to 6 mm [213, 214]. This also reflects the impact of addition transformation on the material research.

9.5.2 Definition on Addition Transformation

Definition 9.7 Suppose that $u(t) \in U$ is the object needing error elimination defined under judging rule G on universe of discourse U, if $T(u(t)) = \{u(t), \ldots, v(t)\}$, then T has conducted addition transformation on $u(t)$ under $G(t)$ on U denoted by T_{zj}.

Definition 9.8 Suppose that $(u(t) \in U, (u(t), x(t)) \in C$, C is the error set defined under judging rule G on universe of discourse U, if $T(u(t), x(t) = f(u(t), \vec{p}, G(t)))$ $= \{(u(t), x(t) = f(u(t), \vec{p}, G(t))), \ldots, (v(t), x(t) = f(v(t), \vec{p}, G(t)))$, then T has conducted addition transformation on $(u(t), x(t)) \in C$ under $G(t)$ on U denoted by T_{zj}.

Definition 9.9 Suppose that $A((U, S(t), \vec{p}, T(t), L(t)), x(t) = f((u(t), \vec{p}), G(t)))$ is the object needing error elimination defined under judging rule $G(t)$ on universe of discourse $U(t)$, if $T(A((U, S(t), \vec{p}, T(t), L(t)), x(t) = f((u(t), \vec{p}), G(t)))) = \{A((U, S(t), \vec{p}, T(t), L(t)), x(t) = f((u(t), \vec{p}), G(t))), A_1((U_1, S_1(t), \vec{p_1}, T_1(t), L_1(t)), x_1(t) = f_1((u_1(t), \vec{p_1}), G_1(t))), A_2((U_2, S_2(t), \vec{p_2}, T_2(t), L_2(t)), x_2(t) = f_2((u_2(t), \vec{p_2}), G_2(t))), \ldots, A_n((U_n, S_n(t), \vec{p_n}, T_n(t), L_n(t)), x_n(t) = f_n((u_n(t), \vec{p_n}), G_n(t)))\}$, then T has conducted addition transformation on $A((U, S(t), \vec{p}, T(t), L(t)), x(t) = f(u(t), G(t)))$ under $G(t)$ on U denoted by T_{zj}.

9.5.3 Types of Addition Transformations in Error Elimination

$T_{zj}(A((U, S(t), \vec{p}, T(t), L(t)), x(t) = f((u(t), \vec{p}), G(t)))) = \{A((U, S(t), \vec{p}, T(t), L(t)), x(t) = f((u(t), \vec{p_1}), G(t))), A_1((U_1, S_1(t), \vec{p_1}, T_1(t), L_1(t)), x_1(t) = f_1((u_1(t), \vec{p_1}), G_1(t))), A_2((U_2, S_2(t), \vec{p_2}, T_2(t), L_2(t)), x_2(t) = f_2((u_2(t), \vec{p_2}), G_2(t))), \ldots, A_n((U_n, S_n(t), \vec{p_n}, T_n(t), L_n(t)), x_n(t) = f_n((u_n(t), \vec{p_n}), G_n(t)))\}$, and if

(1) $U(t) \rightarrow U(t) \cup U_1(t) \cup U_2(t) \cup \cdots \cup U_n(t)$, then T_{zj} had conducted addition transformation on domain of $A((U, S(t), \vec{p}, T(t), L(t)), x(t) = f((u(t), \vec{p}), G(t)))$ denoted by T_{zjly}. For $U(t) \rightarrow U(t) \cup U_1(t) \cup U_2(t), \ldots, \cup U_n(t)$, addition transformation is conducted on the universe of discourse U of object $u(t)$ to achieve the expected goal.

(2) $U(t) \rightarrow u(t) \, h \, u_1(t) \, h \, u_2(t) \, h \, \ldots \, h \, u_n(t)$, then T_{zj} has conducted addition transformation on subject of $A((U, S(t), \vec{p}, T(t), L(t)), x(t) = f((u(t), \vec{p}), G(t)))$ under $G(t)$ on U denoted by T_{zjsw}. In this situation, subject addition transformation is carried out on subject $(u(t), \vec{p})$ to attain the expected objective.

(3) $\vec{p} \rightarrow \vec{p} + \vec{p_1} + \vec{p_2} + \cdots + \vec{p_n}$, then T_{zj} has conducted spatial addition transformation with respect to $G(t)$ and $A((U, S(t), \vec{p}, T(t), L(t)), x(t) = f((u(t), \vec{p}), G(t)))$ denoted by T_{zjkj}. In this situation, spatial addition transformation is conducted on the spatial location $\vec{p}$ of object $u(t)$ to attain the expected goal.

(4) $T(t) \rightarrow T(t) \cup T_1(t) \cup T_2(t) \cup \cdots \cup T_n(t)$, then T_{zj} has conducted addition transformation on property of $A((U, S(t), \vec{p}, T(t), L(t)), x(t) = f((u(t), \vec{p}), G(t)))$ under $G(t)$ on U denoted by T_{zjtx}. In this situation, property addition transformation is conducted on the $T(t)$ of object $u(t)$ to reach the expected goal.

(5) quantifier $L(t) \rightarrow L(t) + L_1(t) + L_2(t) + \cdots + L_n(t)$, then T_{zj} has conducted addition transformation on quantifier of $A((U, S(t), \vec{p}, T(t), L(t)), x(t) = f((u(t), \vec{p}), G(t)))$ under $G(t)$ on U denoted by T_{zjlz}. In this situation, addition transformation on quantifier is conducted on the $L(t)$ of object $u(t)$ to attain the expected goal.

(6) error value $x(t) \rightarrow x(t) + x_1(t) + x_2(t) + \cdots + x_n(t)$, then T_{zj} has conducted addition transformation on error value of $A((U, S(t), \vec{p}, T(t), L(t)), x(t) = f((u(t), \vec{p}), G(t)))$ under $G(t)$ on U denoted by T_{zjcz}. In this situation, addition transformation on error value is conducted on the $x(t)$ of object $u(t)$ to achieve the expected goal.

(7) $G(t) \rightarrow G(t) \cup G_1(t) \cup G_2(t) \cup \cdots \cup G_n(t)$, then T_{zj} has conducted addition transformation on rule of $A((U, S(t), \vec{p}, T(t), L(t)), x(t) = f((u(t), \vec{p}), G(t)))$ on U denoted by T_{zjgz}. In this situation, addition transformation on rule is conducted on the $G(t)$ of object $u(t)$ to attain the expected goal.

(8) $f(t) \rightarrow f(t) + f_1(t) + f_2(t) + \cdots + f_n(t)$, then T_{zj} has conducted addition transformation on error function of $A((U, S(t), \vec{p}, T(t), L(t)), x(t) = f((u(t), \vec{p}), G(t)))$ under $G(t)$ on U denoted by T_{zjhs}. In this situation, addition transformation on error function is conducted on the $f(t)$ of object $u(t)$ to achieve the expected objective.

(9) $\{A_1((U_1, S_1(t_1), \vec{p_1}, T_1(t_1), L_1(t_1)), x_1(t_1) = f_1((u_1(t_1), \vec{p_1}), G_1(t_1))), A_2((U_2, S_2(t_2), \vec{p_2}, T_2(t_2), L_2(t_2)), x_2(t_2) = f_2((u_2(t_2), \vec{p_2}), G_2(t_2))), \ldots, A_n((U_n, S_n(t_n), \vec{p_n}, T_n(t_n), L_n(t_n)) x_n(t_n) = f_n((u_n(t_n), \vec{p_n}), G_n(t_n)))\}$, T_{zj} has conducted temporal addition transformation on $A((U, S(t), \vec{p}, T(t), L(t)), x(t) = f((u(t), \vec{p}), G(t)))$ under $G(t)$ denoted by T_{zjsj}. In this situation, temporal addition transformation is conducted on the t of object $u(t)$ to achieve the expected goal.

(10) In $\{A_1((U_1, S_1(t_1), \vec{p_1}, T_1(t_1), L_1(t_1)), x_1(t_1) = f_1((u_1(t_1), \vec{p_1}), G_1(t_1))), A_2((U_2, S_2(t_2), \vec{p_2}, T_2(t_2), L_2(t_2)), x_2(t_2) = f_2((u_2(t_2), \vec{p_2}), G_2(t_2))), \ldots, A_n((U_n, S_n(t_n), \vec{p_n}, T_n(t_n), L_n(t_n)) x_n(t_n) = f_n((u_n(t_n), \vec{p_n}), G_n(t_n)))\}$, addition transformation is simultaneously carried on domain, subject, property, quantifier, error function, time,space, error value, and rules for judging errors. then T_{zj} has conducted comprehensive addition transformation on $A((U, S(t), \vec{p}, T(t), L(t)), x(t) = f((u(t), \vec{p}), G(t)))$ under $G(t)$ denoted by T_{zjq}. In this situation, comprehensive addition transformation is conducted on all the elements in the object $u(t)$ to attain the expected goal.

9.5.4 Examples Illustrating Error Elimination Through Addition Transformation

Example 1: Peeling egg using addition transformation

In previous example, ugly peeled egg is a result of erroneous processing. Another method that can prevent this error from happening is to **add** salt to the water used for boiling eggs. Having done that, it is thereby easy to peel the shell off without taking long time and scratching the egg white. Here, addition transformation is conducted on handling process to eliminate error.

Example 2: Fresh vegetables are better than that of frozen ones in term of preserved nutrients

This statement is correct if the vegetables were just picked from garden. *Fresh* is a relative term because vegetables might be in transit, stores, or refrigerator for some weeks. Research findings told us nutritional contents can be well preserved by quick freezing for vegetables (plus blanching process prior to freezing) even there are some percentage loss in some micronutrients such as Vitamin C and Thianmin (vitamin B1) [50, 181, 202]. Other nutrients such as anthocyanin flavonoids can be well preserved during freezing process [158].

1. Error analysis

   (a) Definition and annotations

   Suppose that the research object $u(t) = A((U, S(t), \vec{p}, T(t), L(t)), x(t) = f((u(t), \vec{p}), G(t)))$ is defined under rule $(G(t)$ on universe of discourse U, where $S(t)$ is the subject; $p(t)$ is location (spatial position); $T(t)$ is the property (or predicate); $L(t)$ is quantifier (predicative); $x(t) = f((u(t), \vec{p}), G(t))$ is truth value function (or error function); and $G(t)$ is a group of rules for judging errors on universe of discourse U. In the above example, the research object $u(t) =$ vegetable, universe of discourse $U =$ all vegetables, $S(t) =$ certain fresh vegetable (confirmed), $p(t) =$ spatial position (earth or a particular country such as China), $T(t) =$ nutrient of vegetable, $L(t) =$ content/concentration of nutrient (such as vitamin, anthocyanin flavonoids), $G(t) =$ standards (rules) for judging nutrient content in vegetables. Error function $x(t) = f$ (relationship (gap) between concentration of nutrients in the chosen vegetable and $G(t)$).

   (b) Identification of errors

   There are two errors in this statement:(1) fuzzy understanding on the micronutrients of a particular vegetable; (2) unscientific understanding on the dynamics of micronutrients over time of a particular vegetable under different preservation conditions.

2. Approach to eliminate error

   Addition transformation was conducted on time-i.e., $t = t_0 + \Delta t$, where $(t_0 + \Delta t) \in [0, T]$, T is the longest time that vegetable can be kept fresh (two conditions must be met here: (a) before vegetable becoming rotten; (b) the nutrients do not reach unacceptable level [183]. The content of micronutrients of the chosen vegetables in the control group and the experimental group are measured at time $t_0 + \Delta t$.

Example 3: Drinking natural spring water is safe

 Many commercial advertisements claimed that bottled natural spring water has rich minerals and therefore benefits the health of users. However, some ads exaggerated the functions and safety of drinking bottled natural spring water and ignored the negative side of it. According to research findings, detectable heavy metals such as lead, nickel, and arsenic were found in some bottled water brands [45, 206, 210]. Pathological microorganisms were also found exceeding acceptable criteria in some bottled mineral water (or natural spring water) even the amount does not directly cause disease, which is still very risky for users with weak immune system [138, 163, 219].

1. Error analysis

   (a) Definition and annotations

      Suppose that the research object u(t) $= A((U, S(t), \overrightarrow{p}, T(t), L(t)), x(t) = f ((u(t), \overrightarrow{p}), G(t)))$, is defined under rule $G(t)$ on universe of discourse U, where $S(t)$ is the subject; $p(t)$ is location (spatial position); $T(t)$ is the property (or predicate); $L(t)$ is quantifier (predicative); $x(t) = f ((u(t), \overrightarrow{p}), G(t))$ is truth value function (or error function); and $G(t)$ is a group of rules for judging errors on universe of discourse U. The research object here $u(t) =$ natural mineral water, universe of discourse $U =$ all brands of natural mineral water, $S(t) =$ certain chosen natural mineral water(confirmed), $p(t) =$ spatial position (earth or a particular country such as China or a particular region), $T(t) =$ minerals contained in the chosen natural mineral water, $L(t) =$ concentration/content of minerals (such as calcium, magnesium), $G(t) =$ standards (rules) for judging mineral content in natural spring water. Error function $x(t) = f$(relationship (gap) between concentration of minerals in the chosen natural spring water and $G(t)$).

   (b) Identification of errors

      There are two errors in this statement: (1) fuzzy understanding on the concentration of minerals in a particular brand of natural spring water; (2) missing check on concentration of pathological microorganisms and heavy metals in the natural spring water.

2. Approach to eliminate error

   Addition transformation can be conducted on property of natural spring water (or the process of confirming the properties of natural spring water), i.e., T_{zjtx} $(T(t)) = T_{zjtx}$ { essential minerals, water, heavy metal, pathological micro, other non-harmful contents } $= (T'(t)) =$ { essential minerals, water, other non-harmful contents }.

9.6   Eliminating Errors in Systems Through Destruction Transformation

9.6.1   Starting Story

Sweeping dust from appliance vents, dismantling dangerous buildings, and amputating organ with malignant tumor are all examples for destruction transformations.

9.6.2   Definition on Destruction Transformation

Definition 9.10 Suppose that $u(t) \in U$ is the object needing error elimination defined under judging rule G on universe of discourse U, if $T(u(t)) = \Phi$, then T has conducted destruction transformation on $u(t)$ under $G(t)$ on U denoted by T_h.

Definition 9.11 Suppose that $(u(t) \in U, (u(t), x(t)) \in C, C$ is the error set defined under judging rule G on universe of discourse U, if $T(u(t), x(t) = f((u(t), \vec{p}), G(t))) = \Phi$, then T has conducted destruction transformation $(u(t), x(t))$ under $G(t)$ on U denoted by T_h.

Definition 9.12 Suppose that $A((U, S(t), \vec{p}, T(t), L(t)), x(t) = f((u(t), \vec{p}) G(t)))$ is the object needing error elimination defined under judging rule $G(t)$ on universe of discourse $U(t)$, if $T_h(A((U, S(t), \vec{p}, T(t), L(t)), x(t) = f(u(t), G(t)))) = \{A((\Phi, \Phi, \Phi, \Phi, \Phi), \Phi = \Phi((\Phi, \Phi), \Phi))$, then T_h has conducted destruction transformation on $A((U, S(t), \vec{p}, T(t), L(t)), x(t) = f((u(t), \vec{p}) G(t)))$ under $G(t)$ on U denoted by T_h. The meaning of destruction is: T_h (destruction transformation connectives) $\rightarrow$ { kill, eradicate, annihilate, disappear, fire, sell out, discard, move away, and leave, ...}.

9.6.3   Types of Destruction Transformations in Error Elimination

Definition 9.13 Suppose that $A((U, S(t), \vec{p}, T(t), L(t)), x(t) = f((u(t), \vec{p}) G(t)))$ is the error logical variable defined under judging rule $G(t)$ on universe of discourse $U(t)$, if $T_h(A((U, S(t), \vec{p}, T(t), L(t)), x(t) = f((u(t), \vec{p}) G(t)))) = A((\Phi, S(t), \vec{p}, T(t), L(t)), x(t) = f((u(t), \vec{p}) G(t)))$, then T_h has conducted destruction transformation on domain of $A((U, S(t), \vec{p}, T(t), L(t)), x(t) = f((u(t), \vec{p}) G(t)))$

under $G(t)$ on U denoted by T_{hly}. The meaning of destruction transformation on domain is: T_{hly} (domain destruction) $\rightarrow$ (domain does not exist) $\rightarrow$ (there is no domain to discuss or there is no need to discuss the object in current domain).

Definition 9.14 Suppose that $A((U, S(t), \overrightarrow{p}, T(t), L(t)), x(t) = f((u(t), \overrightarrow{p})$ $G(t)))$ is the error logical variable defined under judging rule $G(t)$ on universe of discourse $U(t)$, if $T_h(A((U, S(t), \overrightarrow{p}, T(t), L(t)), x(t) = f((u(t),\overrightarrow{p}),$ $G(t)))) = A((U, \Phi, \overrightarrow{p}, T(t), L(t)), x(t) = f((u(t),\overrightarrow{p}), G(t)))$, then T_h has conducted destruction transformation on subject of $A((U, S(t), \overrightarrow{p}, T(t), L(t)), x(t) = f$ $((u(t),\overrightarrow{p}), G(t)))$ under $G(t)$ on U denoted by T_{hsw}. The meaning of destruction transformation on subject is: T_{hsw} (thing destruction) $\rightarrow$ (thing does not exist) $\rightarrow$ (there is no thing to discuss or there is no need to discuss the current thing in the object of interests, or the thing has been removed through action such as "remove, eradicate, annihilate, fire, sell out, discard, move away, and leave").

Definition 9.15 Suppose that $A((U, S(t), \overrightarrow{p}, T(t), L(t)), x(t) = f((u(t), \overrightarrow{p}),$ $G(t)))$ is the error logical variable defined under judging rule $G(t)$ on universe of discourse $U(t)$, if $T_h(A((U, S(t), \overrightarrow{p}, T(t), L(t)), x(t) = f((u(t),\overrightarrow{p}),$ $G(t)))) = A((U, S(t), \Phi, T(t), L(t)), x(t) = f((u(t), \overrightarrow{p}), G(t)))$, then T_h has conducted spatial destruction transformation on $A((U, S(t), \overrightarrow{p}, T(t), L(t)), x(t) = f$ $((u(t), \overrightarrow{p}), G(t)))$ under $G(t)$ on U denoted by T_{hkj}. The meaning of spatial destruction transformation is: T_{hkj} (spatial destruction) $\rightarrow$ (space does not exist) $\rightarrow$ (there is no space to discuss or there is no need to discuss the object in current space).

Definition 9.16 Suppose that $A((U, S(t), \overrightarrow{p}, T(t), L(t)), x(t) = f((u(t),\overrightarrow{p}),$ $G(t)))$ is the error logical variable defined under judging rule $G(t)$ on universe of discourse $U(t)$, if $T_h(A((U, S(t), \overrightarrow{p}, T(t), L(t)), x(t) = f((u(t), \overrightarrow{p}),$ $G(t)))) = A((U, S(t), \overrightarrow{p}, \Phi, L(t)), x(t) = f((u(t), \overrightarrow{p}), G(t)))$, then T_h has conducted destruction transformation on property of $A((U, S(t), \overrightarrow{p}, T(t), L(t)),$ $x(t) = f((u(t), \overrightarrow{p}), G(t)))$ under $G(t)$ on U denoted by T_{htx}. The meaning of destruction transformation on property is: T_{htx} (property destruction) $\rightarrow$ (property does not exist) $\rightarrow$ (there is no property to discuss or there is no need to discuss the property in the current object; or this object does not possess the property).

Definition 9.17 Suppose that $A((U, S(t), \overrightarrow{p}, T(t), L(t)), x(t) = f((u(t), \overrightarrow{p}),$ $G(t)))$ is the error logical variable defined under judging rule $G(t)$ on universe of discourse $U(t)$, if $T_h(A((U, S(t), \overrightarrow{p}, T(t), L(t)), x(t) = f((u(t), \overrightarrow{p}),$ $G(t)))) = A((U, S(t), \overrightarrow{p}, T(t), \Phi), x(t) = f((u(t), \overrightarrow{p}), G(t)))$, then T_h has conducted destruction transformation on quantifier of $A((U, S(t), \overrightarrow{p}, T(t), L(t)),$ $x(t) = f((u(t), \overrightarrow{p}), G(t)))$ under $G(t)$ on U denoted by T_{hlz}. The meaning of destruction transformation on quantifier is: T_{hlz} (quantifier destruction) $\rightarrow$ (quantifier does not exist) $\rightarrow$ (there is no quantifier to discuss or there is no need to discuss the quantifier in the current object; or this object does not include the quantifier).

Definition 9.18 Suppose that $A((U, S(t), \vec{p}, T(t), L(t)), x(t) = f((u(t), \vec{p}), G(t)))$ is the error logical variable defined under judging rule $G(t)$ on universe of discourse $U(t)$, if $T_h(A((U, S(t), \vec{p}, T(t), L(t)), x(t) = f((u(t), \vec{p}), G(t)))) = A((U, S(t), \vec{p}, T(t), L(t)), \Phi = f((u(t), \vec{p}), G(t)))$, then T_h has conducted destruction transformation on error value of $A((U, S(t), \vec{p}, T(t), L(t)), x(t) = f((u(t), \vec{p}), G(t)))$ under $G(t)$ on U denoted by T_{hcz}. The meaning of destruction transformation on error value is: T_{hcz} (error value destruction) $\rightarrow$ (error value does not exist) $\rightarrow$ (there is no error value to discuss or there is no error in the object of interests).

Definition 9.19 Suppose that $A((U, S(t), \vec{p}, T(t), L(t)), x(t) = f((u(t), \vec{p}), G(t)))$ is the error logical variable defined under judging rule $G(t)$ on universe of discourse $U(t)$, if $T_h(A((U, S(t), \vec{p}, T(t), L(t)), x(t) = f((u(t), \vec{p}), G(t)))) = A((U, S(t), \vec{p}, T(t), L(t)), x(t) = \Phi((u(t), \vec{p}), G(t)))$, then T_h has conducted destruction transformation on error function of $A((U, S(t), \vec{p}, T(t), L(t)), x(t) = f((u(t), \vec{p}), G(t)))$ under $G(t)$ on U denoted by T_{hhs}. The meaning of destruction transformation on error function is: T_{hhs} (error function destruction) $\rightarrow$ (error function does not exist) $\rightarrow$ (there is no error function to discuss or there is no need to discuss current error function; or the error function for this object has not been built).

Definition 9.20 Suppose that $A((U, S(t), \vec{p}, T(t), L(t)), x(t) = f((u(t), \vec{p}), G(t)))$ is the error logical variable defined under judging rule $G(t)$ on universe of discourse $U(t)$, if $T_h(A((U, S(t), \vec{p}, T(t), L(t)), x(t) = f((u(t), \vec{p}), G(t)))) = A((U, S(t), \vec{p}, T(t), L(t)), x(t) = f((u(t), \vec{p}), \Phi))$, then T_h has conducted destruction transformation on rule of $A((U, S(t), \vec{p}, T(t), L(t)), x(t) = f((u(t), \vec{p}), G(t)))$ under $G(t)$ on U denoted by T_{hgz}. The meaning of destruction transformation on rule is: T_{hgz} (rule destruction) $\rightarrow$ (rule does not exist) $\rightarrow$ (there is no rule to discuss or there is no need to discuss current rule, or the current rule for judging errors has been abolished).

Definition 9.21 Suppose that $A((U, S(t), \vec{p}, T(t), L(t)), x(t) = f((u(t), \vec{p}), G(t)))$ is the error logical variable defined under judging rule $G(t)$ on universe of discourse $U(t)$, if $T_h(A((U, S(t), \vec{p}, T(t), L(t)), x(t) = f((u(t), \vec{p}), G(t)))) = A((U, S(\Phi), \vec{p}, T(\Phi), L(\Phi)), x(\Phi) = f((u(\Phi), \vec{p}), G(\Phi)))$, then T_h has conducted temporal destruction transformation on $A((U, S(t), \vec{p}, T(t), L(t)), x(t) = f((u(t), \vec{p}), G(t)))$ under $G(t)$ on U denoted by T_{hsj}. The meaning of temporal destruction transformation is: T_{hsj} (temporal destruction) $\rightarrow$ (time does not play roles in current object) $\rightarrow$ (there is no time element to discuss or current time interval is not appropriate for the object of interest).

Definition 9.22 Suppose that $A((U, S(t), \vec{p}, T(t), L(t)), x(t) = f((u(t), \vec{p}),$ $G(t)))$ is the error logical variable defined under judging rule $G(t)$ on universe of discourse $U(t)$, if $T_h(A((U, S(t), \vec{p}, T(t), L(t)), x(t) = f((u(t), \vec{p}), G(t)))) \in$ $\{A((\Phi, \Phi, \vec{p}, T(t), L(t)), x(t) = f((u(t), \vec{p}), G(t)))), A((U, \Phi, \Phi, T(t), L(t)),$ $x(t) = f((u(t), \vec{p}), G(t)))), A((\Phi, S(t), \vec{p}, T(t), L(t)), x(t) = f(u(t), \vec{p}),$ $\Phi)),\ldots, A((\Phi, \Phi, \Phi, T(t), \Phi), \Phi = f((\Phi, \Phi), \Phi))\}$, then T_h has conducted partial or comprehensive destruction transformation on $A((U, S(t), \vec{p}, T(t), L(t)),$ $x(t) = f((u(t), \vec{p}), G(t))) G(t)))$ under $G(t)$ on U denoted by T_{hqb}. The meaning of partial or comprehensive destruction transformation is: T_{hqb} (more than two or all elements are destroyed) $\rightarrow$ (more than two or all elements do not exist in current object) $\rightarrow$ (more than two or all elements are not necessary to discuss in this object; partial or all elements have been removed through action such as "remove, eradicate, annihilate, disappear, fire, sell out, discard, move away, and leave").

9.6.4 Examples Illustrating Error Elimination Through Destruction Transformation

Example 1: Wiring error in new apartment
Miss Ma just moved into her new bought apartment. One circuit switch breaker was tripped when she attempted to turn on the light in one of her bedrooms. She called an electrician to investigate the issue. Have done a thorough checkup, the electrician found that the wiring of that room caused short connection of live wire and neutral wire. The electrician had to tear down the underground tiles and replace the poorly installed wires after he got permission from Miss Ma.

1. Error analysis
   Root cause: short connection of live wire and neutral wire
2. Approach to eliminate error

   (1) Destruction transformation
       Remove live wire from the terminal which is supposed to connect neutral wire.
   (2) Displacement transformation
       Buy two clearly labeled terminals and connect the live and neutral wires accordingly.

Example 2: Successful demolition of the Nanxi bridge at Yunyang in Chongqing
Nanxi bridge (referring to Fig. 9.8), a 147 m long masonry bridge with open circular arch, was built in 1976. It was successfully demolished using explosive dismantlement on 28th October 2008. This was an explosive removal of snag for the purpose of making Yunxi river become tertiary waterway (channel) when Three Gorges reservoir hit its 175-m water level (full capacity).

Fig. 9.8 Explosive dismantle on the bridge

1. Error analysis
   Yunxi bridge was built on the Tangxi river which is one of the branches of Yangtze river. With the impounding of Three Gorges Reservior to its full capacity of 175 m, the Tangxi river became a tertiary waterway. The Yunxi bridge thus became a snag which must be removed to guarantee the navigation safety.
2. Approach to eliminate error
   Destruction transformation was conducted on the subject by using explosive demolition.

9.7 Case Study on Eliminating Errors in Enterprise Management

9.7.1 Case Background

In October 2004, two engineers Mr. Zhu and Mr. Zi (abbreviated as **Zhu** and **Zi**) acquired certain shares of Old Forest Technology Co., Ltd.(abbreviated as Old Forest) using Intellectual Properties (abbreviated as **IPs**) illegally obtained from their former employer Longdong Digital Technology Co., Ltd. (abbreviated as Longdong Digital). In December 2004, **Zhu** and **Zi**, using fake names, attended New River Plastic Injection Co., Ltd. (abbreviated as New River) in Shenzhen as hardware engineer and software engineer, respectively. In April 2005, **Zhu** and **Zi** developed new set-top box models using the DVB (Digital Video Broadcasting) technical materials illegally obtained from Longdong Digital. In June 2005, Longdong Digital filed a lawsuit against **Zhu** and **Zi** in court of Shenzhen.

In 2002, Longdong Digital started to organize its R & D engineers to develop technologies related to set-top box models DVB-S, DVB-C, and DVB-T, etc. The company purchased related software platforms to support the development process. The development was successfully concluded at the end of 2004. Longdong Digital had registered its software at Science, Technology and Innovation Commission of Shenzhen Municipality (previously named: Shenzhen Bureau of Technology and Information).

Zhu worked as a hardware engineer from May through December 2004 and was in charge of development on set-top box model DVB-C. **Zi** worked as a software engineer from October through December 2004 and was responsible for the development of set-top box model DVB-S. During **Zhu**'s tenure of working at Longdong Digital, he signed *Employment Agreement* and *Non-disclosure Agreement* with the company. **Zi** also signed *Probationary Employment Agreement* and *Non-disclosure Agreement* with Longdong Digital. Both of them agreed not to disclose company's confidential information to any receiving party for any purpose without written consent of an authorized representative of Longdong Digital.

In October 2004, Mr. Wang (abbreviated as **Wang**) in New River approached **Zhu** and **Zi** and came up with a business plan to build a startup. In November 2004, they registered Old Forest to develop DVB products. Wang acquired 83% shares of the company using cash capital and **Zhu** and **Zi**, using **IPs**, acquired 7% and 6% shares of the company, respectively. **Zhu** took advantage of his position that can have access to technical materials of DVB and illegally copied technical materials of *DVB-S, DVB-T*, and *DVB-C*. On December 2nd 2004, Longdong Digital found what **Zhu** had done and thereby fired him. **Zi** also illegally copied technical materials of *DVB-S* and he resigned from Longdong Digital on December 14th 2004.

Zhu and **Zi**, using fake names, went to New River to work as hardware and software engineers, respectively. They started to develop set-top box models *DVB-S* and *DVB-T* by referring to material illegally acquired from Longdong Digital. A sample product was successfully made in April 2005 and then taken to attend Hong Kong Electronics Fair (Spring Ed). In July 2005, they produced 50 sample products and shipped them to more than 20 potential customers. They got an order of 500 pieces of product from a company in Saudi Arabia. Moreover, on May 30th 2005, **Zhu** sold the illegally copied technical materials of *DVB-S* to another party with a price of RMB 120,000. **Zhu** also admitted that he received a service fee of RMB 2,000 for providing Shenzhen Shanchuan Tech with technical support using technical materials of *DVB-T*.

1. Error identification

   In this case, Longdong Digital had put significant efforts in the protection its **IPs**. Nevertheless, the company made a mistake too, which was over confidence on its management system and binding forces of agreements. This case was even not uncommon in well-organized enterprises with standardized procedures. It was hard to guarantee that company had disloyal employees. It was even harder to prevent a company's business partners, negotiating opponents, important customers, and service providers from committing IPs infringement by taking advantage of legal gray areas in business. Under such circumstance, the binding forces of law, agreements, and business regulations may have very limited powers.

2. Error in calculating loss

   In June 2005, Longdong Digital filed a lawsuit against **Zhu** and **Zi** in court of Shenzhen. Municipal People's Procuratorate in Shenzhen investigated the case and recognized that DVB series software developed by Longdong Digital belonged to business secret because: (1) they were not known by the public; (2) they had practical use and can meet actual needs of potential customers; (3) they can generate revenue for the company; (4) the company had taken necessary measures to avoid infringement. Therefore, Municipal People's Procuratorate in Shenzhen prosecuted **Zhu** and **Zi** because: (1) they violated the requirements listed in the non-disclosure agreement; (2) they had caused significant loss for Longdong Digital since they acquired shares from their newly founded company (Old Forest) by using the illegally obtained technical materials of *DVB*.

   The attorney of the defendants stated that it was unreasonable to calculate the economic loss in the **IPs** infringement based on the actual value of **IPs** since the defendants did not actually sell products and made profit out of it. Therefore, what they had done did not incur significant economic loss for the creator of **IPs**.

   How to calculate the loss? The court held that the calculation on the value of loss should be based on the actual investment during development for the subject matter. With evidence regarding relevant investment on the subject matter, the court thereby ruled that **Zhu** and **Zi** had caused economic loss in the amount of RMB 1,525,000 for Longdong Digital and found them guilty on the offence of infringing business secret. **Zhu** and **Zi** were sentenced to 1 year and 10 months in prison and a fine of RMB 20,000 and 10,000, respectively.

   Business secret and **IPs**, with their tangible and especially intangible value, come to play critical roles in the furiously competitive market. Bringing business secrets to new employers by job-hopping employees, selling technical materials to competitor by internal moles of a company, building a new rival company by key members or key team of a company, and stealing a company's core business secrets using "Trojan" code in the software have been widely reported by the media, which have become major factors to discourage innovation in private companies all over the world.

3. Error analysis

In this case, the fine on the **IPs** infringement was far too low although Long-dong Digital won the lawsuit and two defendants received their due punishment. Regarding the amount of fine in lawsuit related to **IPs** infringement, there is no terms regulating the definite calculation of fines in Criminal Law of the People's Republic of China, and judicial interpretations from The Supreme People's Court of The People's Republic of China and The Supreme People's Procuratorate of the People's Republic of China. Therefore, courts in different regions had different standards and practices. The relatively light punishment indulged some people's fluke mind. Infringers undoubtedly want to have a try if the benefit is $1 million while fine is $100,000. It has no much difference with famous description on features of **Capital** by Dunning *"Capital is said by a Quarterly Reviewer to fly turbulence and strife, and to be timid, which is very true; but this is very incompletely stating the question. Capital eschews no profit, or very small profit, just as Nature was formerly said to abhor a vacuum. With adequate profit, capital is very bold. A certain 10% will ensure its employment anywhere; 20% certain will produce eagerness; 50%, positive audacity; 100%. will make it ready to trample on all human laws; 300%, and there is not a crime at which it will scruple, nor a risk it will not run, even to the chance of its owner being hanged. If turbulence and strife will bring a profit, it will freely encourage both. Smuggling and the slave-trade have amply proved all that is here stated [46, 166]."*

4. Approach to eliminate error

As far as the case is concerned, company should take as many measures as possible to enhance the identification and protection of business secrets before the loss actually happens.

(1) Define universe of discourse for business secrets and categorize their scopes
There are 4 steps:
   (a) Define the scope for a company's business secrets.
   (b) Categorize secrets into different classes-i.e., vital business secrets, important business secrets, and generic business secrets, which can help reasonably distribute resources on the protection of different business secrets.
   (c) Define the period of confidentiality for different business secrets.
   (d) Label and log the status of different business secrets based on the aforementioned requirements.
(2) Addition transformation on regulations
By referring to the law of a country on guarding state secret and procedures practiced by relevant guarding departments, companies should develop a set of regulations to guard against various channels/outlets for divulgence of business secrets.
(3) Addition transformation on management system
A dedicated department or standing committee with member(s) coming from key managers in relevant departments could be established to administer and coordinate the management of business secrets or **IPs**, which should directly report to the owner or CEO of the company. The committee should be granted

the **veto power** by the board of directors or CEO on issues which might lead to divulgence of business secrets. Meanwhile, non-compete clause (NCC) should be clearly defined and communicated to employees who have opportunity to have access to business secrets. In a general NCC, employees must agree not to be employed by a competitor or start a similar business (using similar business models or technologies) or profession for up to certain period of time (e.g. three years) after leaving the company.

(4) Eliminating error through decomposition transformation

For a project involving business secrets, it is necessary when possible to dissect the vital parts into subsections and arrange different engineers and persons to develop and manage them, which makes it impossible for a person to master the whole materials of vital secrets.

(5) Using preventive mechanisms

(a) Scheduling periodic training on regulations and procedures related to business secrets;

(b) Conducting scheduled and unscheduled audit on the manuals, procedures, appraisals, approvals, processes, preventive actions, corrective actions, implementation, and awareness on significance related to management of business secrets;

(c) Searching web, conducting field studies, investigating products produced by competitors (and their patent petition), and identifying the employment of former employees (fired, resigned, retired), etc.;

(d) Updating manuals and procedures for preventive mechanisms.

(6) Delineating the intension and extension of agreements and regulations regarding business secrets

The non-disclosure agreement signed by employees should state the following responsibilities and obligations:

(a) Obligations to protect company's business secrets;

(b) Obligations of legally and reasonably using company's business secrets;

(c) Obligations of timely reporting development outputs and chance of having access to business secrets;

(d) Obligations of agreeing not to start a similar business (using similar business models or technologies) or profession for up to certain period of time after leaving the company or during the tenure with the company;

(e) Obligations of agreeing not to be employed by a competitor for up to certain period of time after leaving the company or during the tenure with the company;

(f) Obligations of maintaining and administering the business secrets if being assigned a relevant job.

(7) Verification of the subject

According the principles in civil law system in China, burden of proof should be borne by claimant. In lawsuit related to **IPs** infringement, **IPs** owner or creator or the party which claims **IPs** should bear the obligation of proof. Let's use the case of **IPs** infringement dispute caused by talent drainage in company **BA** as an example. The company **BA**, as plaintiff (claimant), should testify:

(a) The information (or technical materials) claimed belongs to the scope of the company **BA'**s business secret and should be under the tutelage of law;

(b) The information illegally used or disclosed is the same as or almost identical to the company **BA'**s business secrets;

(c) The defendant had the chance to have access to the the company **BA'**s business secrets and simultaneously had the action of illegally disclosing or using the above-claimed business secrets, or permitting third party or providing access convenience to third party in obtaining the company **BA'**s business secrets.

In brevity, companies must enforce their awareness in preserving evidence for the purpose of taking legal actions to protect their **IPs**. Technical materials must be documented and labeled in different forms such as written, digital (audio or video). The process of creating technical materials should also, if possible, documented for the purpose of testifying in court.

9.8 Analyzing and Eliminating Four Typical Errors in Parenting Children

9.8.1 Four Principles Used in Parenting Children

There are four general principles used in parenting children. We also used them in the following examples as rules of judging errors.

1. Reward good behaviors in time;
2. Do not unintentionally punish good behaviors;
3. Do not "unintentionally" reward delinquent/inappropriate conducts;
4. Do not punish delinquent conducts gently or have no punishment at all.

9.8.2 Four Examples in Parenting Children

Example 1

Wei Li was in grade 4. He showed his mid-term exam score to his dad while his Dad was reading newspaper on the sofa. Wei Li said: "Dad, I got 97 in Chinese and 100 in math. This is my exam sheet." Dad said "That is great. I will take a look after a while."

1. Error analysis
   What Wei Li's parents did violated the first principle mentioned above. They unintentionally ignored kid's expectation (maybe just a hug or oral praise).

2. Approach to eliminate error
   Generation transformation (inverse destruction transformation): Wei Li's parents should give some timely reward (even just oral recognition) for what he had done.

Example 2

Lili, 9 years old, tried to help Mom wash dish plates. Lili said "Mommy, I just washed dish plates." Mom said "It is time for you help Mom do some chores. Have you washed the cooking pot on the table?"

1. Error analysis
   Lili's Mom violated the aforementioned second principle. She could have avoided speaking the sentence of "It is time for you help Mom do some chores."
2. Approach to eliminate error
   Displacement transformation can be conducted to replace the unnecessary sentence with words of praise.

Example 3

Leilei's parents brought him to travel. One day, Mom felt very tired and hungry when they came back from a place of interest. Leilei said "Mom, I want to play in the swimming pool for a while before lunch." Mom said "You can go to the swimming poor after we have had lunch and taken a nap." Leilei replied "I will not eat lunch and will disturb you if you do not let me go over there right now." It seemed that Mom had not other choice. She then responded "I really can't help you. Go ahead."

1. Error analysis
   The action of Leilei's mother went against the above third principle. She unintentionally allowed unreasonable request and encouraged inappropriate behavior.
2. Approach to eliminate error
   Displacement transformation can be conducted to substitute denial for approval on Leilei's unreasonable request.

Example 4

Dongdong, a 7 years old neighborhood kid, came to Qiqi's home to play game with him. Qiqi's parents were sitting in living room talking to each other. It was not long before they had a dispute. Qiqi then started to recklessly kick Dongdong. Mom said to Dad "you must stop Qiqi right now." Dad replied "Leave them alone. They are boys."

1. Error analysis
   Qiqi's father did not stop his bad behavior in time, neither did he punish him.
2. Approach to eliminate error
   Generation transformation can be conducted by giving Qiqi appropriate punishment.

9.9 Applications of Error Theory

Errors theory can be applied in different subjects, fields, social life, time periods, and regions. Due to the length limit, only is the application of error systems theory discussed here. The error system, system structure, objective and intrinsic features of system, rules for judging error, and error functions are major factors that will be considered for building error system model. The objective of building error system model is to holistically investigate the error and judge the error of system from the perspective of system features. With bearing that in mind, what should be considered in the error system model and what properties does the error system model have? In the process of judging error of system, how to use error system model in the practical application is the question we are going to answer hereby.

9.9.1 Establishing Error System Model

1. Composition of error system model

   In order to holistically identify and judge possible errors in a system, a model must be built to depict the system. In literature, scholars in systems science had built various models to illustrate different systems. Bertalanffy (1968) established the model for general system by employing partial differential equations. In this session, we attempt to build models for systems containing error (simplified as error systems), which can not only identify and judge error in system but also help prevent and eliminate errors in systems. Hereby, we first propose basic structures and then models for system functional structure and error structure. This hierarchical model serves two purposes: (1) systematically exhibits the features of the system as a whole; (2) helps to diagnose errors in system and investigate the laws of error transitions and transformations in the system.

   In our research, the model for error system is mainly applied to man-made systems. The error in a system is primarily related to objective features of the system. First of all, the proposed model for error system must include the objective feature function. Secondly, the rules for judging errors must also be included. There are three types of rules for judging system errors: (1) rules G_m for judging if there are errors in objective features; (2) rules G_{gm} for judging if the objective features can be realized; (3) rules for error judgment in system behaviors G_h. The 3 major rules can be further divided into more specific rules, which form a rule system and varies with the change of system hierarchies.

   In order to quantitatively characterize errors in a system, the model for it should consider all possible errors for which corresponding error functions are constructed and error values are calculated according to the rules defined on their associated universe of discourse.

In summary, with the above-mentioned requirements, the model for error system can serve as a useful tool for identifying/judging errors and diagnosing/preventing/eliminating them accordingly.

2. Model of error system

According to the basic structures for general systems and error systems, the models for system features and errors in the system are described as follows.

(1) Feature-based model in a system

    (i) Complete additivity

$$GY_j = \sum_{i=1}^{n} GY_{ji}$$

    (ii) Partial additivity

$$GY_j \neq \sum_{i=1}^{n} GY_{ji}$$

    (iii) No additivity between certain feature GY_j of system S and some features GY_{ji} of subsystem

There exists no additivity between the optimum of feature GY_j of system S and certain features GY_{ji} of subsystem. In this case, it is necessary to know what specific features (GY_{ji}) of the subsystem need to be presented when the the feature GY_j of the system achieves optimum. At the same time, it is also necessary to know the value ranges of the features GY_{ji} of different subsystems. It is represented by $S(GY_i) = S(S_1(GY_{j1}$ $(a_1, b_1)), S_2(GY_{j2} (a_2, b_2)), \ldots \ldots, S_i(GY_{ji} (a_i, b_i)), \ldots \ldots, S_n(GY_{jn}$ $(a_n, b_n)))$. It can be divided into 6 basic types.

(a) Feature model of series system structure

$$S(GY_j) = CS(S_1(GY_{j1} (a_1, b_1)), S_2(GY_{j2} (a_2, b_2)), \ldots \ldots,$$
$$S_i(GY_{ji} (a_i, b_i)), \ldots \ldots, S_n(GY_{jn} (a_n, b_n)))$$

(b) Feature model of parallel system structure

$$S(GY_j) = BS(S_1(GY_{j1} (a_1, b_1)), S_2(GY_{j2} (a_2, b_2)), \ldots \ldots,$$
$$S_i(GY_{ji} (a_i, b_i)), \ldots \ldots, S_n(GY_{jn} (a_n, b_n)))$$

(c) Feature model of scaling system structure

$$S(GY_j) = KS(S_1(GY_{j1} (a_1, b_1)), S_2(GY_{j2} (a_2, b_2)), \ldots \ldots,$$
$$S_i(GY_{ji} (a_i, b_i)), \ldots \ldots, S_n(GY_{jn} (a_n, b_n)))$$

(d) Feature model of inclusion system structure

$$S(GY_j) = YS(S_1(GY_{j1} (a_1, b_1)), S_2(GY_{j2} (a_2, b_2)), \ldots \ldots,$$
$$S_i(GY_{ji} (a_i, b_i)), \ldots \ldots, S_n(GY_{jn} (a_n, b_n)))$$

(e) Feature model of feedback system structure

$$S(GY_j) = FS(S_1(GY_{j1} (a_1, b_1)), S_2(GY_{j2} (a_2, b_2)), \ldots \ldots,$$
$$S_i(GY_{ji} (a_i, b_i)), \ldots \ldots, S_n(GY_{jn} (a_n, b_n)))$$

(f) Feature models of other system structures

$$S(GY_j) = QS(S_1(GY_{j1} (a_1, b_1)), S_2(GY_{j2} (a_2, b_2)), \ldots \ldots,$$
$$S_i(GY_{ji} (a_i, b_i)), \ldots \ldots, S_n(GY_{jn} (a_n, b_n)))$$

(2) Error model of a system

$SC(s_1, s_2, \ldots \ldots, s_n)$ can be divided into 6 types:

    (i) Error model of series system structure

       $CSC(s_1, s_2, \ldots \ldots, s_n)$

    (ii) Error model of parallel system structure

       $BSC(s_1, s_2, \ldots \ldots, s_n)$

    (iii) Error model of shrinking and expanding system structure

       $KSC(s_1, s_2, \ldots \ldots, s_n)$

    (iv) Error model of inclusion system structure

       $YSC(s_1, s_2, \ldots \ldots, s_n)$

    (v) Error model of feedback system structure

       $FSC(s_1, s_2, \ldots \ldots, s_n)$

    (vi) Error model of other system structures

       $QSC(s_1, s_2, \ldots \ldots, s_n)$

(3) Error function of a system

G are rules for judging error, which must be scientifically proven true, reasonable, and operable. In general, G can be a set composed of three subsets, namely rule subset for judging errors in system objective features, rule subset for judging if the objective features have been achieved, and rule subset for judging error in system behaviors, i.e., $G = \{G_m, G_{gm}, G_h\}$. And G_m is also composed of three subsets, i.e., rule subset for judging input error in the system objective features, rule subset for judging efficiency error in the system objective features, and rule subset for judging output error in the system objective features: $G_m = \{G_{mr}, G_{mx}, G_{mc}\}$.

Establishing error function:

$$\overrightarrow{x} = f(G, Z(n), MG) =$$

$$\begin{cases} \overrightarrow{x_m} = f_1(G_m \nRightarrow MG) = \begin{cases} x_{mr} = f_{11}(G_{mr} \nRightarrow (G_{mr} \subseteq MG)) \\ x_{mx} = f_{12}(G_{mx} \nRightarrow (G_{mx} \subseteq MG)) \\ x_{mc} = f_{13}(G_{mc} \nRightarrow (G_{mc} \subseteq MG)) \end{cases} \\ x_{gm} = f_2(G_{gm} \nRightarrow (MG \subseteq GY)) \\ x_h = f_3(G_h \nRightarrow H_z(t)) \end{cases}$$

$\overrightarrow{x}$ stands for the error value of object system, which is a vector. Of course, according to actual needs, error function can be expressed by scalar form where the error value is the sum of all sub-scalars. For example, when initiating the comparison of multiple systems, it is necessary to use scalar functions to find out the optimal system. The error value $\overrightarrow{x}$ is calculated through relevant error functions which is expressed as above.

The specific forms that different error functions can take are determined by principles whether the error function can objectively reflect to what extent the object violates the rules for judging errors.

3. Model hierarchy of error system

Hierarchy is one of the fundamental attribute of system. System is composed of subsystems at different hierarchical levels and system features emerge accordingly. Generally, low level constituent subsystems belong to and support higher-level ones and higher-level subsystems contain and dictate the lower-level ones. The higher-level subsystems have the emergence of features or behaviors. Once the system is de-leveled/dis-aggregated, the emergence disappears. Multi-hierarchy is an intrinsic property of complicated system, which is one of the sources for generating complexity. Except for very simple system, common system is constructed from aggregating elements to subsystems and aggregating subsystems to final entire system. In the hierarchical system, different level of subsystems have distinct emergence and the comparatively large subsystems also possess hierarchical characteristics. In these kinds of systems, hierarchy provides a frame of reference by which people know where is the problem is located. Mispositioning the problem in irrelevant hierarchy may render the process of identifying and eliminating errors invalid or chaotic.

As system has hierarchy, so does the model of error system, which is determined by the hierarchy of the object system. Systems at different hierarchies have different objectives and intrinsic features and they are using different rules for judging corresponding errors, which must also take part in the decomposition and aggregation transformations when those transformations take place in the system. Please refer to Fig. 9.9 for the hierarchical structure of error system model.

In Fig. 9.9, C_i^{k-s} ($(S_i^{k-s}(n_i)$, MG^{k-s}, $G^{k-s})$, $\overrightarrow{x}^{k-s}$) stands for the established model for error subsystem $S_i^{k-s}(n_i)$ at the $(k-s)$th layer, which is built based on the objective features and related rules of corresponding error subsystem. Figure 9.9 also demonstrates that the decomposition and aggregation transformations conducted on the model of error system are contingent on the relevant changes in object system. However, the decomposition and aggregation transformations are not the case of simple summation or subtraction but a process of dynamic and organic integration. With the dynamic integration of the object system, the objective features and rules in the model of error system must also take necessary integration. One more thing worth noting is that the subsystem set derived from system decoupling or decomposition transformation is the context of any obtained subsystem. The context of a particular subsystem is expanded with the proceeding of system decoupling.

4. Implication of building model of error system to the elimination of system errors

As we emphasized at the beginning of this section, the objective of building model for error system is to identify and judge the errors in the object system. The actual implications of using model for error system are more than enough. The processes of using error system include: (1) error is identified in the object system using error system model, (2) error is analyzed based on the error value in the error

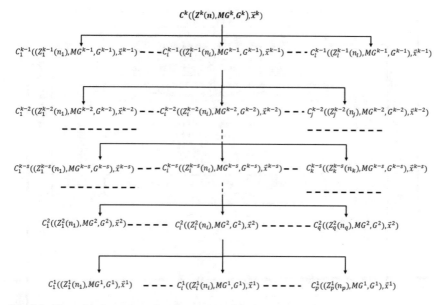

Fig. 9.9 Hierarchical structure of error system model

system model, (3) pertinent measures and methods are chosen to eliminate error, (4) error system model is employed to verify and evaluate the effectiveness of error elimination. Therefore, error system model is very important for diagnosing and eliminating errors. The process of using error system model is demonstrated in Fig. 9.10.

Several points need to be noted in Fig. 9.10:

1. The object system discussed in Fig. 9.10 is changeable error system. In dead error system, error can only be reduced because error in this case can not be fully eliminated.
2. Based on Axiom requirements in Sect. 2.4.1, the judging rules must be scientifically proven true, reasonable, and operable.
3. It is assumed that the function forms adopted in the error system model include classic error function, fuzzy error function, or unilateral error function. Therefore, error occurs when error value is larger than 0.
4. The transformation on the object system dictates the changes not only on system structure or elements of the object system but also on the context of the object system. When error value $x_{gm} > 0$ or $x_h > 0$, error may be generated from the inside of the system or external environment. General system structure must be further analyzed to identify the mechanism in generating errors. As the system error coming from the inside of the system, one must determine which hierarchy does the error come from. The hierarchical structure of error system model provides

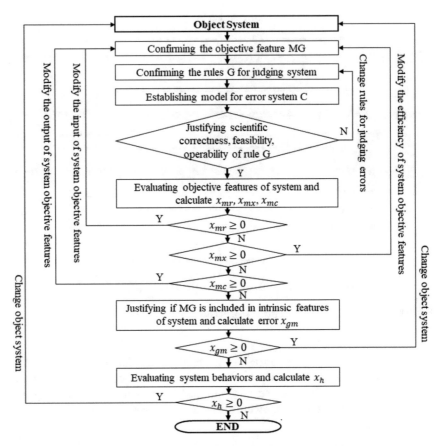

Fig. 9.10 Flowchart representing process of identifying and eliminating errors through error system model

a powerful tool for conducting the analysis. Error system model is decomposed into different levels and corresponding models for error subsystems are built to conduct respective judgment.

9.9.2 Practical Application of Error System Theory

I. Background of the Water Treatment Factory in East Guangzhou Economic and Technological Development Zone

1. Introduction to the water treatment factory

The waste water treatment factory is located at Hongguang road of east Economic and Technological Development Zone in Guangzhou (ETDZG) with an area of 3.5 hectares. As a critical part in the Guangzhou comprehensive waste water treatment and improvement of river system management, two stage constructions had been completed in the east zone. The first stage construction financially supported by fiscal investment of **ETDZG** and government loan from Austria was completed and put into use in May 2004. The designed treatment capacity was 25,000 tons/day. And the second stage construction was finished and put into use in July 2010, which was a regional cooperation project cosponsored by Global Environment Fund (GEF), Guangzhou Water Supplies Authority (GWSA), and government of ETDZG, which had a designed capacity of 75,000 tons/day.

The waste water treatment factory at east ETDZG was responsible for processing the civil and industrial waste water coming from east ETDZG, Yunpu industrial zone, and Nangang town covering an area of 45.12 km^2. The treated water was discharged into Nangang river. In 2010, the average daily waste water discharge in the covered area was totaled 103,000 tons/day where 34,400 tons/day coming from east ETDZG, 37,000 tons/day coming from Yunpu industrial zone, and 28,900 tons/day coming from Nangang town. According to historic data, the basic indicators of the waste water were inspected at the inlet of the treatment factory (Refer to Table 9.1).

The treated water was discharged into Nangang river and finally into East River. Therefore, the Nangang river was regulated as Grade III water system because East River is the drinking water source area. The standard requirement for the discharged water from east waste water treatment factory must meet Grade I-B in GB18918-2002 and Grade I in DB44/26-2001 indicated by Table 9.2.

The waste water, through sewage tubing system in the covered regions, flows to the treatment factory, which then flows through coarse screen, lift station, fine screen, aerated grit chamber, A/O biochemistry tank, UV disinfection tank, and finally discharges into Nangang river. The facilities and equipment used in the treatment process mainly include coarse screen well, lift station, fine screen, aerated grit chamber, anaerobic sludge digester, A/O biochemistry tank, UV disinfection tank, sludge tank, air compressor, dehydrator, and power supply station. The whole system adopted imported equipment and is an automated treatment system.

2. Introduction to the treatment process of the factory

The setup of this treatment process was determined by the major pollutants in the waste water. The need for ammonia-nitrogen removal requires the process to have capability of nitrification; the requirement for phosphorous removal needs measure to

Table 9.1 Indicators of the waste water inspected at the inlet

Pollutants	COD_{Cr}	BOD_5	Suspended solids (SS)	NH_3–N	Total phosphorus (TP)
Average value (mg/L)	400	200	250	25	5

Table 9.2 Indicators of the treated water discharged at the outlet

Pollutants	COD_{Cr}	BOD_5	SS	NH_3–N	TP	E.coli/L
Average value (mg/L)	40	20	20	8	1.0	1000

take care of phosphorous content; the need for removing suspended solids limits the surface load and solid flux of the secondary settling tank. Therefore, due to the quality requirements for the treated water and the limitations in the area of treatment factory, the water treatment factory had adopted consolidated SBR (Sequencing Batch Reactor) mechanism to process the waste water. Please refer to the demonstration of the process (Refer to Fig. 9.11).

In Fig. 9.11, the whole process of treating waste water is divided into three stages described as follows:

(1) Pre-treatment stage

Pre-treatment process is composed of coarse grid screen, waste water lift station, fine grid screen, and aerated grit chamber. The waste water in the designated regions flows into treatment system through tubing system designed according to the topography of terrain. All kinds of suspended solids (SS) with diameter over 20 mm will be blocked by the coarse grid screen and lifted to the landfill or storage center. Thereafter, waste water is pumped up to lift station. Because waste water treatment factory is generally located at low terrain ground which can allow waste water to naturally flow into the treatment system without the help of pump station, further treatment needs to lift waste water to higher ground. Then lifted waste water flows through fine grid, which removes the SS with diameter over 6 mm for purpose of mitigating negative impacts of SS on other processes. And then screened water enters aerated grit chamber. Perforated aeration pipes were installed at the bottom of the aerated grit chamber. The compressed air enters aeration tube and adds oxygen to waste water with a positive inclination angle of 45°. With aeration effect, the newly entered water is well mixed with the previously stored waste water; the grease and SS are separated from waste water; the aeration is also a grit washing mechanism to remove the inorganic particles and makes heavier particles settle down. The sedimentation process takes a while and grit is periodically pumped to the ground grit-water separator.

(2) Biochemical reaction stage

Having done the pretreatment where large-size SS and heavier inorganic particles have been removed, many other undesirable elements need to be processed and removed. In order to qualify for the treatment criteria, biochemical treatment is an indispensable step in the waste water treatment process. There are two parts in the biochemical treatment process, which include anaerobic chamber and modified sequencing batch reactor (MSBR) biochemical treatment chamber. Pretreated waste water first enters anaerobic chamber where newly-entered waste water is mixed with the returned activated sludge from MSBR. MSBR process is composed of water filling-aeration, sedimentation, decantation, and system idle.

(a) *Water filling-aeration*

In this stage, biochemical chamber conducts two functions-i.e., watering and aerating. At the same time, it returns 20% of activated sludge from MSBR to anaerobic

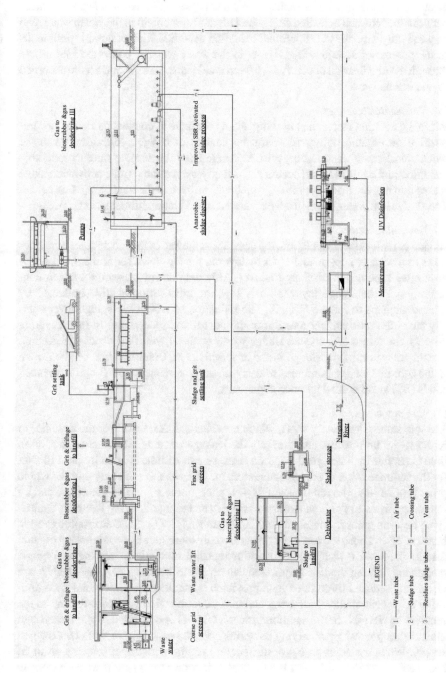

Fig. 9.11 Process flowchart in the waste water treatment factory

chamber. Aerating system provides oxygen to biochemical chamber, which not only meets the oxygen needs of aerobic organism but also facilitates the intensive mixture between the returned activated sludge and organic compound in the waste water, and consequently expedites the microbial oxidation decomposition of organic pollutants. In the process of aeration, the NH_3–N in the waste water is converted into nitrate through the nitrification of nitrifying bacteria and phosphorus P and SS are absorbed to microorganisms.

(b) *Sedimentation stage*
When the aeration stops, the microorganisms continue to conduct oxidation decomposition on organic pollutants. With the decline of oxygen concentration in the biochemical chamber, microorganisms change from the aerobic state to anaerobic condition and denitrification occurs accordingly. At the same time, activated sludge settles down in the static condition, which will be used in the next processing round. The sludge and water are naturally separated in the static sedimentation process.

(c) *Decanting stage*
Having completed sedimentation, the decanter at the end of the biochemical chamber starts to discharge supernatant layer by layer in the top-down sequence. Decanter is reset after supernatant discharge is over. Although, at this stage, the oxygen concentration in the sludge layer is fairly low, the nitrification is still conducted by microorganisms to reduce NH_3–N. The reduction of NH_3–N in sludge layer actually eases the biochemical reaction in the next round processing. In the decanting process, the return of activated sludge works as usual with the purpose of keeping certain concentration of sludge in the anaerobic area. Denitrification on the nitrate in the returned sludge continues to carry on and the phosphorus P and suspended solids (SS) are absorbed by microorganisms.

(d) *Idle stage*
The idle stage is relatively short, which provides a time interval for the decanter to return its initial position and prevents the leakage of sludge. The actual decanting time is shorter than designed duration because part of time is actually used to wait for the sedimentation of sludge and restore the absorbability of sludge. In the whole process of MSBR, the removal of NH_3–N, phosphorus P, and SS are conducted by microorganisms in the biochemical chamber (including nitrifying bacteria, denitrifying bacteria, phosphorus accumulating organisms-PAO, and heterotrophic aerobic bacteria, etc.). Those microorganisms survive and reproduce by decomposing organic matters in the waste water and completing metabolism while processing the organic pollutants. Among the five major elements in waste water, COD_{Cr} and BOD_5 are effectively reduced through oxygen provision by aeration and microbial decomposition of organic matter. Nitrifying and denitrifying bacteria are contributing to the removal of NH_3–N. In the aeration process, NH_3–N is converted to nitrate through nitrification process by nitrifying bacteria under the condition of sufficient oxygen supply. When the aeration stops, nitrification reaction continues to carry on in an ever-decreasing intensity with the gradual reduction of oxygen in waste water. As the oxygen concentration in the waste water decreases to a threshold, denitrifying

bacteria start to initiate denitrification action on the nitrate because the condition in which microorganisms reside change from aerobic state to anaerobic state. The nitrogen in the nitrate is converted to nitrogen N_2 via oxidation-reduction reaction, which at the same time is emitted into air. Moreover, microorganisms also play an important role in the removal of SS and phosphorous (P). In the aeration process, the aerated air drives the adequate contact between microorganisms, SS and P and then they are absorbed by the microorganisms. After the aeration, in the process of sedimentation, microorganisms settle down with sludge and sink to the bottom of the chamber. The average duration of sludge staying in the chamber is 25–30 days. According to the duration of stay, the treatment factory discharges certain amount of sludge everyday and meets the requirements for sludge metabolism. In summary, microorganisms play pivotal role in the waste water treatment. In order to fully take advantages of the function of microorganisms, 20% activated sludge is returned to anaerobic chamber. This process is to improve the reproduction of microorganisms in the eutrophic environment. In the condition of anaerobic state, microorganisms release the absorbed SS and P, which keeps the absorption capability when they enter biochemical chamber.

Having been processed in the biochemical reaction, the waste water has become clean supernatant and the major pollutants in the treated water have reached discharging criteria. However, treated water discharged from biochemical chamber contain all kinds of microorganisms including some pathogenic bacteria and viruses. Therefore, disinfection is a necessary process after previous treatments.

(3) Disinfection and sludge discharging stage
The clean supernatant enters ultraviolet disinfection chamber (UV) when it is discharged from the biochemical treatment chamber. Pathogenic bacteria and viruses are killed in the UV chamber and the treated water is metered and discharged into Nanggan river. The discharged sludge from biochemical treatment chamber is collected by sludge storage tank, which is then transported to sludge dehydrator and the dehydrated sludge (hydration rate below 80%) is shipped to landfill.

II Analysis on the waste water treatment system

1. System structure
Figure 9.12 depicts the system structure of the first-stage waste water treatment project at east ETDZG. The water treatment system consists of series structure (coarse grid screen, lift station, and fine grid screen) and parallel structure (aerated grit chamber A and B, the connection among four biochemical treatment chambers), augmenting and expanding structure (the radiating connection between anaerobic selection chamber and four biochemical treatment chambers, the shrinking connection between four biochemical treatment chambers and UV disinfection chamber), feedback structure (20% activated sludge returns from four biochemical treatment chambers to anaerobic selection chamber), and inclusion structure (air compressor system is composed of three interconnected air compressors A, B, and C). Apparently, the whole system is constructed by interacting constituent subsystems.

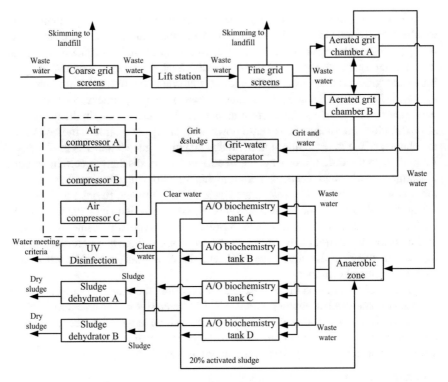

Fig. 9.12 Structure of waste water treatment system

2. Composite functional structure

Figure 9.13 is a map depicting all functions in the waste water treatment system. The ultimate goal and function of waste water treatment system are to process urban civil and industrial waste water to reach national I-B criteria quality. In order to achieve the ultimate goal, the system needs to have the first-order functions such as removal of large size debris, small size grit, and SS, dephosphorization, denitrification, microbial decomposition of organic matter, power supply, and disinfection. The first-order functions need to be divided into second-order functions and then third-order functions, and so on and so forth until all the functions are assigned to specific subsystems. For example, the first-order function "dephosphorization" needs the second order functions including "sludge metabolization" and "sludge activation" combined with the participation of subsystem "air compressor" and "biochemical treatment chamber". And the second-order function "sludge metabolization" can be finished with coordinated actions of biochemical treatment chamber, sludge storage chamber, and sludge dehydrator. Similarly, "improvement in sludge activation" requires the coordination of anaerobic selection chamber, biochemical treatment chamber, and air compressors. Therefore, the "dephosphorization" involves

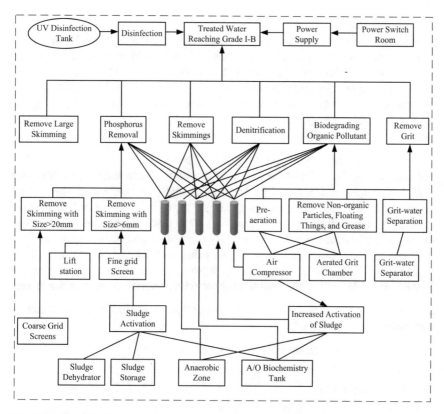

Fig. 9.13 Composite functional structure

participation of subsystems including anaerobic selection chamber, biochemical treatment chamber, air compressor, sludge storage chamber, and sludge dehydrator.

According to the system structure and composite functional structure as well as the processing mechanism of the treatment system, for any function, it is easy to pinpoint its critical subsystem, important subsystem, critical structure, important structure, and critical subsystem connections. For instance, regarding the first-order function "dephosphorization", biochemical treatment chamber is the critical subsystem because the phosphorous in the waste water can not be removed without the work of it. While anaerobic selection chamber, air compressor, sludge storage chamber, and sludge dehydrator are important subsystem where the first two subsystems are contributing to the improvement of activeness of sludge and the rest two subsystems are responsible for the metabolization of sludge. The power supply station is the critical subsystem of the whole system because the whole system can not work without power. The connection between power supply station and biochemical treatment chamber forms critical subsystem interface.

3. Objective features of system

With the approval of Guangzhou municipal government, Luogang district was established in 2005. Luogang district, with mountainous and forest terrains accounting for 50% of the 393km² total area, is located at northeast part of Guangzhou, which serves as a modern eco-friendly new urban area. Therefore, the municipal government developed measures to consolidate water environmental governance in Luogang district. With this strategy, the objective feature of first-stage project at east ETDZG was to build the processing capacity of 25, 000 tons/day.

(1) Inputs for the function of system objective features

The independent variables of the function for system objective features include the state of target environment and the impacts of target environment on the system. Suppose that the state of target environment keeps stable during certain periods, the independent variable of the function for system objective features only contains the impacts of target environment on the system, i.e., objective input of system. The objective inputs of system include many different materials which vary within certain range. That is to say the objective input of system is a vector composed of multiple variables and each variable changes within certain range. For the convenience of analysis and description, the input $(\overrightarrow{X'})$ for the objective feature function was measured using unit of day.

x'_1 (kw·h): electricity consumption, $x'_1 \in [0, 5325]$;
x'_2 (m^3): air, $x'_2 \in [0, +\infty)$;
$\overrightarrow{x'_3}$: waste water, where,
    Waste water (10,000 tons), $x'_{31} \in [0, 2.5]$;
    BOD$_5$ (mg/L), $x'_{32} \in [0, 200]$;
    COD$_{Cr}$ (mg/L), $x'_{33} \in [0, 400]$;
    SS (mg/L), $x'_{34} \in [0, 250]$;
    NH$_3$–N (mg/L), $x'_{35} \in [0, 25]$;
    TP (mg/L), $x'_{36} \in [0, 5]$;
    TN (mg/L), total nitrogen $x'_{37} \in [0, 40]$;
    Cu (mg/L), $x'_{38} \in [0, 2)$;
    Zn (mg/L), $x'_{39} \in [0, 5)$.

x'_1, x'_2, and x'_3 are not independent because the electricity consumption and air volume are determined by the volume and degree of pollution of the waste water to be treated. The more the volume and the worse the level of pollution are, the higher it is in the consumption of electricity and air. So far, we only do a preliminary systematic analysis and we try to use a function to represent the relationship (Eqs. 9.1 and 9.2).

$$x'_1 = f_1(\overrightarrow{x'_3}) = f_1(x'_{31}, x'_{32}, x'_{33}, x''_{34}, x'_{35}, x'_{36}, x'_{37}) \tag{9.1}$$

$$x'_2 = f_2(\overrightarrow{x'_3}) = f_2(x'_{31}, x'_{32}, x'_{33}, x'_{34}, x'_{35}, x'_{36}, x'_{37}) \tag{9.2}$$

Therefore, the input for the objective feature function of system $\overrightarrow{X'}$ can be represented by (Eq. 9.3):

$$\overrightarrow{X'} = \begin{vmatrix} x_1' \\ x_2' \\ x_3' \end{vmatrix} = \begin{vmatrix} f_1(\overrightarrow{x_3'}) \\ f_2(\overrightarrow{x_3'}) \\ x_3' \end{vmatrix} = MD(\overrightarrow{x_3'}) \tag{9.3}$$

(2) Outputs for the function of system objective feature

The outputs for the objective feature function of system is also called the objective behavior of the system. In the first stage, the objective is to discharge less than 25,000 tons of treated water $\overrightarrow{Y'}$ to Nangang river, where,

Treated water (10,000 tons/day), $y_1' \in [0, 2.5]$;

BOD$_5$ (mg/L), $y_2' \in [0, 20]$;
COD$_{Cr}$ (mg/L), $y_3' \in [0, 40]$;
SS (mg/L), $y_4' \in [0, 20]$;
NH$_3$–N (mg/L), $y_5' \in [0, 8]$;
TP (mg/L), $y_6' \in [0, 0.5]$;
TN (mg/L), $y_7' \in [0, 20]$;
No of E. coli/L $y_8' \in [0, 1000]$.

(3) Function for the system objective features

Under normal operation conditions, there must be an output that is corresponding to the input of the system objective features with transformation process. The corresponding relationship between inputs and outputs of system objective is the function for the system objective features. An abstract function can be used to describe this kind of relationship. Although the function for system objective feature varies with the system state, the function for system objective features is relatively stable given that system operates under the objective state.

$$MD_j = MD(\overrightarrow{X'}) = MD(\overrightarrow{x_3'}) \tag{9.4}$$

(4) System intrinsic features

In this research, it is not possible to exhaust all the system intrinsic features since (1) cognitive capability of human being is very limited and sometimes it is prohibitively expensive for collecting all features even it is possible; (2) it is not necessary to have all features investigated because some of them are not directly related to the problem of interests. Therefore, for the intrinsic features of waste water treatment factory at east ETDZG, we only focus on the features which are related to the realization of system objective features or factors that must be considered for other systems in the environment.

(a) *Inputs for the function of system intrinsic features*

Based on the historic data and materials in the waste water treatment system, it is generally assumed that the quality indicators of the waste water are relative stable. However, with the increase in population and enterprises in this region, the volume in waste water increased rapidly over time. The designed capacity in the first-stage was 25,000 tons/day and this was a conservative capacity which can be stretched to 30,000 tons/day in a saturated operation.

Therefore, similar to the function of system objective features, the independent variables for the function of system intrinsic features only represent the input $(\overrightarrow{X})$ of the system in daily base. They are listed as follows:

x_1 (kw·h): electricity consumption, $x_1 \in [0, 6390]$;
$x_2 (m^3)$: air, $x_2 \in [0, +\infty)$;
$\overrightarrow{x_3}$: waste water, where,

Waste water (10,000 tons), $x_{31} \in [0, 3.0]$;
BOD_5 (mg/L), $x_{32} \in [0, 200]$;
COD_{Cr} (mg/L), $x_{33} \in [0, 400]$;
SS (mg/L), $x_{34} \in [0, 250]$;
$NH_3–N$ (mg/L), $x_{35} \in [0, 25]$;
TP (mg/L), $x_{36} \in [0, 5]$;
TN (mg/L), $x_{37} \in [0, 40]$;
Cu (mg/L), $x_{38} \in [0, 2)$;
Zn (mg/L), $x_{39} \in [0, 5)$.

x_4: other system inputs.

Notes: the ranges for the above system inputs are obtained in the condition that system environment is in a relatively stable state. The values in the system inputs might exceed the preset range when system environment has extreme conditions.

(b) *Output for the function of system intrinsic feature*

The output $\overrightarrow{Y}$ for the function of system intrinsic feature is the system behavior. Normally, the outputs are in two forms: the treated water and the discharged grit, sludge, and the offensive gas. Specifically, they are listed as below:

$\overrightarrow{y_1}$: treated water, where,

Treated water in volume (10,000 tons), $y_{11} \in [0, 3.0]$;
BOD_5 (mg/L), $y_{12} \in [0, 20]$;
COD_{Cr} (mg/L), $y_{13} \in [0, 40]$;
SS (mg/L), $y_{14} \in [0, 20]$;
$NH_3–N$ (mg/L), $y_{15} \in [0, 8]$;
TP (mg/L), $y_{16} \in [0, 0.5]$;
TN (mg/L), $y_{17} \in [0, 20]$;
E. coli/L, $y_{18} \in [0, 1000]$.

$y_2(m^3)$, grit/skimming $y_2 \in (0, 1]$;
y_3, : sludge,where,

  Sludge (m^3), $y_{31} \in (0, 4.38]$;
  Water content: $y_{32} \in (0, 80\%]$;

$y_4(m^3)$: offensive gas, unmeasured;
y_5: other outputs.

Notes: the ranges for the above system outputs are derived in the condition that system environment is in a relatively stable state. The values in the system output may exceed the preset range when system environment has extreme conditions.

(c) *Function for the system intrinsic feature*
The function for the system intrinsic feature can be depicted as follows (Eqs. 9.5 and 9.6):

$$GY = GY(\vec{X}) \tag{9.5}$$

$$\begin{vmatrix} \vec{y_1} \\ y_2 \\ \vec{y_3} \\ y_4 \\ y_5 \end{vmatrix} = GY \begin{vmatrix} x_1 \\ x_2 \\ \vec{x_3} \\ x_4 \end{vmatrix} = GY \begin{vmatrix} \vec{x_3} \\ x_4 \end{vmatrix} \tag{9.6}$$

3. Error system model and its error value calculation for water treatment process

(1) Establishing error system model
Based on the analysis on the first stage project of east ETDZG, the error system model is built as follows (Eq. 9.7):

$$C((S(n), MG, G), \vec{x}) \tag{9.7}$$

where, $S(n)$ stands for the waste water treatment system which includes all the inter-acting subsystems and their relationships; the function for system intrinsic features $GY : \vec{Y} = \Psi(\vec{Y})$; the function for system objective features MG is $\vec{Y'} = \Psi'(\vec{X'})$; G is the rule system for judging errors, $G = \{\{G_{mr}, G_{mx}, G_{mc}\}, G_{gm}, G_h\}$, where $G_{mr} = \{$the inputs for the function of system objective features must conform to the industry regulations and law in ruling waste water treatment industry$\}$, $G_{mx} = \{$the efficiency for the function of system objective feature must be equal or higher than that of the average efficiency in the relevant industry$\}$, $G_{mc} = \{$the outputs for the function of system objective feature must meet the water quality criteria required by the Nation and Guangzhou$\}$, $G_{gm} = \{$function of system objective features must be included in the function of system intrinsic features;$\}$, $G_h = \{$except for the output of system objective, other outputs must conform to the industry regulations and law in ruling waste water treatment industry and can not bring harm to other systems in the environment}; The error value vector $\vec{x}$ is derived by calculating error function after judging the system with rules, where $\vec{x} = ((x_{mr}, x_{mx}, x_{mc}), x_{gm}, x_h)$.

(2) Calculation of error value of waste water treatment system
(a) *Calculation of x_{mr}*
Suppose that the inputs for the objective feature function of system are $\vec{X'}$, the judging rules G_{mr} = {the inputs for the objective features function of system must conform to the industry regulations and law in ruling waste water treatment industry}, the classic error function is built as follows (Eq. 9.8):

$$x_{mr} = f(G_{mr}, \vec{X'}) = \begin{cases} 0 & G_{mr} \Rightarrow \vec{X'} \\ 1 & G_{mr} \not\Rightarrow \vec{X'} \end{cases} \tag{9.8}$$

Have analyzed the inputs $\vec{X'}$ of the objective feature function of system, no violations of industry regulations and law G_{mr} have been found-i.e., $G_{mr} \Rightarrow \vec{X'}$, so $x_{mr} = f(G_{mr}, \vec{X'}) = 0$.

(b) *Calculation of x_{mx}*
The system objective features of east ETDZG were proposed after extensive research had been done to compare different mature waste water treatment systems in China. In Table 9.3, it is to compare the efficiency of the function of the current system objective features with that of another system UNITANK in China.

Based on the data, the efficiency of current system objective features is better than that of UNITANK. Therefore under the judging rule G_{mx} = {the efficiency for the function of system objective features must be equal or higher than that of the average efficiency in the relevant industry}, this system does not violate the rules for judging system errors. The established fuzzy error function is (Eq. 9.9):

Table 9.3 Comparison on indicators of the treated water discharged at the outlet

Items to compare	Current system	UNITANK system
Investment ($¥/m^3$ waste water excluding sludge handling)	852.63	883.77
Land area (m^2/m^3 waste water)	0.25	0.25
Power consumption (kwh/ m^3)	0.213	0.213
Unit management cost ($¥/m^3$ waste water)	0.48	0.49
Unit handling cost ($¥/m^3$ waste water)	0.72	0.73
Sludge (tons $DS/10\,km^{3d}$)	1.46	1.46

$$x_{mr} = f(G_{mr}(\mu_0), \mu) = \begin{cases} 0 & \mu \geq \mu_0 \\ \frac{(\mu_0 - \mu)}{\mu_0} & 0 \leq \mu \leq \mu_0 \end{cases} \tag{9.9}$$

The efficiency of this system objective feature is apparently better than that of average level, i.e., $\mu \geq \mu_0$, so the error value $x_{mr} = 0$.

(c) *Calculation of* x_{mc}

The objective output of this system is that the treated water must meet preset criteria, namely GB18918-2002 I-B and DB44/26-2001 I-Enforced Criteria. The quality also meets the requirements designed in "Environmental Impact Evaluation Report for Waste Water Treatment System at Nangang, Guangzhou" and the approval comments for the above report by Guangzhou Environment Protection Bureau. The established error function is (Eq. 9.10):

$$x_{mc} = f(G_{mc}, \vec{Y'}) = \begin{cases} 0 & G_{mc} \Rightarrow \vec{Y'} \\ 1 & G_{mc} \nRightarrow \vec{Y'} \end{cases} \tag{9.10}$$

Apparently, $x_{mc} = 0$.

(d) *Calculation of* x_{gm}

Given that no error exists in objective inputs, objective outputs, and the efficiency of objective features, the system intrinsic features must include system objective features (i.e., rules for judging error G_{mc}) if the system attempts to achieve its objective features. In order words, for any objective inputs, corresponding objective outputs of the system can be derived, and consequently the objective features can be simultaneously achieved. Sometimes, the system outputs possibly include more than just objective outputs. Therefore, system objective features must be contained in the system objective features.

In the waste water treatment system, Eqs. 9.4 and 9.6 represent the functions for system objective features and intrinsic features, respectively. From the perspective of inputs, both functions contain electricity consumption, air, and waste water; while from the angle of outputs, both functions include treated water meeting national and regional criteria. Supposed that Ω represents the vector space composed of all objective inputs $\vec{X'}$ and Λ stands for the vector space composed of all the objective outputs $\vec{Y'}$. For any $\vec{X'_0} = (a, b, (c_1, c_2, \ldots, c_9))^T \in \Omega$, substitute it in the following Eq. (9.11):

$$\begin{vmatrix} (y_{11}, y_{12}, \Lambda, y_{18})^T \\ y_2 \\ \vec{y_3} \\ y_4 \\ y_5 \end{vmatrix} = MD \begin{vmatrix} a \\ b \\ (c_1, c_2, \Lambda, c_9)^T \\ x_4 \end{vmatrix} = MD \left| \vec{X_0} \right| \tag{9.11}$$

If $\vec{y_1} \in \Lambda$, the system's objective features can be achieved under the objective inputs, which does not violate the judging rules G_{gm}; if $\vec{y_1} \notin \Lambda$, the system's objective features can not be achieved under the objective inputs, which violates the judging rules G_{gm} and the error appears. However, for $\vec{y_1} \notin \Lambda$, two reasons exist: (1) the system inputs $\vec{X}$ do not meet the requirements of objective inputs when there are errors in system inputs; (2) there might be error in the function of system intrinsic features, i.e., the issues in the system itself. Thereby, the calculation of x_{gm} in waste water treatment system can be conducted in two stages with first one calculating the error values x'_{gm} of system inputs and the second stage computing the error values x''_{gm} of system outputs given the condition when system inputs meet the requirements of system's objective inputs. In the waste water treatment system, the judging rule G_{gm} emphasizes whether system outputs include system objective outputs-i.e., x''_{gm}; while error value x'_{gm} is generally discussed under the condition that x''_{gm} is larger than 0. Establishing error function as follows (Eq. 9.12):

$$x''_{gm} = f_2(G_{gm}, MG, GY) = \begin{cases} \frac{|\vec{y'_1} - \vec{y_1}|}{|\vec{y'_1}|} & \vec{y'_1} \neq \vec{y_1} \\ 0 & \vec{y'_1} = \vec{y_1} \end{cases} \qquad (9.12)$$

where, $\vec{y_1}$ and $\vec{y'_1}$ are normalized by $\Psi(\vec{X_0})$ and $\Psi'(\vec{X_0})$, respectively.

Error 1: Suppose that the measurement for the outputs of the waste water treatment is $\vec{y_1} = (2.5, 20, 40, 20, 8, 3, 20, 1000)^T$. By analyzing the outputs, except for the TP component, the values of other properties in $\vec{y_1}$ were all within the range of objective outputs-i.e., $\vec{y_1} \notin \Lambda$ where the system violated judging rule G_{gm} and there existed error in this system. The calculation of error value is conducted through (Eq. 9.13):

$$x''_{gm} = f_2(G_{gm}, MG, GY) = \frac{3-1}{1} = 2 \qquad (9.13)$$

Error 2: Suppose that the measurement for the outputs of the waste water treatment is $\vec{y_1} = (2.5, 100, 200, 120, 15, 3, 30, 1000)^T$. By analyzing the outputs, except for the total volume of waste water and number of E. Coli, the values of other properties in $\vec{y_1}$ were not within the range of objective outputs-i.e., $\vec{y_1} \notin \Lambda$ where the system violated judging rule G_{gm} and there exists error in this system.

The first stage project at east ETDZG was constructed based on the objective features proposed in 2004. Therefore, the system should normally be able to realize the objective features. However, with the increase of population and enterprises in this affected region, the emitted civil and industrial waste water had increased significantly by 2010. The waste water needing treatment had increased to 100,000 tons/day.

Error 3: Suppose that the waste water treatment capacity designed in system objective features was changed to the criteria designed in 2010 (i.e.,100,000 tons/day), then this objective features can not be included in the system intrinsic features of this waste water treatment system. That is to say this system can not achieve the objective features and error appears. The error value is computed through (Eq. 9.14):

$$x''_{gm} = f_2(G_{gm}, MG, GY) = \frac{10 - 3}{3} \approx 2.33 \qquad (9.14)$$

(e) *Calculation of x_h*

The rule used for calculating x_h is G_h = {except for the system objective outputs, other outputs must conform to the industry regulations and law in ruling waste water treatment industry and can not bring harm to other systems in the environment}. This rule needs to be further clarified when dealing with different system outputs. In the waste water treatment system, system outputs contain not only objective outputs but also grit/skimming y_2, sludge y_3, and offensive gas y_4 which brought negative impacts on the surrounding environment. Therefore, Guangzhou Environment Protection Bureau (GEPB) had put strict ordinances and requirements on the handling of those wastes. Specifically, G_h = {skimming and grit should be shipped to landfill or local trash handling center G_{h1}; the hydration rate in sludge is less than 80% and should be shipped to Guangzhou sludge handling center G_{h2}; the offensive gas generated in the treatment process should be deodorized and emitted back to environment G_{h3}}. In actual operation, the skimming and grit had been shipped to landfill or local trash handling center; the sludge with hydration rate 75–80% had shipped to Guangzhou sludge handling center; the offensive gas generated in the treatment process had not been processed in the first-stage project. Therefore, the offensive gas had to be emitted to environment without being processed. According to rule G_h, the error function is established as follows (Eqs. 9.15, 9.16, and 9.17):

$$x_{h1} = f_{h1}(y_2, G_{h1}) = \begin{cases} 1 & G_{h1} \nRightarrow y_2 \\ 0 & G_{h1} \Rightarrow y_2 \end{cases} \qquad (9.15)$$

$$\overrightarrow{x_{h2}} = f_{h2}(\overrightarrow{y_3}, G_{h2}) = \begin{cases} x'_{h2} = \frac{y_{32} - 80\%}{80\%} \\ x''_{h2} = \begin{cases} 1 \\ 0 \end{cases} \end{cases} \qquad (9.16)$$

where **1** means that sludge was not transported to treatment center and **0** means that sludge was transported to treatment center.

$$x_{h3} = f_{h3}(y_4, G_{h4}) = \begin{cases} 1 & G_{h3} \nRightarrow y_4 \\ 0 & G_{h3} \Rightarrow y_4 \end{cases} \qquad (9.17)$$

By comparing the actual outputs with the judging rules, we can easily obtain x_{h1} = 0, and $\overrightarrow{x_{h2}} = (\frac{(75\% - 80\%)}{80\%}, 0) = (-0.4, 0)$. In this situation, $x'_{h2} = -0.4$ indicates that the hydration rate of the sludge is blew the lower limit of the rule. The error appears when we calculate the x_{h3}.

Error 4: Because the processing shop floor was not a closed design (sealed), the offensive gas generated during the process can not be collected and deodorized, which apparently violated rule G_{h3}-i.e., $x_{h3} = 1$.

Error 5: According to the newly issued requirements by Guangzhou Environment Protection Bureau, the hydration rate requirement in G_{h2} must be decreased from 80% to the level below 60%-i.e., G'_{h2}. Under this new rule, the system output y_{32} did not meet the requirement. The error value is calculated as follows:

$$\vec{x_{h2}} = f_{h2}(\vec{y_3}, G'_{h2}) =$$

$$\begin{cases} x'_{h2} = \frac{y_{32}-80\%}{80\%} \\ x''_{h2} = \begin{cases} 1, & sludge\ was\ not\ transported\ to\ treatment\ center \\ 0, & sludge\ was\ transported\ to\ treatment\ center \end{cases} \end{cases} = \begin{vmatrix} 0.25 \\ 0 \end{vmatrix}$$

In the above session, errors have been identified in the waste water treatment system according to the judging rules. Among the 5 errors, errors 1 and 2 were haphazard; errors 3 and 5 occurred when judging rules were updated; error 4 had been that way since the system came into being. The values of the errors only reflected the extent to which the system behaviors were away from the judging rules, which had nothing to with the loss or impacts caused by the error.

4. Error elimination in the water treatment system
In previous session, error system model has been employed to identify the errors in the waste water treatment system. Given that the errors have been discovered, it is necessary to investigate the root cause for generating the errors and eliminate these errors accordingly. In the following session, by understanding the processing operation of the waste water treatment system, we use the basic transformations within the error system theory to discuss how to eliminate the identified errors.

(1) Analysis and removal of error 1
In error 1, the problem resides in the fact that the TP exceeded design criteria. In the judging rule, it is required that the total phosphorus can not exceed 1 mg/L and the actual phosphorus concentration in the treated water exceeded 3 mg/L, where the particular system features cannot be achieved. As discussed in previous session, there were two reasons that rendered the system objective features unreachable: (1) actual system inputs did not meet the requirements of objective inputs; (2) the function of system intrinsic features had been changed, which caused the objective features unattainable. Therefore, we need to make sure if the actual system inputs meet the requirements of objective inputs. Suppose that when error 1 emerges, the actual system inputs are:

$$\vec{x_3} = (2.5, 200, 400, 250, 25, 10, 40, 0.5, 1.2)^T$$

By observing the actual system inputs, one can see that, except for TP, the values of all other property variables were within the required range of objective input. Error arises in the actual system inputs and its error value is:

$$x'_{gm} = \begin{cases} \frac{x_{36} - x'_{36}}{x'_{36}}, & x_{36} \geq x'_{36} \\ 0, & x_{36} < x'_{36} \end{cases}$$

$$x'_{gm} = \frac{10-5}{5} = 1$$

In order to eliminate the errors in the system inputs, two aspects could be considered: (1) system inputs are changed to make it meet the requirements of the objective inputs; (2) given that the system inputs can not produce the desired system outputs, the function of system features is modified to guarantee that the objective outputs can be attained. First of all, we start from changing the system inputs:

$$T_{zj}^{-1}(\vec{x_3}, x'_{gm}) = T_{zj}^{-1}((2.5, 200, 400, 250, 25, 10, 40, 0.5, 1.2)^T, 1)$$
$$= T_{zj}^{-1}((2.5, 200, 400, 250, 25, 5, 40, 0.5, 1.2)^T, 0)$$

By reducing the TP concentration in the system inputs, the actual system inputs meet the requirements of objective inputs. The actual meaning of this transformation is to enhance the compliance monitoring on the source of waste water and prohibit the noncompliance with standards. In the other way around, one can change the system intrinsic features. Specifically, aluminum sulfate $Al_2(SO_4)_3$ can be added to the anaerobic tank to reduce the TP in the waste water. Here, addition transformation in the system was employed to eliminate error.

$$T_{zj}(S(n), MG, G, x''_{gm}) = T_{zj}(S(n), MG, G, 2) = (S'(n), MG, G, 0)$$

(2) Analysis and removal of error 2

Error 2 has similar characteristics as error 1, where some contents' measurements in the treated water exceeded the requirements of objective outputs. In error 2, the BOD_5, COD_{Cr}, SS, NH_3–N, TP, and TN all exceeded the required range. In order to eliminate the errors in the system outputs, solutions can be considered from two angles: (1) system inputs are changed to make it meet the requirements of the objective inputs; (2) given that the system input can not produce the desired system outputs, the function of system feature is modified to guarantee that the objective outputs can be attained. The approach used in eliminating error 1 can be employed to eliminate error 2 if this error is caused by system inputs. Otherwise, if the error is caused by system structure, hierarchical structural analysis will be conducted to identify the particular structure that is responsible for producing the error. In the functional diagram of waste water treatment system (Fig. 9.12), it is easy to find that the above-mentioned contents and their measurements are directly related to A/O biochemical tank and air compressor in which A/O biochemical tank is more critical in determining the performance of the treatment. Using system hierarchical analysis:

$$T_f(C((S(n), MG, G), \overrightarrow{x})) = \{C_1^1((S_1^1, MG_1^1, G_1^1), \overrightarrow{x}_1^1), \ldots, C_n^1((S_n^1, MG_n^1, G_n^1), \overrightarrow{x}_n^1)\}$$

So, it is assumed that the error system model for the subsystem of air compressor is expressed as $C_i^1((S_i^1, MG_i^1, G_i^1), \overrightarrow{x}_i^1)$, where S_i^1 stands for the subsystem of air compressor, MG_i^1 represents the objective feature of the subsystem of air compressor (the subsystem of air compressor can pump $6399.8\,\mathrm{m}^3$ air with wind pressure of 69.6 KPa under the condition of electricity consumption of 180 kw/h), G_i^1 denotes the rule for judging errors in the subsystem of air compressor and $G_i^1 = \{\{G_{mr}, G_{mx}, G_{mc}\}, G_{gm}, G_h\}$. Having gone through error identification, it was found that no error arose in the subsystem of air compressor.

Suppose that the error system model for the subsystem of A/O biochemical tank is expressed as $C_j^1((S_j^1, MG_j^1, G_j^1), \overrightarrow{x}_j^1)$, where the objective features for the A/O biochemical tank is to process the COD_{Cr}, BOD_5, SS, NH_3–N, TP, and TN to achieve the water standard. In error 2, with the current intrinsic features of the subsystem, the objective features can not be realized because the objective features are not completely included in the intrinsic features. From error system model, in $\overrightarrow{x}_j^1 = (x_m, x_{gm}, x_h)$, the calculated error values of x_m and x_h are all 0s. In this case, it is necessary to identify the reasons why the objective features can not be achieved under the current intrinsic features. With further analysis, it is found that there exist no errors in the subsystems providing inputs to the system. We can rule out the case that actual inputs did not meet system objective inputs. Therefore, we can confirm that the subsystem structure had defects. In order to fully understand the causes for the defects, decomposition transformation is conducted:

$$T_f(C_j^1((S_j^1, MG_j^1, G_j^1), \overrightarrow{x}_j^1)) = \{C_{j1}^2((S_{j1}^2, MG_{j1}^2, G_{j1}^2), \overrightarrow{x}_{j1}^2), \ldots, C_{jm}^2((S_{jm}^2, MG_{jm}^2, G_{jm}^2), \overrightarrow{x}_{jm}^2)\}$$

Having done decomposition, the lower hierarchy subsystems were analyzed one by one. In the analysis, the error was found at the aerator which was blocked and malfunctioned. As indicated by the working mechanism of A/O biochemical tank, the broken or blocked aerator was definitively the primary cause for generating error 2. It is very easy to eliminate this error by replacing the old aerator and restore the A/O biochemical tank to normal operation.

$$T_z(C_{ji}^2((S_{ji}^2, MG_{ji}^2, G_{ji}^2), \overrightarrow{x}_{ji}^2 > 0)) = C_{ji}^2((S_{ji}^2, MG_{ji}^2, G_{ji}^2), \overrightarrow{x}_{ji}^2 = 0)$$

(3) Analysis and removal of error 3

In the process of investigating errors, we have been emphasizing the temporal characteristics of error occurrence. Error must arise at certain time point or during certain time periods (It can not exist without time and space). From the variable definition in the error system model, system, subsystem, and rules for judging error all vary with time. In error 3, due to the dynamics of system objective features, the error tended to arise because the former non-erroneous system can not meet the requirements of changed system objective features.

By comparing the objective features between current and old systems, it was found that the requirements for daily treatment volume had been changed in the new objective features. The maximum daily processing capacity of the old system was 30, 000 tons while the daily processing capacity of 100,000 tons was demanded in the new system objective features. The new requirements in the system objective features were proposed based on rigorous evaluation on the ever-increasing volume in surrounding regions and the goal of establishing eco-friendly city in Guangzhou. The newly added system objective features were reasonable and feasible.

Under the newly added objective features, the error value is 3.33. The error can be eliminated as long as the daily waste water processing capacity is increased to 4 times as large as the original one. Specifically, similarity transformation was conducted on the waste water treatment system and the capacity was expanded to 4 times as large as the original one.

$$T_x(S(n), MG', G, x''_{gm}) = T_x(S(n), MG', G, 3.33) = (4 * S(n), MG', G, 0)$$

(4) Analysis and removal of error 4

Errors 4 and 5 arose because the system created negative impact on the surrounding environment. In order to realize system objective features, undesirable outputs such as offensive gas, grits, and sludge cannot be avoided in the waste water treatment process. As far as the emitted offensive gas is concerned, the emission not only generated unfavorable smell in the environment but also produced negative impacts on health of the residents in surrounding communities and employees working in this treatment factory. It is not acceptable no matter which environment it is affecting-i.e., internal or external. It is not surprising to add the requirements for regulating the emission of offensive gas.

Because offensive gas emission is the by-product in the process of realizing system objective features, it is impossible to eliminate error 4 by just changing current system structure. On the contrary, it must guarantee the realization of system objective features. Therefore, addition transformation conducted on current system is a good measure of addressing current issues. First of all, a system used for purification and deodorization of offensive gas was added to current waste water treatment system.

$$T_{zj}(S(n), MG, G, x_{h3}) = T_{zj}(S(n), MG, G, 1) = \{(S(n), MG, G, 1), (S'(n'), MG', G', 0)\}$$

Integration of the added system into the current system forms a complete new system.

$$T_f^{-1}\{(S(n), MG, G, 1), (S'(n'), MG', G', 0)\} = (S_0(n+1), MG, G, 0)$$

Have done the integration, the updated system can not only achieve the original system objective features but also realize the removal of offensive gas by using the added processing system S'. Error 4 was eliminated as offensive gas in the system output has been removed.

(5) Analysis and removal of error 5

Similar to error 3, error 5 arose because the former non-erroneous system was not in conformance with the updated rules for judging error. According to the new requirements in 2011 by Guangzhou Environment Protection Bureau, the hydration rate of sludge from the waste water treatment system must be below 60%. However, the minimum hydration rate of the sludge from the treatment system was 75% and reduction in the hydration rate was ultimate goal of eliminating this error.

In the diagram composed of process flow and system features (Fig. 9.12), the hydration rate was related to the sludge dehydrator, which became the object for error elimination by conducting system transformation. Three types of transformations can be undertaken: (1) addition transformation is initiated to the current subsystem to improve the effectiveness of current dehydrators; (2) displacement transformation is done on the current subsystem to replace current dehydrators with more advanced and effective dehydrators; (3) combination transformation is employed, where newly developed dehydration accessories is installed on the current dehydrators to lower the hydration rate in the sludge. Although all three methods could be used to eliminate errors, there is very large difference in the costs of achieving the same goal. Have done comprehensive evaluation and analysis, the third method was adopted and its specific process is demonstrated as follows:

$$T_h^{-1}(\Phi) = (S', MG', G', 0)$$

$$T_f^{-1}\{(S(n), MG, G, 0.25), (S', MG', G', 0)\} = (S_0(n+1), MG, G, 0)$$

In this chapter, the general form of error system model is established and relevant properties are discussed. Thereafter, the method of employing error system model to identify system error is discussed. The general error system model is further explored in the error diagnosis and elimination. Actual applications are exhibited to demonstrate the specific procedures of using error system model. The concrete foundation of error system theory and methodologies have foreseen their future wide applications including but are not limited to feasibility analysis of project, business re-engineering, decision making, project evaluation, diagnosis of mechanical and electronic equipment, system design, system management, and system optimization, etc.

Glossary

Acting force of system elements refers to the value w of certain property of system elements; w_{ij} denotes the acting force of jth property of ith system element.

Axiomatic set theory is the research using modern axiomatized method to reestablish Cantor set theory.

Cost indicators refer to the case that higher cost mean larger error.

Efficiency indicators refer to the case that smaller value means larger error.

Error one aspect in the two aspects of contradiction, which is non-conforming to objective facts and coexists with correctness or truth.

Error function deterministic relationship between error object set and real number set $\Re$.

Error matrix developed based on matrix theory, is a powerful instrument used to study transformation and transfer of errors.

Error set Suppose that $U(t)$ is object set, $G(t)$ is a set of rules for judging error, if $C = \{((U(t), S(t), \overrightarrow{p(t)}, T(t), L(t)), x(t) = f(G \not\Rightarrow u(t))) \mid (U(t), S(t), \overrightarrow{p(t)}, T(t), L(t)) = u(t) \in U(t), f \subseteq \times \mathbb{R}, x(t) = f(G \not\Rightarrow u(t))\}$ then C is called an error set defined on $U(t)$ under the rule of judging errors $G(t)$.

Error theory is to investigate the causes and mechanism that produce errors with the objective of ultimately eliminating errors.

Features of system indicate the functionality of a system. We use this word "feature" instead of "function", although it is not semantically matching the exact meaning of functionality, for preventing generating ambiguity because function is often used to indicate the mathematical function.

Fixed value indicates the phenomenon that an particular value will be located (or optimal) to meet the requirements.

© Springer Nature Switzerland AG 2020
K. Guo and S. Liu, *Error Systems: Concepts, Theory and Applications*,
Studies in Systems, Decision and Control 275,
https://doi.org/10.1007/978-3-030-40760-5

Fixed value approximation indicator the indicator that approaches to a fixed value where the error value is the maximum.

Formal language of error predicate logic is the single-meaning synthetic language that uses various logical forms in error predicate logic.

Function is a mapping for one set to the other, which also depicts the relationship between two sets.

Judging rules in reality, are requirements and conditions for evaluating and judging the objects of interest.

Mathematical error predicate logic is the subject that uses mathematics and semantics to investigate the forms and laws of error proposition reasoning which contain quantifier, domain, thing, space, property, dimension value, function, time, error value, and rules for judging error.

Mathematical error propositional logic is to use mathematics and semantics to investigate different types of reasoning and laws of error compound propositions. It is a kind of thinking that employs object and judging rules to evaluate and justify the erroneous structure of certain phenomenon and/or problem of interests.

Model theory mainly explores the relationship between formalism system and the mathematical model.

Numeric type is used to store the data having types of *int* and *float*.

Proposition is the sentence for being the primary bearers of truth and falsity.

Propositional calculus is the branch of logic for studying more advanced and complicated propositions (whether they are true or false) through the use of logical connectives.

Range indicators refer to the case that the chosen values must be in certain predefined range.

Range approximation indicator the indicator that approaches to certain range within which the maximum error value can be found.

Recursion theory focuses on the study of computability, which has close relationship with the development of computer science.

Rulebase is used to store rules. Two dimensions are used to profile the rules with the first one describing the practical meaning of rules, i.e., "qualitative description" and the second one describing the mathematical meaning of rules,i.e., "logical expression".

Similarity transformation on system structure The morph that scale of a structure is augmented or lessened (miniature or giant) while shape is kept unchanged is called similarity transformation on system structure.

System structural acting force dictates the strength, direction, and sequence of mutual actions of system's elements, which is a vector used to quantify the effect of one element's action on another one denoted by ω.

System Structure the totality for type and order of connections, organizations, and interactions of elements within a system.

Total acting force of one structural chain is the "sum" of acting forces of all edges of that chain denoted by $w = \sum w_i$ (if there exists additivity) or other aggregation mechanism (if there exists no additivity).

References

1. Afsar, B., Ali, Z., Dost, M., Safdar, U.: Linking error management practices with call center employees? helping behaviors and service recovery performance. Pakistan J. Commerce Soc. Sci. **11**(1), 185–205 (2017)
2. Anonymous: Analysis on the error in locating a restaurant (2012). http://blog.sina.com.cn/s/blog_900e50d601012569.html. Cited on 10 May 2019
3. Anonymous: Finding an enhanced implementation in SAP (2018). http://progra-mmersought.com/article/35291363781/;jsessionid=5AE70A7479F3CD9386899B-693D36E228. Cited on 10 October 2019
4. Arecchi FT: Towards a Post-Bertalanffy Systemics. Martino Fine Books (2016)
5. Ashby, W.R.: An Introduction to Cybernetics. Chapman & Hall, London (1956)
6. Ashby, W.R.: An Introduction to Cybernetics. Springer International Publishing, Verlag (2015)
7. Banerjee, S., Rondoni, L.: Applications of Chaos and Nonlinear Dynamics in Science and Engineering, Vol. 4 (Understanding Complex Systems). Springer (2015)
8. Barwise, J. (ed.): Handbook of Mathematical Logic. p. 1–11. North–Holand Publishing Company, Amastdam (1977)
9. Bell, B.S., Kozlowski, S.W.J.: Collective failure: the emergence, consequences, and management of errors in teams. In: Hofmann, D.A., Frese, M. (eds.) Errors in organizations, pp. 1–44. Routledge, New York (2011)
10. Belyakov, R.A., Marmain, J.: MiG: Fifty Years of Secret Aircraft Design. Airlife Publishing, Shrewsbury, UK (1994)
11. Bian, Y., Guo, K.: Optimization of the non-additive feature systems. Math. Pract. Theory **41**(9), 90–96 (2011)
12. Bian, Y., Guo, K.: A definition on basic structure of enterprise management system. In: Proceedings of 2011 2nd International Conference on Artificial Intelligence, Management Science and Electronic Commerce. Zhengzhou, China, AIMSEC 2011, 8–10 August 2011, pp. 907–910 (2011)
13. Bian, Y., Guo, K.: Mathematical definition on basic structure of enterprise management systems. Math. Pract. Theory **41**(18), 104–111 (2012)
14. Blokdyk, G.: Error Management Theory the Ultimate Step-By-Step Guide. Emereo Publishing (2018)
15. Bogner, M.S.: Human Error in Medicine (Human Error And Safety). CRC Press (1994)
16. Boniface, D.E., Bea, R.G.: Assessing the risk of and countermeasures for human and organizational error. SHAME Trans. **104**(1996), 157–177 (1996)

© Springer Nature Switzerland AG 2020
K. Guo and S. Liu, *Error Systems: Concepts, Theory and Applications*,
Studies in Systems, Decision and Control 275,
https://doi.org/10.1007/978-3-030-40760-5

17. Bostock, L., Bairstow, S., Fish, S., Macleod, F.: Managing Risk and Minimizing Mistakes in Services to Children and Families. SCIE, London (2005)
18. Brain, W., Salt, D., Reid, W.: Resilience Thinking: Sustaining Ecosystems and People in a Changing World, 1st edn. Island Press (2006)
19. Bregman, O.C., White, C.M.: Bringing Systems Thinking to life: Expanding the horizons for Bowen Family Systems Theory. Routledge, New York (2011)
20. Brodbeck, F.C., Zapf, D., Prümper, J., Frese, M.: Error handling in office work with computers: a field study. J. Occup. Organ. Psychol. **66**, 303–317 (1993)
21. 1% error will result in 100% failure. http://my.icxo.com/. Cited on 26 Sept 2007
22. Cacciabue, P.C.: Human error risk management for engineering systems: a methodology for design, safety assessment, accident investigation and training. Reliab. Eng. Syst. Saf. **83**(2), 229–240 (2004)
23. Campbell, D., Draper, D., Huffington, C. (eds.): Teaching Systemic Thinking. Routledge, London (1988)
24. Carroll, J., Christianson, K.M., Frese, M., Lei, Z., Naveh, E., Vogus, T.: Errors in organizations: exploring new frontiers, and developing new opportunities for theory, research and impact. Paper presented at AMD Errors in Organization session at the 2018 Academy of Management Conference, Chicago, 13 August 2018 (2018)
25. CCPS (Center for Chemical Process Safety): Guidelines for preventing human error in process safety, 1st Ed. Wiley-AIChE (2004)
26. Checkland, P.: Systems Thinking, Systems Practice: Includes a 30-Year Retrospective, 1st edn. Wiley (1999)
27. Chen, S.: Measuring the weight of an elephant. Records of the Three Kingdoms · Book of Wei · Biographies of Cao Chong (3rd century)
28. Chamberlain, R.M.: Lies, damned lies and counterinsurgency (2008). http://armedforcesjournal.com/lies-damned-lies-and-counterinsurgency/. Cited on 18 September 2019
29. Charles, M.C.: Dynamic Systems 3e. Wiley (2001)
30. Clarke, B. (Ed.): Hansen MBN(Editor), Smith BH (Series Editor), Weintraub ER (Series Editor). Emergence and embodiment: New essays on second-order systems theory (Science and Cultural Theory). Duke University Press (2009)
31. Clarke, D.: Descartes: A Biography. Cambridge University Press, England, Cambridge (2006)
32. Cooper, J., Newbower, R., Kitz, R.: An analysis of major errors and equipment failures in anesthesia management: conditions for prevention and detection. Anasthesiology **60**, 42–43 (1984)
33. Dahrendorf, R.: the modern social conflict: an essay on the politics of liberty. University of California Press (2007)
34. Dahlin, B.K., Chuang, Y., Roulet, T.J.: Opportunity, motivation and ability to learn from failures and errors: review, synthesis and ways to move forward. Acad. Manag. Ann. **12**, 252–277 (2017)
35. Dekker, S.: The Field Guide to Human Error Investigations. Ashgate Publishing (2002)
36. Dekker, S.: Ten Questions About Human Error: A New View of Human Factors and System Safety (Human Factors in Transportation), 1st edn. CRC Press (2004)
37. Dhillon, B.S.: Human Reliability and Error in Transportation Systems (Springer Series in Reliability Engineering). Springer, London (2007)
38. Dhillon, B.S.: Human reliability, error, and human factors in engineering maintenance: with reference to aviation and power generation, 1st edn. CRC Press (2009)
39. Dhillon, B.S.: Safety and Human Error in Engineering Systems, 1st edn. CRC Press (2012)
40. Dhillon, B.S.: Human Reliability, Error, and Human Factors in Power Generation (Springer Series in Reliability Engineering). Springer International Publishing (2014)
41. Dimitrova, N.G., Van Dyck, C., van Hooft, E.A.J., Groenewegen, P.: Error prevention, error management, or both? In: Annual Meeting Proceedings of Academy of Management. Philadelphia, Pennsylvania United States, 1–5 August 2014, 2014(1), 15702–15702 (2014)
42. Dismukes, R.K.: Human Error in Aviation. CRC Press, Taylor & Francis Group (2009)

43. Dormann, T., Frese, M.: Error training: replication and the function of exploratory behavior. Int. J. Hum. Comput. Interact. **6**(4), 365–372 (1994)
44. Dorner, D.: The Logic of Failure: Recognizing and Avoiding Error in Complex Situations, Revised Ed. Basic Books (1997)
45. DR Pepper Snapple Group: Arsenic found in 11 bottled water brands, Consumer Reports says (2019). https://www.fooddive.com/news/arsenic-found-in-11-bottled-water-brands-consumer-reports-says/553071/. Cited on 10 August 2019
46. Dunning, T.J.: Trades' Unions and Strikes: Their Philosophy and Intention, pp. 35–36 (1860). https://ia802908.us.archive.org/30/items/tradesunionsstri00dunnrich/tradesunionsstri00dunnrich.pdf. Cited on 10 July 2019
47. Eden, B.: Encyclopaedia Britannica CD 99 (Multimedia version). Electron. Resour. Rev. **3**(1), 9–10 (1999)
48. Elzer, P.F., Kluwe, R.H., Boussoffara, B.: Human Error and System Design And Management (Lecture Notes in Control and Information Sciences). Springer, London (2000)
49. Engels, F.: Dialectics of Nature (Translator, Clements Dutt). New York: International Publishers (Original work published 1940) (1925)
50. Favell, D.J.: A comparison of the vitamin C content of fresh and frozen vegetables. Food Chem. **62**(1), 59–64 (1998)
51. Fieguth, P.: An Introduction to Complex Systems: Society, Ecology, and Nonlinear Dynamics, 1st edn. Springer (2017)
52. Fischer, S., Frese, M., Mertins, J.C., Hardt-Gawron, J.V.: The role of error management culture for firm and individual innovativeness. Appl. Psychol. **67**(3), 428–453 (2018)
53. Fitts, P., Jones, R.: Analysis of factors contributing to 460 'pilot error' experiences in operating aircraft controls. Memorandum Rep. Ohio, Aero Medical Laboratory (1947)
54. Fraser, J., Smith, P., Smith, J.: A catalog of errors. Int. J. Man-Mach. Stud. **37**, 265–393 (1992)
55. Frese, M.: To err is human. Review of Human Factors in Hazardous situations by D.E. Broadbent, A. Baddeley & J.T. Reason (Eds.). The Psychologist, 14, 8, 341 (1991)
56. Frese, M.: Error management or error prevention: two strategies to deal with errors in software design. In: Bullinger, H.-J. (Ed.), Human Aspects in Computing: Design and use of Interactive Systems and Work with Terminals, Elsevier Science Publisher, pp. 776–782 (1991)
57. Frese, M.: Error management in training: conceptual and empirical results. In: Zuccermaglio, C., Bagnara, S., Stucky, S.U. (eds.) Organizational Learning and Technological Change, pp. 112–124. Springer, Berlin, Heidelberg, New York (1995)
58. Frese, M.: Error management: an alternative concept to error prevention in organizations and in technical system design. In: Sheridan, T.B. (Ed.), Proceedings of ManMachine Systems (MMS'95). Cambridge, Massachusetts, June 1995. Elsevier Science Publisher, Amsterdam (1995)
59. Frese, M., Altmann, A.: The treatment of errors in learning and training. In: Bainbridge, L., Quintanilla, S.A.R. (eds.) Developing Skills with New Technology, pp. 65–86. Wiley, Chichester (1989)
60. Frese, M., Brodbeck, F.C., Zapf, D., Prümper, J.: Users' errors and error handling: its relationships with task structure and social support. SIGCHI Bull. **23**(2), 59–62 (1991)
61. Frese, M., Brodbeck, F.C., Heinbokel, T., Mooser, C., Schleiffenbaum, E., Thiemann, P.: Errors in training computer skills: on the positive function of errors. Human Comput. Interact. **6**, 77–93 (1991)
62. Frese, M., Keith, N.: Action errors, error management, and learning in organizations. Ann. Rev. Psychol. **66**, 661–687 (2014)
63. Frese, M., van Dyck, C.: Error management: Concept to error prevention in technical system design. Int. J. Psychol. Abstracts of the XXVI. International Congress of Psychology, Montréal, Canada, 16–21 August 1996, **31**, 147 (1996)
64. Gall, J., Blechman, R.O.: Systemantics: how systems work and especially how they fail, 1st Ed. Quadrangle (1977)
65. Gell-Mann, M.: The Quark and the jaguar: Adventures in the Simple and the Complex. W.H. Freeman, San Francisco (1994)

66. Georgiou, I.: Thinking Through Systems Thinking. Routledge, London (2007)
67. German Federal Bureau of Aircraft Accident Investigation(2004). Investigation Report:AX001-1-2/02 May 2004. http://www.bfu-web.de/DE/Publikationen/ Untersuchungsberichte/2002/Bericht_02_AX001-1-2.pdf?_blob=publicationFile. Cited on 18 June 2019
68. Gharajedaghi, J.: Systems Thinking: Managing Chaos and Complexity: a Platform for Designing Business Architecture, 3rd edn. Morgan Kaufmann (2011)
69. Ghosh, A.: Dynamic Systems for Everyone: Understanding how our World Works. Springer (2015)
70. Gleik, J.: Chaos: Making a New Science. Penguin Books (2008)
71. Goodman, P.S., Ramanujam, R., Carroll, J.S., Edmondson, A.C., Hofmann, D.A., Sutcliffe, K.M.: Organizational errors: directions for future research. Res. Organ. Behav. **31**, 151–176 (2011)
72. Gordon, Y.: Mikoyan MiG-25 Foxbat: Guardian of the Soviet Borders (Red Star Vol. 34). Midland Publishing Ltd., Hinckley, UK (2008)
73. Gray, N.: Maintenance error management in the ADF, Touchdown (Royal Australia Navy), December 2004, p. 1–4 (2004). http://www.navy.gov.au/publications/touchdown/dec.04/ maintrr.html. Cited on 10 July 2019
74. Greif, S., Janikowski, A.: Aktives Lernen durch systematische Fehlerexploration oder programmiertes Lernen durch Tutorials? [Active learning through systematic error exploration or programmed learning through tutorials?]. Zeitschrift für Arbeits- und Organistionspsychologie **31**, 94–99 (1987)
75. Gros, C.: Complex and Adaptive Dynamical Systems: A Primer. Springer (2015)
76. Grout, J.R.: Mistake proofing: changing designs to reduce error. Quality Saf. Health Care **15**(S1), i44–i49 (2006)
77. Gully, S.M., Payne, S.C., Koles, K.L.K., Whiteman, J.A.K.: The impact of error training and individual differences on training outcomes: an attribute-treatment interaction perspective. J. Appl. Psychol. **87**, 143–155 (2002)
78. Guo, K.: Preliminary analysis on error systems. J. Guangdong Univ, Technol (1986)
79. Guo, K.: Research of eliminating error in system. In: International Conference of AMSE System, Control Information Methodologies and Applications, Wuhan, China, Oct 1994 (1994)
80. Guo, K.: Research of eliminating error of system transformation. Adv. Model. Anal. D. **19**(4), 9–14 (1992)
81. Guo, K.: "Fifteen-six-three" methods of eliminating error of system. Adv. Model. Anal. D **26**(2), 59–63 (1995)
82. Guo, K.: Logic of error elimination. J. Guangdong Univ. Technol. **4**, (1997)
83. Guo, K.: Study on the global optimization. Sci. Technol. Assoc. Forum. **3**, 51–52 (2007)
84. Guo, K.: Error Systems. China Science Press, Beijing (2012)
85. Guo, K.: How to Avoid and Correct Errors. China Science Press, Beijing (2015)
86. Guo, Q., Guo, K.: Error-eliminating theory of intelligent system: the optimization of systems, subsystems and elements. In: 2009 International Workshop on Intelligent Systems and Applications,ISA, Wuhan, China, 23–24 May 2009 (2009)
87. Guo, Q., Guo, K., Wang, Q.: The impact of decision-making during environmental mutation-substantial change of fuzzy error system. In: Proceedings of 2008 IEEE International Conference on Industrial Engineering and Engineering Management. IEEM, Singapore, 8–11 December 2008, pp. 695–698 (2008)
88. Guo, K., Huang, J.: 1% error leading to 100% system failure: investigation on the mechanism of error generation and methods for eliminating and avoiding errors. J. Guangdong Univ. Technol. **25**(2), 1–5 (2008)
89. Guo, K., Huang, J., Xiong, H.: Feature additivity-based system optimization. J. Guangdong Univ. Technol. **25**(4), 1–4 (2008)
90. Guo, K., Liu, S.: Research on laws of security risk-error logic system with critical point. Model. Measurement Control D. **22**(1–2), 1–10 (2001)

91. Guo, K., Liu, S.: Fundamentals of Error Theory (Studies in Systems Decision and Control). Springer International Publishing (2019)
92. Guo, K., Min, X.: Study on the solution-finding of inclusive error matrix equation. J. Guangdong Univ. Technol. **64**(3), 12–15 (2010)
93. Guo, K., Shi, J.: Equality-type fuzzy error-eliminating programming-based system optimization. Presented at 3rd Workshop on Management Studies in China, Lanzhou, China, 13–15 August 2010 (2010)
94. Guo, Q., Wang, Q., Guo, K.: Exploration of fuzzy system in decision making—relationship of fuzzy error logic decomposition word T^f and connotative antithesis $+^{nhd}$. In: Proceedings of International Conference on Industrial Engineering and Engineering Management. IEEM, Singapore, 8–11 December 2008, p 450–454 (2008)
95. Guo, K., Xiong, H.: Research on thing transformation connectives in fuzzy error logic. Fuzzy Syst. Math. **20**(2), 34–39 (2006)
96. Guo, K., Yao, J.: Application of system structure analysis to the system of decision-making objects. Wuhan, China 25–26 November 2009 (2009)
97. Guo, K., Zhang, S.: Fuzzy error set. Fuzzy Syst. Math. **5**(2), 67–75 (1991)
98. Guo, K., Zhang, S.: Decomposition of error system. Adv. Model. Anal. D. **17**(4), 25–32 (1993)
99. Guo, K., Zhang, S.: Introduction to Error Elimination. Press of South China University of Technology, Guangzhou (1995)
100. Guo, K., Zhang, S.: Theory and Method for judging Decision-Making Errors in Enterprise Capital Asset Investment. Press of South China University of Technology, Guangzhou (1995)
101. Guo, K., Zhang, S.: Theory of Error Set. Press of Central China University of Technology, Hunan (2001)
102. Hagen JU(2010). Why we all make big mistakes? 21 July 2010, 03:50pm https://www.forbes.com/2010/07/21/mistakes-error-prevention-leadership-managing-crm.html.Cited on 15 October 2018
103. Hagen, J.U.: How the lack of error management contributed to today?s problems in the financial services industry (2013). 31 January 2013 https://www.globalbankingandfinance.com/how-the-lack-of-error-management-contributed-to-todays-problems-in-the-financial-services-industry/. Cited on May 19th, 2019
104. Hagen, J.U.: Confronting Mistakes: Lessons from the Aviation Industry when Dealing with Errors. Palgrave Macmillan (2013)
105. Hagen, J.U: Error management: Not just a wing and a prayer. EFMD Global Focus **8**(2), 52–55 (2014)
106. Hagen, J.U.: How Could this Happen?-Managing Errors in Organizations. Palgrave Macmillan (Ed.) (2018)
107. Hanson, B.G.: General Systems Theory-Beginning with Wholes, 1st edn. Taylor & Francis (1995)
108. Harrald, J.R., Mazzuchi, T.A., Spahna, J., Van Dorp, R., Merrick, J., Shrestha, S., Grabowski, M.: Using system simulation to model the impact of human error in a maritime system. Saf. Sci. **30**, 235–247 (1998)
109. Hasegawa, T., Kemeda, A.: Analysis and evaluation of human error events in nuclear power plants. Presented at the meeting of IAEA?s CRP on Collection and Classification of Human Reliability Data for Use in Probabilistic Safety Assessments: Available from the Institute of Human Factors, Nuclear Power Engineering Corporation 3–17-1 Toranomon. Minato-Ku, Tokyo (May 1998)
110. Haselton, M.G., Buss, D.M.: Error management theory: a new perspective on biases in cross-sex mind reading. J. Personal. Soc. Psychol. **78**(1), 81–91 (2000)
111. Haselton, M.G., Galperin, A.: Error management in relationships. In: Simpson, J., Campbell, L. (eds.) The Oxford Handbook of Close Relationships (2013)
112. Hegel, G.W.F.: Science of logic, §716ff-718ff, (p 335 in the Miller edition) (1812). https://www.marxists.org/reference/archive/hegel/works/hl/hl333.htm#0719. Retrieved 21 Sept 2017

113. Heimbeck, D., Sonnentag, S., Frese, M., Keith, N.: Integrating errors into the training process: the function of error management instructions and the role of goal orientation. Pers. Psychol. **56**(2), 333–361 (2003)

114. Helmreich, R.L.: On error management: lessons from aviation. BMJ **320**(7237), 781–785 (2000)

115. Henningsen, D.D., Henningsen, M.L.M.: Testing error management theory: exploring the commitment skepticism bias and the sexual overperception bias. Human Commun. Res. **36**(4), 618–634 (2010)

116. Highway Capacity Manual: Sixth Edition: A Guide for Multimodal Mobility Analysis. Transportation Research Board (2016). http://www.trb.org/Main/Blurbs/175169.aspx. Cited on 1 Aug 2018

117. Hofmann, D.A., Frese, M. (eds.): Error in Organizations (SIOP Organizational Frontiers Series), 1st edn. Routledge (2011)

118. Holland, J.H.: Outline for a logical theory of adaptive systems. J. ACM **9**(3), 297–314 (1962)

119. Holland, J.H.: Adaptation in natural and artificial systems: an introductory analysis with applications to biology, control, and artificial intelligence. University of Michigan Press, Ann Arbor, MI (1975)

120. Holland, J.H.: Hidden Order: How Adaptation Builds Complexity. Perseus Books, New York (1995)

121. Holland, J.H.: Emergence: From Chaos to Order. Oxford University Press, Oxford, U.K. (2000)

122. Holland, J.H.: Signals and Boundaries: Building Blocks for Complex Adaptive Systems. MIT Press, Cambridge, MA (2012)

123. Holland and Reitman: Cognitive systems based on adaptive algorithms. ACM SIGART Bull. **63**, 49 (1977)

124. Hollnagel, E.: Cognitive Reliability and Error Analysis Method. Elsevier, Oxford (1998)

125. Huang, J., Xiong, H., Guo, K.: The transformation of fuzzy error set in a system intelligent decision making. In: Proceedings of International Conference on Advanced Computer Control(ICACC). Singapore, 22–24 January 2009, p 518–522 (2009)

126. Huang, J., Xiong, H., Guo, K.: System optimizing based on function additivity. Presented at International Workshop on Intelligent Systems and Applications (ISA), Wuhan, China, 23–24 May 2009 (2009)

127. International Atomic Energy Agency (IAEA): Human Error Classification and Data Collection. Report of a technical committee meeting organized by the IAEA and held in Vienna, 20–24 February 1989

128. Ivancic, K., Hesketh, B.: Learning from errors in a driving simulation: Effects on driving skill and self-confidence. Ergonomics **43**, 1966–1984 (2000)

129. Jackson, M.C.: Critical Systems Thinking and the Management of Complexity. Wiley (2019)

130. Jantsch, E.: The self-organizing universe: Scientific and human implications of the emerging paradigm of evolution (Systems Science and World Order Library. Innovations in Systems Science), 1st Edn. Pergamon (1980)

131. Johns Hopkins Medical Institutions (JHMI): Fewer airline crashes linked to "pilot error"; Inclement weather still major factor (2001). https://www.sciencedaily.com/releases/2001/01/010109083707.htm. Retrieved on 15 November 2018

132. Johnson, C.W.: Human error, interaction and the development of safety-critical systems. In: Boy, G.A. (ed.) The Handbook of Human-Machine Interaction: a Human-centered Design Approach, pp. 91–106. Ashgate, Farnham (2011)

133. Johnson, D.D.P., Blumstein, D.T., Fowler, J.H., Haselton, M.G.: The evolution of error: error management, cognitive constraints, and adaptive decision-making biases. Trends Ecol. Evolut. **28**(8), 474–481 (2013)

134. Keith, N., Frese, M.: Self-regulation in error management training: emotion control and metacognition as mediators of performance effects. J. Appl. Psychol. **90**(4), 677–691 (2005)

135. Laszlo, E.: The Relevance of General Systems Theory. G. Braziller, New York (1972)

136. Laszlo, E.: The systems view of the world: A holistic vision for our time (Advances in Systems Theory, Complexity, and the Human Sciences). Hampton Press (1996)
137. Lazonder, A.W., Van der Meij, H.: Effect of error information in tutorial documentation. Interact. Comput. **6**(1), 23–40 (1994)
138. Leclerc, H., Moreau, A.: Microbiological safety of natural mineral water. FEMS Microbiol. Rev. **26**(2002), 207–222 (2002)
139. Lei, Z., Naveh, E., Novikov, Z.: Errors in organizations: an integrative review via level of analysis, temporal dynamism, and priority lenses. J. Manag. **42**(5), 1315–1343 (2016)
140. Leo, M.A., Lieber, C.S.: Alcohol, vitamin A, and beta-carotene: adverse interactions, including hepatotoxicity and carcinogenicity. Am. J. Clin. Nutrition **69**(6), 1071–1085 (1999)
141. Li, M.: The profile of mathematical logic in 20th century. Stud. Dialect. Nature. **S12**(9) (2002)
142. Li, M., Guo, K.: Research on decomposition of fuzzy error set. Adv. Model. Anal. A: Gener. Math. Comput. Tools **43**(26), 15–26 (2006)
143. Lin, L., Guo, K.: Studies on the similarity transformation connectives of error domain: investigating errors in management system. Theoret. Investig. **213–216**, (2008)
144. Lin, L., Guo, K.: Examination on the rules for judging error systems. Statist. Decis. **10**, 41–43 (2009)
145. Liu, Y.: Theory and method for understanding conflicts and errors in large-scale complicated systems. Press of South China University of Technology, Guangzhou (2000)
146. Liu, S., Guo, K: Introduction for the theory of error-eliminating. Adv. Model. Anal. A? Gener. Math. **39**(2), 39–66 (2002)
147. Liu, S., Guo, K.: Extreme value of fuzzy error system with change of time and space-zero faults trend of significant decision. Adv. Model. Anal.: B. **45**(3), 39–49 (2002)
148. Liu, S., Guo, K.: Fuzzy error system with change of time and space-the effect of change of time and space on decision making. Adv. Model. Anal.: B. **45**(3), 49–61 (2002)
149. Liu, S., Guo, K.: Decision making under condition of uncertainty: entropy change of fuzzy error system. Adv. Model. Anal.: B **39**(2), 53–62 (2002)
150. Liu, S., Guo, K.: Exploration and application of redundancy system in decision making-relation between fuzzy error logic increase transformation word and connotative model implication word. Adv. Model. Anal. A: Gener. Math. Comput. Tools **39**(4), 17–29 (2002)
151. Liu, S., Guo, K.: Substantial change of decision-making environment-mutation of fuzzy error system. Adv. Model. Anal. A: Gener. Math. Comput. Tools **39**(4), 29–39 (2002)
152. Liu, H., Guo, K.: Discussion about risk-control of investment: research on transformation system of fuzzy error set. Adv. Model. Anal. D (2006)
153. Liu, H., Guo, K.: One-element fuzzy error-matrix. Adv. Model. Anal. D **27**(2), 33–42 (2006)
154. Liu, H., Guo, K: Fuzzy multi-element error-matrix and its operation. Adv. Model. Anal. A: Gener. Math. Comput. Tools **47**(1–2), 21–32 (2010)
155. Liu, S., Guo, K: Employing substitution connectives of error-elimination logic in decision making of social system. Adv. Model. Anal.: A Gener. Math. **47**(1–2), 33–45 (2010)
156. Liu, S., Guo, K., Sun, D.: Progress and prospect of research on theory of error-elimination: a new thrust in management science. Chin. J. Manag. **7**(12), 1749–1758 (2010)
157. Liu, J.J., Zhang, Y., Zhao, F.: Robust distributed node localization with error management. In: 7th ACM International Symposium on Mobile Ad Hoc Networking and Computing (MobiHoc 2006), Florence; Italy 22–25 2006 May, pp. 250–261 (2006)
158. Lohachoompol, V., Srzednicki, G., Craske, J.: The change of total anthocyanins in blueberries and their antioxidant effect after drying and freezing. Biomed. Biotechnol. **2004**(5), 248–252 (2004)
159. Lorenz, E.N.: Deterministic nonperiodic flow. J. Atmos. Sci. **20**(2), 130–141 (1963)
160. Lorenz, E.N: The predictability of hydrodynamic flow. Trans. New York Acad. Sci. **25**(4), 409–432 (1963)
161. Lorenz, E.N.: The essence of chaos (Jessie and John Danz Lectures), 1st edn. University of Washington Press (1995)
162. Love, P.E.D., Edwards, D.J., Han, S.: Bad apple theory of human error and building information modelling: A systemic model for BIM implementation. In: 2011 Proceedings of the 28th ISARC, Seoul, Korea, June 29–July 2 2011, pp. 349–354 (2011)

163. Loy, A., Beisker, W., Meier, H.: Diversity of bacteria growing in natural mineral water after bottling. Appl. Environ. Microbiol. **71**(7), 3624–3632 (2005)
164. Luhmann, N., Gilgen, P.: Introduction to Systems Theory. Polity (2012)
165. 100 World-wide proverbs in management. http://www.glswpx.com/. Cited on 27 July 2007
166. Marx, K.: Capital: A Critique of Political Economy—The Process of Capitalist Production, p. 834 (1867)
167. Mao, Z.: On practice and contradiction (Revolutions). Reissue edition, Presented by Slavoj Žižek Verso (1937)
168. Marx, D.A.: Graeber, R.C: Human error in maintenance. In: Johnston, N., McDonald, N., Fuller, R. (eds.) Aviation Psychology in Practice, pp. 87–104. Ashgate Publishing, London (1994)
169. McKay, R., Efferson, C.: The subtleties of error management. Evolut. Human Behav. **31**(5), 309–319 (2010)
170. Meadows, D.H.: Thinking in Systems: A Primer. Chelsea Green Publishing (2008)
171. Miller, J.H., Page, S.: Complex Adaptive Systems: An Introduction to Computational Models of Social Life (Princeton Studies in Complexity). Princeton University Press (2007)
172. Min, X., Guo, K.: A study regarding solution of a knowledge based on the containing-type error matrix equation. In: 2nd International Conference on Computer and Network Technology FSKD. **4**, 1971–1975 (2011)
173. Min, X., Guo, K.: The modeling and solution of knowledge reason for intelligent traffic management systems based on error matrix equation. In: Proceedings of 2011 International Conference on Computer Science and Service System(CSSS), Nanjing, China, 27–29 June 2011, pp. 1827–1830 (2011)
174. Min, X., Guo, K.: Knowledge model based on error logic for intelligent system. In: 2009 WRI World Congress on Computer Science and Information Engineering. Los Angeles, California USA, March 31–April 02 2009 (2009)
175. Min, X., Huang, J., Qi, J., Guo, K.: rror matrix equation based error-elimination and error-avoidance expert system. J. Guangdong Univer. Technology. **29**(2), 21–27 (2012)
176. Mitchell, M.: Complexity: A Guided Tour, 1st edn. Oxford University Press (2009)
177. Mobus, G.E., Kalton, M.C.: Principles of Systems Science (Understanding Complex Systems). Springer (2015)
178. Munro, E.: Common errors of reasoning in child protection. Child Abuse & Neglect **23**, 745–758 (1999)
179. National People's Congress: The Food Hygiene Law of the People's Republic of China (1995). http://www.npc.gov.cn/wxzl/gongbao/1995-10/30/content_1481321.htm. Cited on 10 March 2019
180. Nordstrom, C.R., Wendland, D., Williams, K.B.: To err is human: an examination of the effectiveness of error management training. J. Bus. Psychol. **12**, 269–282 (1998)
181. Nursal, B., Yucecan, S.: Vitamin C losses in some frozen vegetables due to various cooking methods. Nahrung **44**(6), 451–453 (2000)
182. Oostinga, M.S.D: Breaking (the) ice: Communication error management in law enforcement interactions. Enschede: University of Twente (2018)
183. Office of Disease Prevention and Health Promotion: Dietary Guidelines 2015–2020 (2019). https://health.gov/dietaryguidelines/2015/guidelines/chapter-1/a-closer-look-inside-healthy-eating-patterns/. Cited on 15 October, 2019
184. Palmer, D.: Normal organizational wrongdoing: A critical analysis of theories of misconduct in and by organizations. Oxford Scholarship Online (2012)
185. Parents should prevent http://blog.sina.com.cn/s/blog_6ca116c30100u9uf.html. Cited on 20 March 2019
186. Portillo, R.A.: Theory of interaction: The foundations. ydor.org, Toulouse, France (2015). https://www.ydor.org/uploads/YDOR/TheoryOfInteraction093-Excerpt.pdf. Cited on 10 March 2019
187. Prümper, J., Zapf, D., Brodbeck, F., Frese, M.: Errors in computerized office work: differences between novice and expert users (1991). SIGCHI Bull. **23**(2), 63–66

188. Prümper, J., Zapf, D., Brodbeck, F.C., Frese, M.: Errors of novices and experts: Some surprising differences in computerized office work. Behav. Inf. Technol. **11**, 319–328 (1992)
189. Qiao, D.: Rand Decision mMaking: Opportunity Predication and Business Decision Making. Tiandi Press, Chengdu (1998)
190. Rabøl, L..I, Andersen, M.L., Østergaard, D., Bjørn, B., Lilja, B., Mogensen, T.: Republished error management: descriptions of verbal communication errors between staff. An analysis of 84 root cause analysis-reports from Danish hospitals. Postgraduate Med. J. **87**(1033), 783–789 (2011)
191. Rasmussen, J., Gunter, S.K., Leplat, J. (eds.): New Technology and Human Error (New Technologies and Work: A Wiley Series), 1st edn. Wiley (1987)
192. Rasmussen, J.: Human error and the problem of causality in analysis of accidents. Philosoph. Trans. Royal Soc. Lond. **327**, 449–460 (1990)
193. Reason, P.: Human Error. Cambridge University Press, Cambridge (1990)
194. Reason, J., Hobbs, A.: Managing Maintenance Error: A Practical Guide. CRC Press (2003)
195. Reinach, S., Viale, A.: Application of a human error framework to conduction train accident/incident investigation. Accid. Anal. Prev. **38**, 396–406 (2006). https://www.sciencedaily.com/releases/2001/01/010109083707.htm. Retrieved on 12 January 2019
196. Rybowiak, V., Garst, H., Frese, M., Batinic, B.: Error orientation questionnaire (EOQ): reliability, validity, and different language equivalence. J. Organ. Behav. **20**, 527–547 (1999)
197. Rzepnicki, T., Johnson, P.: Examining decision errors in child protection: a new application of root cause analysis. Child. Youth Serv. Rev. **27**, 393–407 (2005)
198. Sasou, K., Reason, J.: Team errors: definition and taxonomy. Reliab. Eng. Syst. Saf. **65**, 1–9 (1999)
199. Sayama, H.: Introduction to the modeling and analysis of complex systems. Binghamton University, Open SUNY Textbooks (2015)
200. Senders, J.W., Moray, N.P.: Human Error: Cause, Prediction, and Reduction (Applied Psychology Series), 1st edn. CRC Press (1991)
201. Senge, P.M.: The Fifth Discipline. Doubleday/Random House, New York (2006)
202. Severi, S., Bedogni, G., Manzieri, A.M., et al.: Effects of cooking and storage methods on the micronutrient content of foods. Eur. J. Cancer Prev. **6**(S1)), S21–S24 (1997)
203. Sharma, K.D., Karki, S., Thakur, N.S., Attri, S.: Chemical composition, functional properties and processing of carrot-a review. J. Food Sci. Technol. **49**(1), 22–32 (2012). https://doi.org/10.1007/s13197-011-0310-7
204. Sheridan, T.B.: Risk, human error, and system resilience: fundamental ideas. Hum. Fact. **50**(3), 418–426 (2008)
205. Shi, W., Jiang, F., Zheng, Q., Cui, J.: Analysis and control of human error. Proced. Eng. **26**, 2126–2132 (2011)
206. Shotyk, W., Krachler, M.: Lead in bottled waters:? Contamination from glass and comparison with Pristine groundwater. Environ. Sci. Technol. **41**(10), 3508–3513 (2007)
207. Skyttner, L.: General Systems Theory: Problems, Perspectives, Practice, 2nd edn. World Scientific Publishing (2006)
208. Smith, T.A.: Why originalism won't die-common mistakes in competing theories of judicial interpretation. Duke J. Const. Law Public Policy **2**(1), 159–216 (2017)
209. Sterman, J.: Business Dynamics: Systems Thinking and Modeling for a Complex World. McGraw-Hill Education (2000)
210. Sullivan, M.J., Leavey, S.: Heavy metals in bottled natural spring water. J. Environ. Health. **73**(10), 8–13 (2011)
211. Ross, J., Arkin, A.P.: Complex systems: From chemistry to systems biology. Proc. Natl. Acad. Sci. **106**(16), 6433–6434 (2009)
212. Strogatz, S.H.: Nonlinear Dynamics and Chaos: with Applications to Physics, Biology, Chemistry, and Engineering (Studies In Nonlinearity), 1st edn. CRC Press (2000)
213. Thomason, J.L., Vlug, M.A: Influence of fibre length and concentration on the properties of glass fibre-reinforced polypropylene: 1. Tensile and flexural modulus. Composites Part A: Appl. Sci. Manuf. **27**(6), 477–484 (1996)

214. Thomason, J.L., Vlug, M.A.: Influence of fibre length and concentration on the properties of glass fibre-reinforced polypropylene: 4. Impact properties. Compos. Part A: Appl. Sci. Manuf. **28**(3), 277–288 (1997)
215. Valacich, J.S., George, J.F., Hoffer, J.A: Essentials of Systems Analysis and Design (6th Edition). Pearson (2014)
216. Van der Linden, D., Sonnentag, S., Frese, M., Van Dyck, C.: Exploration strategies, performance, and error consequences when learning a complex computer task. Behavi. Inf. Technol. **20**, 189–198 (2001)
217. Van der Schaaf, T.W., Frese, M., Heimbeck, D.: Human recovery and error management. In: Proceedings of the XV. European Annual Conference on Human Decision Making and Manual Control. Soesterberg, June 1996, 5.2-1–5.2-10 (1996)
218. Van Dyck, ., Frese, M., Baer, M., Sonnentag, S.: Organizational error management culture and its impact on performance: a two-study replication. J. Appl. Psychol. **90**(6), 1228–1240 (2005)
219. Varga, L.: Bacteriological quality of bottled natural mineral waters commercialized in Hungary. Food Control **22**(2011), 591–595 (2011)
220. Veldkamp, C.: The human fallibility of scientists: Dealing with error and bias in academic research. Doctoral Thesis. Department of Methodology and Statistics, Tilburg University https://research.tilburguniversity.edu/en/publications/the-human-fallibility-of-scientists-dealing-with-error-and-bias-i. Cited on 10 April 2019 (2017)
221. Von Bertalanffy, L.: General System Theory: Foundations. Development. Penguin University Books, Applications (1969)
222. Wallace, B., Ross, A.: Beyond Human Error: Taxonomies and Safety Science. Taylor & Francis, Boca Raton (2006)
223. Wang, C.: Introduction to Fuzzy Mathematics. Press of Beijing University of Technology, Beijing (1998)
224. Wasson, C.S.: System analysis, design, and development concepts, principles, and practices. Wiley-Interscience (2005)
225. Wasson, C.S.: System Engineering Analysis, Design, and Development: Concepts, Principles, and Practices (Wiley Series in Systems Engineering and Management), 2nd edn. Wiley (2015)
226. Weinberg, G.M.: An Introduction to General Systems Thinking (Silver Anniversary Edition). Dorset House (2001)
227. Weinberg, G.M., Weinberg, D.: General Principles of Systems Design. Dorset House (1988)
228. Wen, Q.: Discussion on Error. Press of Liaoning University, Shenyang (1995)
229. Whittingham, R.: The Blame Machine: Why Human Error Causes Accidents. Elsevier Butterworth-Heinemann, Oxford (2004)
230. Wiener, N.: Cybernetics: Or the Control and Communication in the Animal and the Machine. Paris, France: Librairie Hermann & Cie, and Cambridge, MA: MIT Press. Cambridge, MA: MIT Press (1948)
231. Wilson, J.Q., Kelling, G.L.: Broken windows: The police and neighborhood safety (1982). The Atlantic. https://www.manhattan-institute.org/pdf-atlantic-monthly-broken-windows.pdf. Cited on 20 August 2017
232. Wilson, B.A., Baddeley, A., Evans, J.: Errorless learning in the rehabilitation of memory impaired people. Neuropsychol. Rehabil. **4**, 307–326 (1994)
233. Woods, D., Cook, R.: Nine steps to move forward from error. Cognit. Technol. Work **4**, 137–144 (2002)
234. Woods, D., Johannesen, L., Cook, L., Sarter, N.: Behind human Error: cognitive systems, computers and hindsight. Wright-Patterson Air Force Base, Ohio, CSERIAC (1994)
235. Xie, B., Guo, K., Wu, W.: Destruction transformation on the structure of error system. China Intelligence and Automation Conference (CIAC). Nanjing, Jiangsu, China, 27–29 September 2009 (2009)
236. Xinhuanet: Successful demolition of Nanxi bridge in Yunyang, Chongqing (2008). http://news.ifeng.com/photo/society/detail_2008_10/29/549010_0.shtml. Cited on 15 December 2018

237. Xiong, H., Guo, K.: Research on domain transformation connectives in fuzzy error logic. Fuzzy Syst. Math. **20**(1), 24–29 (2006)
238. Xiong, H., Huang, J., Guo, K.: Research on the application of fuzzy error logical in system optimization. In: 2008 IEEE International Symposium on Knowledge Acquisition and Modeling Workshop Proceedings. Wuhan, China, 21–22 December 2008, pp. 147–149 (2008)
239. Yao, J., Guo, K.: Investigation on the domain transformation of error systems. In: China Intelligence and Automation Conference (CIAC). Nanjing, Jiangsu, China, 27–29 September 2009 (2009)
240. Ye, Q., Guo, Q., Guo, K.: Methods for the error-eliminating in system decision-making. In: Proceedings of The 2010 International Conference on Measuring Technology and Mechatronics Automation (ICMTMA). Changsha, China, 13–14 March 2010, **2**, 652–655 (2010)
241. Zapf, D., Brodbeck, F., Frese, M., Peters, H., Prümper, J.: Errors in working with office computers. A first validation of a taxonomy for observed errors in a field setting. Int. J. Hum. Comput. Interact. **4**, 311–339 (1992)
242. Zhang, M., Sun, X.: 1% errors can lead to 100% of the failure: a case from Falklands War between Argentina and UK. Liberation Army Daily. 16 March 2005 (2005)
243. Zhao, Z.: Introduction to Mathematical and Dialectic Logic. Press of Renmin University, Beijing (1998)
244. Zhou, P., Jiang, C., Guo, K.: Judgment of error system category of decision-making error. Adv. Model. Anal. A:Gener. Math. **40**(1), 19–30 (2003)
245. Zhou, X., Guo, K.: The optimization of independent and absolutely additive system. In: Proceedings of International Conference on E-Business and E-Government. Shanghai, China, 6–8 May 2011 ICEE2011, pp. 1782–1785 (2011)
246. Zimmermann, H.J.: Fuzzy Set Theory and it Application, 4th edn. Kluwer Academic Publishers, Boston/Dordrecht/London (2001)
247. Zimmermann, J., Efferson, C.: One-shot reciprocity under error management is unbiased and fragile. Evolut. Hum. Behav. **38**(1), 39–47 (2017)

Printed in the United States
By Bookmasters